TRIBOLOGY
OF
INTERFACE
LAYERS

TRIBOLOGY OF INTERFACE LAYERS

HOOSHANG HESHMAT

CRC Press
Taylor & Francis Group
Boca Raton London New York

CRC Press is an imprint of the
Taylor & Francis Group, an **informa** business

CRC Press
Taylor & Francis Group
6000 Broken Sound Parkway NW, Suite 300
Boca Raton, FL 33487-2742

First issued in paperback 2019

No claim to original U.S. Government works

ISBN-13: 978-0-367-45233-9 (pbk)
ISBN-13: 978-0-8247-5832-5 (hbk)

Visit the Taylor & Francis Web site at
http://www.taylorandfrancis.com

and the CRC Press Web site at
http://www.crcpress.com

Library of Congress Cataloging-in-Publication Data

Heshmat, Hooshang.
    Tribology of interface layers / Hooshang Heshmat.
        p. cm.
    "A CRC title."
    Includes bibliographical references and index.
    ISBN 978-0-8247-5832-5 (hardcover : alk. paper)
    1.˝ Tribology. 2.˝ Interfaces (Physical sciences)˝ I. Title.

TJ1075.H48 2010
621.8'9--dc22   2009048848

Dedication

The book is dedicated to my friend and reliable sounding board Melissa D. Heshmat

Contents

Nomenclature

A	Area
B	Damping constant
B_e	Equivalent damping
C	Bearing radial clearance
D_o	Einstein's diffusion coefficient
E	Energy
F	Force
F_r	Viscous shear force
F_t	Tangential force
H	Power loss, energy
K	Stiffness coefficients
L	Length of bearing
M	Mass
N	Rotational speed
P	Fluid film pressure
Q	Volumetric oil flow rate
R	Bearing or journal radius
R_t	Torque arm radius
S	Surface area
S_L	Shear strength of lubricant film
S_m	Shear strength of metallic junction
T	Temperature
U	Velocity
V	Volume
W	Load
W_T	Tractive force
W_n	Normal load
d	Diameter
d_a	Projected area diameter
d_m	Martin diameter
d_p	Perimeter area diameter
d_S	Surface diameter
d_{Si}	Sieve diameter
d_{SV}	Surface volume diameter
d_v	Volume diameter
e	Eccentricity
f	Frequencies
h	Film thickness
h_N	Nominal film thickness
k_n	Constant coefficient
p	Fluid film pressure

p_a	Ambient pressure
u	Velocity
x	Displacement
z	Coordinate in axial direction
α	Temperature-viscosity
β	Bearing arc
f	Coefficient of Friction
f_T	Traction coefficient
γ	Shear rate
λ	Film width
μ	Viscosity
μ_L	Viscosity of liquid
μ_R	Relative viscosity
ρ	Mass density of lubricant
ω	Angular velocity
τ_ℓ	Limiting shear stress
τ_o	Yield shear stress

Preface

Two sets of experimental evidence, one old and the other rather recent, were the impetus to write *Tribology of Interface Layers*. The old evidence, which has long plagued lubrication engineers and analysts, is the fact observed both in the laboratory and in industrial applications that many tribological devices, such as flat surface and centrally pivoted sliders, can act as viable bearings in contradiction to basic hydrodynamic theory. The recent evidence in the same direction is the documented observation that minute particles or powder films, whether introduced deliberately or generated spontaneously as a result of normal wear, exhibit quasi-hydrodynamic features that enable them to act much as conventional fluid film bearings do. This led, in the first instance, to a redefinition of a lubricant as any intermediate layer that does not constitute or remain an integral part of the two mating surfaces. This would then embrace anything from molecular layers to coatings, gases, and liquids, as well as powders and other solid particles small enough to form a tribological film.

The present volume on interface layers in tribology then has two objectives. One is to demonstrate the presence of a continuum across the spectrum of lubrication regimes ranging from dry friction to full fluid film lubrication, which offers a most striking example of the above continuum; next, bearings, seals, dampers, and similar devices using submicron powder lubricants are shown to possess quasi-hydrodynamic characteristics akin to conventional fluid films. Thus, powder-lubricated tribological devices make excellent candidates for operation in extreme environments.

The field of tribology has traditionally been dealt with in terms of disparate regimes such as dry friction, the area of coatings and dry lubrications, boundary or mixed lubrication, and on to elastohydrodynamic and full film lubrication. Even while adhering to this classification, workers in the field were aware of the fact that this approach lacks rigor, not only in fact but even in their very definitions. Thus, it was realized that all surfaces carry oxide layers or at least some molecular monolayer that imparts to them the characteristics of a coated surface. On the other end of the spectrum, it was far from clear when a mixed lubrication regime becomes a full fluid film operation. An attempt is made here to show that, indeed, the various lubrication modes—from dry to hydrodynamic and on to powder lubrication—all overlap to such a degree that one can talk of a continuum from one end of the tribological spectrum to the other. The essence of this postulate is that specific operational mechanisms characteristic of one regime are also present in most others, and that it is only their relative weights that determine the prevalence of one regime over another. Ultimately, the characterization and treatment of any one regime may require the simultaneous accounting of a number of tribological modalities present in a particular mode of operation. In particular, this phenomenon is elaborated with respect to the new technology of powder lubrication. Based on a wide range of experimental and theoretical work by the author and others, the interdependence of powder and hydrodynamic lubrication will be shown to constitute the most striking example of the above postulated continuum in tribological processes.

Acknowledgments

It is my duty and obligation to use this opportunity to record my deep gratitude to the educators of tribology, those who have come before me and those who have guided me throughout my career for the past three decades. They have helped and inspired me with their tireless teachings and encouragement to remain true to the field of science and the engineering discipline. They have provided me with the essential ingredients for laying out my thoughts and findings through experiments in this book. I thank them profusely, in particular the late professors Maurice Godet of INSA, France, for his unblemished devotion to the understanding and finding of a path through the jumbles of the "third body's nature," Frederick F. Ling, and Christopher M. McC. Ettles.

I am truly grateful to my friend and mentor, the late Dr. Jed Walowit, for cultivating in me mathematical simplicity in modeling thought experiments.

I could never find a place for this line of thought here or elsewhere, or a way to say thank you enough to my friend, Melissa, for teaching me to be patient and believing in my outreaching visions.

Without doubt, the work presented in this book could not have been accomplished without the hard work and dedication of my students, co-workers and colleagues— the list goes practically forever but to mention a few names for the sake of sampling: Jim Walton, Oscar Pinkus, Leo Hoogenboom, Mike Tomaszewski, Dave Slezak, Ashley P. Heshmat, Christina Jewell, Said Jahanmir, Crystal A. Heshmat, Rita Kaur, C. Fred Higgs, and Harold Elrod. I am lucky to have had my path cross with these fine people in this life.

It is my humble opinion that without the support of the DOD (Department of Defense), DOE (Department of Energy), NSF (National Science Foundation), and ARL (Army Research Laboratory) in the United States, my work would have been impossible. I thank these organizations for their encouragement, sustained interest, and the flourishing of the field. Noteworthy scientists and management of these organizations are Drs. Alan Garscadden, Nelson Forster, Tom Jackson, Michael N. Gardos, Bob Bill, Jim Dill, and Messrs. David Brew, Ron Dayton, and Theodore Fecke.

Hooshang Heshmat
New York

About The Author

Dr. Hooshang Heshmat is Lead-Founder and President/CEO & Technical Director of Mohawk Innovative Technology, Inc. (MiTi), an internationally renowned applied research and product development company dedicated to green technology. His company specializes in the design, integration and development of mechanical components such as advanced oil-free foil bearings, seals, and dampers, including powder, solid film coatings and dry and fluid film bearings into high-speed oil-free rotating machinery systems. Developed rotating machines employing his novel tribological devices include oil-free compressors, high-speed motors and generators, and gas turbine engines down to Mesoscopic sizes. He has developed special tribological testing methods and measurement techniques and has designed and fabricated the corresponding component test equipment. Dr. Heshmat has established MiTi*Heart*, one of the premier biotechnology companies in the Albany area based on their unique heart pump system, which uses magnetic levitation.

Dr. Heshmat received his B.S. degree from Pennsylvania State University and his M.S. and Ph.D. degrees in mechanical engineering from Rensselaer Polytechnic Institute. He is the principal investigator for a variety of programs concerned with applied tribology, including hydrodynamics, solid lubrication, novel backup bearings for magnetically suspended rotor systems, tribochemistry, and surface morphology. Dr. Heshmat's background in tribology encompasses bearings, seals, dampers, and piston rings, including both fluid film and dry lubricant systems. In the area of industrial liquid lubrication, Dr. Heshmat developed a new class of industrial oil ring bearing and novel lubrication system for a major manufacturer.

While employed at Reliance Electric Company, Dr. Heshmat was responsible for developing high-performance and cost-effective industrial fluid film bearings ranging from 2 to 17 inches in diameter. There, he developed novel bearings and lubrication systems for starved (partially lubricated) bearings.

Dr. Heshmat has also played a primary role in the development of compliant foil bearings and has been responsible for major advances in this field, including analytical and experimental research for bearing/seal design and application. Much of his work has been related to applying compliant foil bearings to high-speed turbomachinery, including advanced turbine engines (cruise missiles and liquid rocket engines), automotive gas turbines, air-cycle machinery for aircraft, turbochargers, turboexpanders, cryocoolers, pumps, compressors, PM/IM motors, and generators and refrigerant systems. Dr. Heshmat has been investigating the interaction of hybrid foil/magnetic bearing and developed computer codes to permit integrated hybrid bearing designs for advanced gas turbine engine applications.

He developed the principle of quasi-hydrodynamic lubrication with dry triboparticulates, i.e., friction- and wear-control theory for application to extreme environments. This theory was developed based upon his having conducted a wide range of experimental test efforts characterizing fundamental aspects of powder lubrication. Additionally, he has investigated the practical application of this promising new technology to high temperature coatings (Korolon™), dampers and bearings for high-speed gas turbine engines, as well as vacuum environments and damping of structural members including turbine blades.

Dr. Heshmat is listed in *Who's Who in America* and *Who's Who in the World*, holds 28 patents, has authored more than 150 papers, coauthored two chapters (Principles of Bearing Design for the *Compressor Handbook* and Principles of Gas Turbine Bearing Lubrication and Design for the *Theory of Hydrodynamic Lubrication*), and authored the book *The Tribology of Interface Layers*. He has received the Society of Tribologists and Lubrication Engineers (STLE) 1983 Wilbur Deutsch Memorial Award; the American Society of Mechanical Engineers (ASME) 1985 Burt L. Newkirk Award; the Mechanical Technology Incorporated (MTI) 1990 Technical Creativity Award for his pioneering work in powder lubrication; and the STLE 1993 Captain Alfred E. Hunt Award for the best paper published by the society, "The Quasi-Hydrodynamic Mechanism of Powder Lubrication: Part II—Lubricant Film Pressure Profile." He is an active member of STLE/ASME, was chairman of the 1994 International Joint ASME/STLE Tribology Conference, chairman of the Research Committee on Tribology of ASME (1999–2000), a member of the ASME Tribology Division Executive Committee, and an invited speaker for the January 1997 ASME Satellite Broadcast on "The Selection, Design, and Performance of Bearings and Seals," and is an ASME/STLE Fellow. For his pioneering work in the field of compliant surface hydrodynamic foil bearings, he received the prestigious ASME/RCT 1995 Creative Research Award. He was the recipient of STLE's 1996 Al Sonntag Award for the outstanding paper written on solid lubrication and published in the society journal ("The Quasi-Hydrodynamic Mechanism of Powder Lubrication: Part III—On Theory and Rheology of Triboparticulates," *Tribology Transactions*, Vol. 38 (1995), No. 2, pp. 269–276) and ASME/IGTI 1999 Best Paper Award, "Application of Foil Bearings to Turbomachinery Including Vertical Operation," ASME Paper 99-GT-391. He was chosen to receive the ASME Board on Research and Technology Development's 2002 Thomas A. Edison Patent Award and STLE 2003 Frank P. Bussick Award for authoring the best paper on sealing systems technology, "Performance of a Compliant Foil Seal in a Small Gas Turbine Engine Simulator Employing a Hybrid Foil/Ball Bearing Support System," published by the society. Dr. Heshmat also received the ASME 2005 Microturbine and Small Turbomachinery Committee Best Paper Award for authoring "Demonstration of a Turbojet Engine Using an Air Foil Bearing," ASME Paper GT2005-68404.

Recently, the following awards were also given to Dr. Heshmat: The ASME 2007 Mayo D. Hersey award for significant contributions to the fundamental science of powder lubrication and advanced compliant foil bearings and the STLE 2008 International Award in recognition of his outstanding contribution to the field of tribology. Dr. Heshmat was also presented the Pennsylvania State University 2009 Alumni Achievement Award for his extraordinary professional accomplishments and the ASME 2009 Propulsion Best Paper Award for his technical paper entitled "Innovative High Temperature Compliant Surface Foil Face Seal Development."

Introduction: A Brief Synopsis

The book is aimed at providing the engineering profession with a unified approach to the discipline of lubrication. Whereas, up to the present, the field consisted of several noncontiguous modes of operation—boundary, fluid-film, and dry and solid lubrication—the book introduces a new concept wherein all these disparate modes are shown to be but particular phases of a tribological continuum. In this approach, bearings, seals, gear teeth, dampers, and other tribological devices operate at all times on a syndrome of confluence of all of the above-mentioned mechanisms. The understanding and practical application of this tribological continuum approach should enable the engineer to design and calculate the performance of seals, dampers, tilting pad, bearings, piston rings, and other parallel surface systems for which previous theory was incapable of providing solutions.

A particularly important aspect of the book is the space allotted to the newest branch of interface media—powder lubrication. There are two important aspects to this new technology. One is that the debris produced by wear at the mating surfaces, instead of being harmful and something to be avoided, can actually be made to act as a lubricant. Given the small size of wear particles, they can be made to form a quasi-hydrodynamic film at the interface, facilitating the smooth performance of the machine in question. The other aspect of this new technology is that of deliberately feeding into tribological devices a supply of powder lubricant. This is a field in which the author has been its foremost proponent over the last dozen years, both as a researcher and product designer. It is a technology of tremendous potential throughout industry.

Throughout the history of conventional lubrication theory and experience, it has been known to workers in the field that many phenomena did not conform to the postulates of hydrodynamic theory, as indicated by the Reynolds equation. In particular, the ability of parallel surfaces, such as thrust bearings with centrally pivoted pads, flat plate sliders and seals, and other parallel surface devices, were known to be capable of supporting a load, in contradiction of the requirement of a convergent film, as postulated by hydrodynamic theory. Many attempts were made to explain or "justify" this behavior, most of them imparting to such surfaces some semblance of convergence, either via surface deformation, thermal distortion, the action of surface roughness, or of such perturbations as variations in lubricant viscosity or density. However, they all proved unsatisfactory, either because the calculated effects were minute or because nondeformable, polished surfaces still showed themselves capable of producing a load capacity.

In a 1988 paper, the author and two of his colleagues for the first time offered the hypothesis that there is an overlap or a continuum between boundary lubrication and hydrodynamic in the sense that both mechanisms contribute to load capacity through the weights carried by the two regimes depend on the geometric and operating conditions of the tribological device. A tribological continuum was proposed for the parallel convergent films or the boundary hydrodynamic action with the total

load capacity and tribological performance compounded of the simultaneous interaction of the two mechanisms of surface interaction. This understanding was later
extended with the introduction of powder lubrication. This form of lubrication can
arise either from the unintended accumulation of debris due to normal wear between
two surfaces or as a deliberate mode of facilitating relative motion by feeding a properly selected layer of triboparticulates between interacting surfaces. In the course of
extensive experimental work with such triboparticulates, it was ascertained that such
films exhibit all the basic characteristics of hydrodynamic lubrication. The main difference of powder films stems from the different rheological properties, similar to
the differences with any other viscoelastic or pseudoplastic substance. These experimentally verified characteristics extended the previously noted overlap of boundary
and hydrodynamic regimes to that of powder lubrication.

The above experimentally observed relations provide the basis for the postulate
of an overall tribological continuum spanning a wide array of material lubricants.
This synthesis is schematically portrayed in Figure I.1, which can be labeled as a
neo-Stribeck curve in which the various regimes of lubrication from boundary to
gas–liquid–powder interfaces are seen to overlap, providing conjointly for the generation of load capacity at surfaces moving at some relative motion with respect to
each other. It is only necessary to assign to each material lubricant—whether a gas,

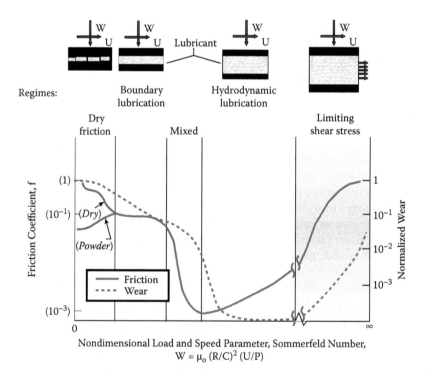

FIGURE I.1   Neo-Stribeck curve, including limiting shear stress regime for the third body.

liquid, powder, or even discrete particles—the proper rheological model to be able to employ the generalized form of equations of motion to obtain the quasi-viscous forces produced by these various kinds of lubricants or, as they are often referred to, "third bodies."

The present book is concerned with this new interdisciplinary approach to the subject of tribological interaction of various third-body lubricants. In addition to postulating the general theory of interdisciplinary tribology, the text provides specific methods of solving powder lubrication problems encountered in bearings, seals, dampers, and a host of other machine components in which debris particles provide a lubricant layer considerably affecting the dynamics of the machine. In addition to the conceptual and theoretical explication of this new approach to tribology, the book offers experimental data on the rheology of triboparticulate films, as well as design guidelines based on extensive experience derived from the use of powder lubrication in a wide array of turbomachinery and other advanced systems.

1 Historical Perspective

As remarked in the preface, the subject of tribology is here treated from the perspective of the lubricant layers separating the two interacting surfaces. Regardless of the modes and means employed—and this includes so-called dry contacts—it will be seen that there is always an intermediate substance, referred to by some as a "third body," to separate the mating surfaces. Approached from that viewpoint, it will be shown that there is a common mechanism in the seemingly disparate processes occurring in friction and lubrication. This remains true whether one deals with liquid, gaseous, powder, sputtered or dry lubricants, or no lubricant at all. Consequently, this chapter is not a history of tribology as such, which would have to include a multitude of subjects. It is, rather, a focused look at the means used over the ages to reduce friction in the then-existing bearings. In brief, it is a history of lubricants as they are here defined (see Chapter 2). These lubricants do not have to be, as conventionally understood, supplied from an external source but can be generic to the mating surfaces. Sintered bearings and lignum vitae are common examples. Perhaps, not unexpectedly, civilization's progress in this area seems to have been out of step with the rest of tribology.

1.1 PREHISTORY TO THE RENAISSANCE (5000 BC–1450 AD)

The earliest recorded use of materials to facilitate relative motion comes to us from Mesopotamia and Egypt. These were bearings made of wood, bone, ivory, or coconut shell. Potters' wheels were mounted in pivot seats, sometimes lubricated with bitumen obtained from petroleum that seeped to the surface. The most urgent need for lubricated bearings arose in connection with the appearance of wheeled vehicles, which started replacing heretofore used sledges around 3500 BC. One of the unsettled questions about these early vehicles is whether it was the axles or the wheels that were equipped with bearings. An argument for the latter is that it would have taken much skill to produce circular axles. Lubricants, if they were used at all, were vegetable oil or animal fats—mostly beef and mutton tallow, used extensively on the Egyptian chariots of the day.

In the Near East, much effort was expended to reduce friction in connection with the erection of various megalithic structures—pyramids, temples, and other colossi—when the need arose to haul massive building blocks or monuments to the construction site. It is certain that wooden planks were laid under these masses in a relay system but it remains unclear whether these were staves or rollers, which would have constituted the first use of linear roller bearings. Some of the bas-reliefs show men pouring a substance in front of the planks, though here, too, it is unclear whether it was some sludge, water, sand, or crushed rocks abounding in those regions. If so, it

FIGURE 1.1   Laborer pouring lubricant in front of wooden sledge (ca. 2500 BC) (Steinhoff, 1913).

would represent the first use of powder lubrication. A frieze from 2400 BC, portraying a laborer lubricating the staves, is shown in Figure 1.1.

Throughout the Middle Ages, most machines used the same lubricants as in antiquity: vegetable oil and fats. Rapeseed and poppy seed oils were also used wherever these were grown. The years 1300–1400 did not advance much beyond ancient tribological practices and, in many ways, even regressed. While in antiquity some bearings made of iron or copper were already in use, in the thirteenth and fourteenth centuries these were mostly wood-on-wood. Thus, in the field of tribology, the Middle Ages represent a dark period, much as they did in most human endeavors.

1.2   THE RENAISSANCE (1450–1600)

From a tribological viewpoint, the Renaissance years were dominated by a single man: Leonardo da Vinci (1452–1519). Celebrated painter, architect, student of anatomy, scientist, and inventor, he conceived, drafted, and improved upon many basic engineering devices. His contributions to the science of tribology and the conceptualization of low-friction machines can, in part, be summarized as follows:

- Nature of friction. He was the first to state that frictional resistance does not depend on the area of contact but on the magnitude of the normal force and the morphology of the mating materials.
- Coefficient of friction. For perfectly "smooth and polished" surfaces, he gave the formula for the coefficient of friction as $f = (\frac{F}{W})$, where F is the

frictional force and W is the normal load. From experiments he deduced its value to be 0.25. Considering that at the time most bearings were wood-on-wood, it was remarkably accurate. Recent figures for wood, as given by Bowden and Tabor, are as follows: for wet surfaces $f = 0.2$ and for dry surfaces $f = 0.2–0.6$.

- Bearing materials. At one time, da Vinci proposed the use of glass and gun-metal for a bearing surface. Eventually he came up with a low-friction bearing alloy, which consisted of 30% copper and 70% tin. This is remarkably close to tin babbitt, which was not invented until 1839 and would become the most widely used and most satisfactory bearing material.
- Rolling friction bearings. There is a wide range of devices he proposed, which relied on rolling in place of sliding friction, having understood fully the advantages from such a conversion. Figure 1.2 shows some of them.

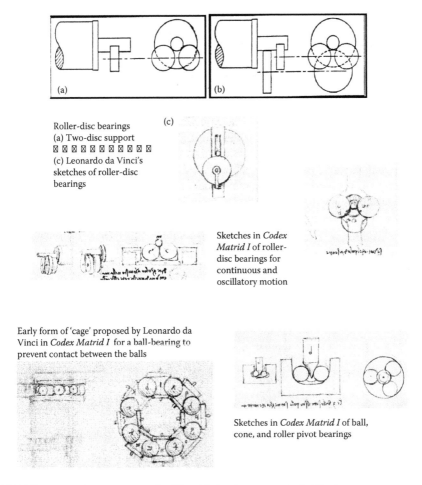

(a)

(b)

Roller-disc bearings
(a) Two-disc support
☒ ☒ ☒ ☒ ☒ ☒ ☒ ☒ ☒ ☒
(c) Leonardo da Vinci's
sketches of roller-disc
bearings

(c)

Sketches in *Codex Matrid I* of roller-disc bearings for continuous and oscillatory motion

Early form of 'cage' proposed by Leonardo da Vinci in *Codex Matrid I* for a ball-bearing to prevent contact between the balls

Sketches in *Codex Matrid I* of ball, cone, and roller pivot bearings

FIGURE 1.2   Various concepts of rolling friction by Leonardo da Vinci.

These include sleeve bearings mounted on rotating rollers; full-fledged ball bearings, including separators or what is known today as cages, to avoid interball friction; and partial bearings for oscillatory movements.

- Solid lubricants. da Vinci proposed the use of leather as an intermediate layer in the sliding surfaces of pump rods. He later improved upon this by suggesting that the leather be soaked in oil, something equivalent to a sintered bearing or wick lubrication.
- Granular interface layers. Most unusually, da Vinci seems to have anticipated the possibility of powder lubrication—a prominent subject in the present volume. In one of his elucidations, da Vinci says:

> All things and everything whatsoever however thin it be which is interposed in the middle between objects that rub together lighten the difficulty of this friction ... And so increasing tiny grains such as millet make it better and easier... (3, p. 103)

1.3   THE PREINDUSTRIAL AND INDUSTRIAL ERAS (1600–1850)

1.3.1   THE PREINDUSTRIAL ERA (1600–1750)

As has been seen, the art of developing proper lubricants lagged conspicuously behind the considerable progress in the conceptualization of new bearing design during the Renaissance, much of it due to Leonardo da Vinci. This holds true for the seventeenth century as well. The parallel is even more pronounced when one realizes that this century produced the greatest scientist of all time: Isaac Newton (1642–1721). The boost that the physical sciences received from this man is truly dazzling and the physical and chemical implication of generating new lubricants continues.

A methodical experimental program to determine the frictional characteristics of various materials was undertaken by Guillaume Amantos in France and the results presented in 1699. The materials he tested included copper, iron, lead, and wood in various combinations. The surfaces were lubricated with pork fat. His conclusions were

- Frictional resistance depends only on the normal load and not on the area of contact—a corroboration of da Vinci's law. Amantos explained the foregoing results in terms of the existence of surface asperities: the higher the load, the more the asperities interlock, producing resistance to motion; hence, a larger area for the same load would produce less intimate interlocking over a larger number of asperities without changing the total resistance to motion.
- By using fat as a lubricant, friction becomes independent of the materials used—the first hint of the presence of a lubricant film.
- The value of the friction coefficient is of the order of 0.33, as compared to da Vinci's 0.25.

Subsequently, Belidor in 1737 conducted experiments with layers of spherical balls to imitate the surface asperities. By setting up arrays of balls, as shown in Figure 1.3, he then measured the force required to pull them apart. He found the force to be

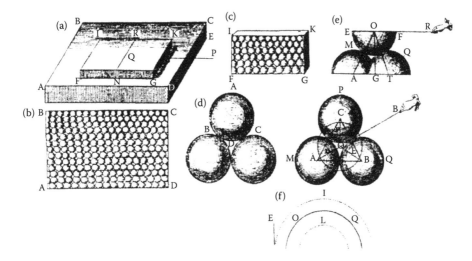

FIGURE 1.3 Belidor's 1737 model of bearing asperities.

independent of the number of asperities, thus confirming Amantos' hypothesis. The value of the coefficient of friction Belidor obtained was 0.35. Interestingly, he saw the role of a lubricant as filling the interspaces between the particles or asperities, thereby minimizing friction. This is, perhaps, the first experimental model for our prime topic of powder lubrication.

While no great progress was made in lubricant technology, a most notable event, though unrealized at the time, was the introduction of the concept of viscosity—a lubricant's all-important property. This was due to Newton.*

Newton's statement about viscosity reads as follows: "The resistance arising from want of lubricity in the parts of a fluid is, other things being equal, proportional to the velocity with which the parts of the fluid are separated from one another." Newton's term "want of lubricity" corresponds to our present notion of internal friction; that is, viscosity. Etymologically, the term derives from the sticky substance of mistletoe berries called viscin. While Newton's work on viscosity remained in the realm of pure science, it laid the foundations of the discipline of hydrodynamic lubrication some 200 years later.

1.3.2 INDUSTRIAL REVOLUTION (1750–1850)

During this important era, wood and brass remained the basic bearing materials while tallow and water remained the common lubricants, which were fed manually. Tin and lead were tried, but they were too soft and had too low a melting point to be practicable. It was not until 1839 when Isaac Babbitt proposed the addition of small

---

* It should, perhaps, interest tribologists that the two greatest scientists of history concerned themselves with the nature of viscosity. See Albert Einstein's "Investigation of the Theory of Brownian Movement," ed. R. Fuerth, Dover Publ. [from Annalen der Physik, (6), 19, 1906].

amounts of antimony and copper that viable bearing materials, known in the United States as babbitts and in England as white metals, became the most successful bearing alloys widely used to this day.

In the struggle against excessive friction, Varlo in Italy came up with an arrangement in 1772 whereby, unlike in trunnions, the balls or rollers are free to rotate, setting the basis for modern rolling element bearings. But his understanding of the nature of rolling friction was deficient in that he attributed all resistance to the friction between individual balls or rollers, assuming that rolling, per se, is frictionless. Therein, perhaps, lies the absurd terminology, which persists to the present day, of calling them antifriction bearings.

A much more comprehensive study of the nature of friction, including the role of intermediate layers, was conducted by C. A. Coulomb (1736–1806). For lubricants, he used tallow and axle grease, testing both smooth and rough surfaces. He concentrated on several crucial aspects, that is:

1. The nature of the materials, including coatings
2. The extent of surface areas
3. The normal load
4. Time of repose (i.e., length of time the surfaces remained in contact)
5. The effects of temperature, humidity, and even vacuum

For dry sliding of wood-on-wood, he found the following values for the friction coefficient:

Oak on oak, $f = 0.43$
Elm on elm, $f = 0.46$
Pine on pine, $f = 0.56$
Oak on pine, $f = 0.67$

They are, thus, close to the values cited by Bowden and Tabor given previously.

A more complete listing of the coefficient of friction for a range of materials in use, or contemplated as bearing materials in the eighteenth century, was provided in England by G. Rennie in tests conducted in 1829. These are given in Table 1.1, including the effect of pressure on some of these materials.

The lubricants used during the Industrial Revolution were simple variations on the familiar substances used since ancient times. These were

- Sperm oil. This was a sought-after lubricant because of its stability, but it was also the most expensive. It was obtained from the head of the sperm whale. The melted blubber was sometimes used but avoided because of its odor, though it was used in the manufacture of soaps.
- Lard oil. This molten pig fat was inexpensive and most widely used.
- Neat's foot oil. This was obtained from melting the hooves of cattle.
- Olive oil. It was as highly rated and had an even higher viscosity than sperm oil and was thus used for heavy-duty bearings.

TABLE 1.1
Rennie's Table of Values of Dry Friction

Materials	Mean Pressure (lbf/in.^{2})	Coefficient of Friction
Hardwood on hardwood	28–728	0.129
Brass on wrought iron	6.09	0.136
Brass on steel	5.33	0.141
Brass on cast iron	6.10	0.139
Soft steel on soft steel	6.10	0.146
Cast iron on steel	6.10	0.151
Wrought iron on wrought iron	6.10	0.160
Cast iron on cast iron	5.33	0.163
Hard brass on cast iron	4.64	0.167
Cast iron on wrought iron	6.10	0.170
Brass on brass	6.10	0.175
Tin on cast iron	5.33	0.179
Tin on wrought iron	6.10	0.181
Soft steel on wrought iron	6.09	0.189
Leather on iron	0.66–28.44	0.25
Tin on tin	6.10	0.265

- Groundnut and castor oils. The first was obtained most commonly from peanuts, the latter from the plant Ricinus, and was preferred to nut oil. Both of these tended to gel when exposed to the atmosphere.
- Mineral oils. Pitch, tar, and a liquid oil were obtained in those days from some stones. The first usage was also made of whale deposits for obtaining oil, mostly, however, for lighting purposes. The usage of these mineral products was small compared to vegetable and animal products.
- Greases and semifluid lubricants. These were still mostly beef or mutton fats melted down so that the organic membranes could be eliminated. These were sometimes composed of a mixture of suet and graphite, representing a transition to solid lubricants.

The Industrial Revolution did bring some progress in solid lubricants. Rennie, in his experiments, used black-lead, also known as plumbago. This was actually graphite, an allotrope of carbon, but was called black-lead because it left traces similar to the black-lead used as ink in the Middle Ages. The other solid lubricant was soapstone, or talc. These powdered substances were administered by the machinists in small quantities as a form of medicine whenever the bearings showed signs of "distress." Correspondingly, these men were referred to as mill doctors. In 1835, the Englishman Booth patented bona fide axle grease. It consisted of common washing soda "mixed with three pounds of clean tallow and six pounds of palm oil." The mixture was heated to 200°F, agitated, and allowed to cool, following which it had the consistency of butter. A competing grease was composed of olive oil and a solution of lime in water, sometimes thickened with carbonaceous matter such as plumbago or soot. These were all brave attempts to provide satisfactory carbon-based lubricants before the advent of petroleum, which followed soon thereafter.

As noted earlier, lubricants were usually fed by hand. Following the appearance of the steam engine and the railroad in the mid-1800s, the need arose for a continuous-lubrication system. This, at first, consisted of a tin box placed atop the bearing with a wick feeding oil to the bearing. In 1835, the aforementioned Henry Booth patented the first grease lubricator. Both of these systems yielded, at best, only partially lubricated bearings. It is true that the machines of the Industrial Revolution—slow and lightly loaded—were not very demanding. Still, it remains a fact that, babbitt bearings excepted, there had been, during the two centuries considered here, no revolution or anything like it in the field of lubricant technology.

1.4   THE SCIENTIFIC ERA (1850–1925)

This era consists of the decades spanning the nineteenth and twentieth centuries that brought about tribology's technological revolution. This holds true for most aspects of this discipline, whether they are the theoretical and experimental discoveries of the basic mechanism of lubrication or the introduction of new lubricants and additives. It is only now that studies of the physical and chemical aspects of interface layers have caught up with the mechanical aspects of bearing technology. The true significance of this new stage in tribology is that, from its primitive role of merely reducing friction, the function of the lubricant was discovered to be that of physically separating the mating surfaces.

1.4.1   THE FORMULATION OF A LUBRICANT FILM

Three men, within a few years and independently of each other, discovered and formulated the mechanism of hydrodynamic lubrication. They were a Russian, N. P. Petrov (1836–1920), and two Britons, B. Tower (1845–1904) and O. Reynolds (1842–1912). All three perceived the process of lubrication as being due to the dynamics of the intervening lubricant. Petrov, whose main interest was, like his predecessors, mainly friction, postulated two cardinal hypotheses: that the important lubricant property is not density, as assumed by his contemporaries, but viscosity; and the lubrication process is due mainly to the shearing of such a viscous substance. Petrov was a true tribologist in that his interests also embraced the properties of materials, on which he wrote some 80 papers during his tenure as professor in St. Petersburg's Technical Institute.

The fundamental discovery of the presence of hydrodynamic pressures in the lubricant is due to Tower, which he discovered serendipitously in the course of experiments on bearing friction. In one of his tests, he had a plug inserted in the hole at the bottom of a journal bearing. During operation, he found that the plug was ejected from the hole, repeatedly so. With keen insight, he surmised that the oil in the bearing must be pressurized. He installed a series of pressure gauges along the bottom of the bearing and, when he integrated the obtained pressures, found them to add up to the external load on the bearing. Thus were the nature of hydrodynamic lubrication and the true role of the lubricant experimentally established. This took place in 1883–1884.

Almost simultaneously, Osborne Reynolds formulated and presented the basic differential equation of hydrodynamic lubrication, which bears his name. This he presented at the Royal Society in London on 1886. One of its basic parameters is the viscosity as well as the requirement that there be a convergent film in the interspace between the two surfaces. This Reynolds equation remains the basic mathematical tool in much of fluid film tribology.

The next important phase was the discovery that the fluid film does not have to be an oil or even a liquid, but can be a gas. This discovery was made by A. Kingsbury (1863–1943), again, serendipitously. As professor of mechanical engineering, in 1892 he had built a torsion-compression machine, which contained a cylinder-piston arrangement. With the cylinder in a vertical position, he one day twirled the piston and found that a minimal effort would make it spin. He tried the same in a horizontal position with the same result, even though there was now a load on the piston. The same thought occurred to him that came to Tower: that a pressurized air film was responsible for the observation. Kingsbury then went on to run special bearings on both air and hydrogen and established the presence of a pressure field in the gas. Thus was the phenomenon of hydrodynamic lubrication extended to compressible fluids and the entrance of gases as a new family of lubricants.

The third major tribological area, next to liquid and gas lubrication, is that of boundary lubrication. A scientific approach to the nature of boundary lubrication did not come about until the beginning of the twentieth century with the original studies of H. B. Hardy in England. He found that, even in vanishingly thin films, there is always present a lubricant layer—as tiny as millionths of a millimeter—due to molecular adsorption to the mating surfaces. In 1920, Hardy postulated that in boundary lubrication, friction depends not only on the lubricant but also on the chemical nature of the solid boundaries. A report prepared by Britain's Department of Scientific and Industrial Research offered a comparison of frictional resistance in various modes of lubrication (Table 1.2).

TABLE 1.2
Values of the Friction Coefficient for Various Modes of Lubrication

Mode of Lubrication	Designation	f
Unlubricated surfaces	Dry friction	0.1–0.4
Partially lubricated surfaces	Greasy friction	0.01–0.1
Fully lubricated surfaces	Viscous friction	0.001–0.01

Source: Dowson, 1998, p. 366.

1.4.2 LUBRICANTS AND MATERIALS

The second most important contributor to tribology's renaissance was the discovery of petroleum and its widespread utilization in everyday life and industry. Although there were scattered instances of oil usage that had seeped to the surface in antiquity, it was not until the 1850s that oil drilling and a petrochemical industry came into being. Its effect on society and industry was profound and immediate, and so too its effect on the development of new lubricants and a host of additives. Within a short time, it completely replaced the vegetable oils and animal fats in usage for the past five millennia.

One of the most remarkable bearing materials used as late as the twentieth century was lignum vitae. This is a hard resinous wood of the family Zygophyllaceae, native to the West Indies and the northern coast of South America, and one of the most durable woods in existence. Its color ranges from greenish-brown to near black, while the sapwood is thin and yellow. The timber's specific gravity is 1.1 to 1.3. The main quality of this wood is, of course, that it is self-lubricating. The wood contains an oily resin, which amounts to 25% of its weight.

Bearings made of this wood were first tried in 1717 as roller pinions by a clockmaker who remarked that it can "move so freely as never to need any Oyl" (3, p. 372). But the wood was not really used extensively until the appearance

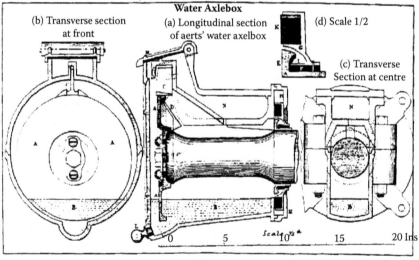

(A) Cast-iron disc dipping into water
(B) Water reservoir
(C) Brass scraper
(D) Spout
(E) Cast-iron collar
(F) Leather collar
(G) Annular cast-iron disc

(H) Back plate
(I) India-rubber band or steel spring
(K) India-rubber ring
(L) Plug
(M) Lid

FIGURE 1.4   Water axlebox.

of steamships, which suffered corrosion problems in the stern tube bearings, whether they were made of iron or brass. The construction of steamship bearings exposed to seawater consisted of brassy cylinders lined with dovetailed lignum vitae staves; thrust bearings had a metal base lined with lignum vitae pads. Not only did these prove to be resistant to seawater and suffer less wear, but they also carried higher loads. Their employment extended into World War II when, in the United States alone, 1000 Liberty and Victory ships were equipped with lignum vitae bearings.

The first continuous-lubrication system appeared at the end of the century when the railroads in England introduced what they called a *water box*. This was to provide a more satisfactory hydrodynamic film. As shown in Figure 1.4, a water box consisted of a disk mounted on the shaft, the lower part of which dipped into a pool of water. A scraper installed on top of the disk transferred the water to the bearing clearance. We, thus, have here not only a forerunner of oil-ring lubrication but the first instance of the use of water as a lubricant.

1.5 THE SPACE AGE (1945–2000)

In the period spanning the two World Wars (1914–1945), tribology can be said to have slipped into the doldrums, perhaps due to the wars. What pulled it out of that moribund state was the appearance of high-speed electronic computers and the advent of space age technology. In the area of petroleum and synthetic oils, process fluids, lubricant additives, and bearing materials, it was a period of full and varied bloom, sprung from the seeds sown during the prewar decades.

1.5.1 LUBRICANT ADDITIVES

- High- and low-temperature oils. Successful strides were made in making petroleum oils more amenable at both very low and very high temperatures. It was found that the addition of polymers led to an increase in viscosity at high temperatures. To prevent lubricants from gelling at low temperatures, polymeric depressants of the pour point are added in tiny amounts, ranging from 0.01% to 0.3%.
- Oxidation inhibitors. Oil oxidation produces acid and corrosive products, which leads to the formation of varnishes and sludge. Antioxidation agents usually consist of sulfur, phosphorous compounds, or of oil soluble amines and phenols in amounts ranging from 0.25% to 5%.
- Detergents. These are oil-soluble, surface-active additives or polymeric dispersants introduced to prevent the formation of sludge or solid combustion deposits. They are added in amounts ranging from 1% to 5%.
- Antiwear compounds. The two kinds of additives added to reduce wear are those for boundary lubrication, which act by chemical adsorption, and extreme pressure (EP) additives based on a number of chemical substances, among which are chlorine, sulfur, and phosphorus.

1.5.2 PROCESS AND SYNTHETIC FLUIDS

- Water. This is used extensively in boiler feed pumps, some marine applications, and nuclear reactors. For water lubrication, nonmetallic bearings are required, such as lignum vitae, resin-bonded materials, or rubber. When high temperatures are encountered, steam lubrication is, in rare cases, resorted to.
- Cryogenic fluids. In liquid oxygen and liquid hydrogen turbopumps, such as those used in space vehicles, rolling element bearings have an extremely short life, sometimes of the order of minutes. Hydrodynamic bearings must often be resorted to, which use the cryogenic fluid as a lubricant.
- Synthetic lubricants. In advanced aviation or space vehicles, the danger of fire or a very low-temperature environment dictates the use of synthetic fuels. These include phosphate and silicate esters, polyglycols, polybutenes, silicons, fluorocarbons, and others. Some of these can lower the pour point to −70°F and are capable of operating in an environment of 500°F.
- Solid lubricants. For extremely high temperatures, solid lubricants with layer-lattice structures are used. The best-known solid lubricants are molybdenum disulfite and graphite. They can be dispersed in liquids, used as loose powders, or impregnated into metal to form sintered bearings.

1.5.3 BEARING MATERIALS

The impetus behind the development of new bearing materials comes from modern machinery's need to run at high speed and high temperatures while in their frequently unusual environments, such as radiation or vacuum. Some of the more recently introduced materials are

- Ceramics and cermet. These materials have good wear resistance, high thermal conductivity, and can operate in temperatures up to 1000°C. Ceramic bearings have been considered for adiabatic internal combustion engines and are in use in some rocket engines.
- Natural and synthetic rubber. They are being used extensively in stern tube bearings on ships, where they are replacing the heretofore used lignum vitae staves.

1.5.4 THE LUBRICANT FILM

It was pointed out previously that, with Hardy's studies, the thickness of the lubricating film has shrunk to almost molecular dimensions. Figure 1.5 shows the progressive shrinking of bearing film thickness with time. From grease layers measured in parts of an inch, tribology has ventured into synovial films on the order of one to several microns and in elastohydrodynamic lubrication (EHL) to fractions of a micron. Eventually, thicknesses of the order of several nanometers were recorded. All this has carried the tribologist in ever more intimate parts of the interface layer, where the lubricant, the surface asperities, and the chemical and molecular activities all participate in whatever the film produces in its final performance. This is the

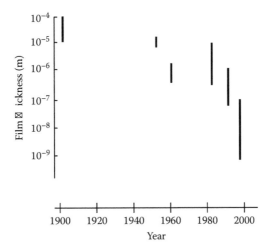

FIGURE 1.5 Variation of bearing film thickness over the last century.

background to the stated objective of this work: that there is a continuum and an overlapping mechanism across the various tribological processes from full lubrication to "dry" contacts.

The perceived interaction between the lubricant and solid surfaces led Godet of France to introduce, in the early 1980s, the notion of a "third body," soon to be taken up vigorously by H. Heshmat in the United States under the category of powder lubrication. Because bearing surfaces always produce wear, the detached particles must somehow participate in transmitting the imposed load. Furthermore, should there be no externally delivered lubricant, these particles, like miniature balls, would act to enable the bearing to somehow carry a load. Regardless of wear, outside debris, and the shedding from oxides or coatings, enough of these particles could undergo shear and flow in a quasi-hydrodynamic fashion, producing a film of powder lubrication.

In view of the foregoing, tribologists had to venture beyond the boundary layer theory propounded by Hardy into the realm of molecular dynamics. Newton's equations of motion are often applied to an immense number of molecules, and such calculations would be inconceivable without the prodigious capacity of supercomputers. Studies showed that the coefficient of friction, for example, for materials such as glass, steel, and bismuth lubricated with either paraffins, alcohols, or acids, shows a dependence on the molecular weight, given by the expression:

$$f = b - aM$$

where M is the molecular weight, a is a function dependent on the lubricant alone, and b is dependent on both the lubricant and the solid surfaces. Figure 1.6 shows a plot of this dependence for the three solids mentioned with alcohols as a lubricant. Work in this area is continuing and will, no doubt, expand into the next millennium.

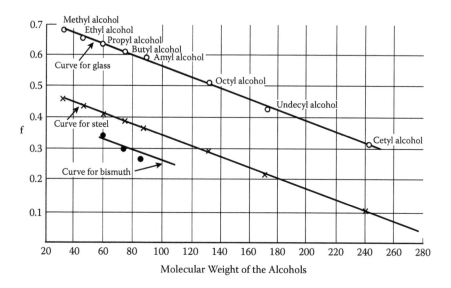

FIGURE 1.6   Effect of molecular weight of alcohol lubricants on static friction in various materials.

1.5.5   MATERIALS

The major advance in materials in the period under discussion was in the field of nonmetallic bearings covering a wide range of thermoplastic, thermosetting, and elastomeric materials. Equally new are trimetal compositions, which may consist of a very thin layer of babbitt on a matrix of steel, brass, copper, or silver, thus combining good surface properties with a rugged base. As mentioned earlier, ceramics and hybrid ceramics, which are highly resistant to heat and wear, are employed in applications where no other materials could possibly serve.

REFERENCES

1. Archibald, F.R., *Men of Lubrication*, ASLE Publication, Chicago, 1957.
2. Cameron, A., *The Principles of Lubrication*, Chapter 11, John Wiley & Sons, New York, 1966.
3. Dowson, D., *History of Tribology*, 2nd ed., Professional Engineering Publication, London, 1998.
4. *Fluid Film Lubrication—A Century of Progress*, ed. Rohde, S.M., ASME Publication 100162, New York, 1983.
5. Harris, T.A. and Wedevan, L.D., Rolling bearing tribology—past, present and future, *Lubrication Engineering*, November 1989.
6. Pinkus, O., The Reynolds centennial: A brief history of hydrodynamic lubrication, *ASME J. Tribology*, Volume 109, New York, January 1987.
7. Reti, L., Leonardo on bearings and gears, *Scientific American*, February 1971.
8. West, G.H. and Rathier, G.I., Some important past and present developments in tribology, *Wear* 20, 1972.

2 The Rheology of Interface Layers

This chapter is concerned with the rheology of a number of substances, an extensive and complex topic. It will be necessary here to restrict the treatment to substances that function as interface layers in tribological devices. On the other hand, some properties that do not strictly qualify as rheological in nature, such as density, will be included because quite often, as in gas lubrication, density plays a prominent role. Thus, while rigor of the presentation may suffer, it should correspond more closely with the theme of this book.

As used in the present volume, an interface layer, or in common parlance, a lubricant, is defined as any substance that does not remain integral to the mating surfaces. This lubricant can be introduced to the interspace either deliberately or can be a by-product of the tribological interaction. An alternate definition would be a layer that has a different chemical composition or a different morphology from that of the rubbing surfaces. The reasons why tribologists over the centuries shied away from such a generalization were twofold: one reason was that the external supply of lubricant constitutes the routine mode of operation; the other reason was that lubrication was mostly the provenance of mechanical engineers, who preferred to deal with mathematical approaches, whereas to generalize the subject to the other modes—dry, boundary, mixed, etc.—would have involved chemical and morphological elements much less amenable to analytical treatment.

2.1 GENERAL CONSIDERATIONS

The general definition of rheology is that it deals with the deformation and flow of materials. A substance is deformed when an applied force alters its shape or size; a body flows when the degree of this deformation changes with time. With simple fluids, the smallest unbalanced force will produce flow, whereas in solids and some other substances, this is not always the case. The science of rheology deals with the relationship between the applied force and the resulting flow or deformation. The quantities that bear on this discipline are as follows:

- Strain rate versus shear rate. This represents the ease with which a substance yields to an applied shear force. Within this category fall such concepts as yield point, limiting shear stress, and relaxation time.
- Viscosity. This is clearly a factor in determining the response to an applied shear stress. It is, to various degrees, a function of the pressure, temperature, rate of shear, and other factors.

- Density. This, too, is a function of pressure and temperature; in the case of powder films and slurries, it also depends on the solids fraction and mode of operation.
- Morphological changes. Whereas, initially, a material may have fixed rheological characteristics, these may not remain constant with time. Exposure to the operational environment may cause substantial changes due to chemical reactions and mechanical intrusions by wear debris from the adjacent surfaces.

2.1.1 CLASSIFICATION

From a rheological standpoint, substances that are used in tribological applications can be divided into several categories. Of direct interest here are viscous fluids, viscoelastic materials, and Bingham plastics. Their place in the spectrum of fluid and solid substances is sketched in Figure 2.1. Their response to an imposed shear stress can be illustrated by a comparison to the mechanical systems shown in Figure 2.2. Following a diagonal from upper left to lower right, one goes from a fluid that can support no shear to some solids that cannot be deformed at all. The model in Figure 2.2a corresponds to a purely viscous fluid, where the rate of dashpot travel

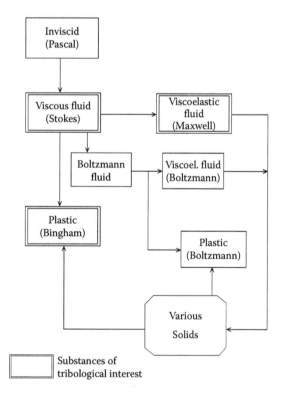

FIGURE 2.1   Classification of rheodynamic substances.

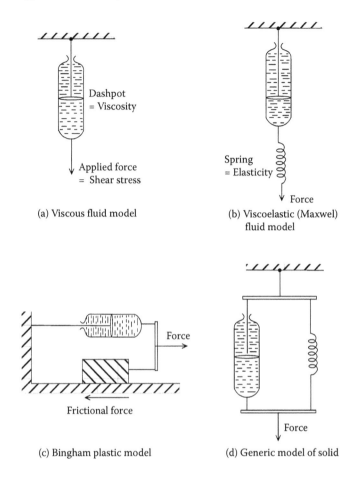

FIGURE 2.2 Mechanical models for rheodynamic substances: (a) Viscous fluid model;
(b) Viscoelastic (Maxwel) fluid model; (c) Bingham plastic model and (d) Generic model of
solid.

is a function of the applied force. In our parlance, this expresses the notion that the
rate of strain (travel) is a function of the shearing stress. In the mechanical model,
the force applied is related only to the rate of dashpot travel and not to the absolute
distance traveled; likewise, in a viscous fluid, the shear stress is related to rate of
shear and not to the total magnitude of the strain.

A viscoelastic fluid is simulated by a spring and dashpot assembly in Figure 2.2b.
Should the model be subjected to a sudden force, the spring will elongate but the
dashpot, unable to respond instantaneously, will not move. With time, the dashpot
will move, and the spring will contract to its equilibrium position, which corresponds
to the relaxation process in a viscoelastic fluid. Should the applied force be suddenly
removed before spring equilibrium is reached, the system will recoil. Such relaxation and recoil do not occur in a purely viscous fluid. Consequently, unsteady state
experiments are required to distinguish viscous from viscoelastic fluids.

Figure 2.2c is a model for fluid exhibiting properties of a Bingham plastic. Here, before the viscous fluid can be sheared, a threshold stress has to be exceeded. This threshold is represented by the static friction under the block, which the applied force must overcome before the dashpot will move. Subsequently, the dashpot will travel at a rate corresponding to the applied force in excess of the static frictional force. This we shall later call the yield stress of the substance. In certain materials, the yield point is clear and definite; in others, such as rubbers or organic plastics, it is vague. According to one source (14), there seems to be a reciprocal relationship between the yield stress and viscosity, as is suggested by the following tabulation:

Substance	Viscosity (Poises)	Yield Point (dynes/cm^2)
Phenol and formaldehyde resins	10^5–10^7	0
Plasticized rubber	10^6	0
Clay suspensions paints	0.1–10	10^3
Oleomargarine	7	510

A Boltzmann fluid, appearing in the diagram of Figure 2.1, is one in which the properties are a function of time. In general, solids can be represented by a spring and dashpot arranged in parallel, as shown in Figure 2.2d. Here, the movement of the system depends only on the extension of the spring, which depends only on the applied force; once equilibrium is reached, no further extension will take place.

These rheological characteristics must be incorporated in the relevant equations of continuity, momentum, and energy to obtain the motion of a finite mass of gas, liquid, or solid.

2.1.2 STABILITY PROBLEMS

In addition to their inherent characteristics, interface layers suffer from the effects of time, prolonged usage, or exposure to the operational environment. These morphological transformations are either chemical in nature or result from mechanical intrusions, both of which degrade the inherent qualities of the product.

- Oxidation. In general, the degradation of lubricants by oxidation starts with the formation of peroxides. From this reaction, secondary processes are set in motion, which may lead to the formation of sludges and varnish-like deposits. The rate of oxidation depends on the prevailing temperature, availability of oxygen, and the kind and purity of the lubricant.
- Metallic debris. The presence of metallic particles has a direct effect on the oxidation process. Thus, while steel will accelerate the rate of oxidation, copper and brass will help to slow it down, and aluminum and lead will have no effect.
- Thermal stability. This is related primarily to the molecular structure of the material. In common lubricants, thermal stability is not a problem

until temperatures in excess of 300°F are encountered. In some cases, high temperatures can lead to the formation of acidic components and the breakdown of the original molecular structure. Gelling or thixotropy, too, can sometimes be attributed to extremes in the thermal environment.

2.2 NEWTONIAN FLUIDS

Newtonian fluids, be they gases or liquids, are characterized by a linear relationship between shear stress and shear strain. The coefficient relating them is the viscosity or

$$\tau = \mu \gamma' \tag{2.1}$$

Although it is a function of neither shear stress nor strain, the viscosity does depend on the thermodynamic state of the fluid, its pressure, and temperature. In thick films, the dependence is primarily on temperature; in thin films, such as in elastohydrodynamic operation or in rolling element bearings, the viscosity is a strong function of pressure. Much of the work in Newtonian fluid rheology is invested in establishing the function $\mu = \mu(p,T)$ that would hold over a sufficiently wide range of pressures and temperatures.

2.2.1 STANDARD CONDITIONS

One theoretical method of finding the $\mu(p, T)$ function is by the free volume molecular approach. According to Reference 15, this results in an expression of the form

$$\mu(p,T_1) = \mu(0,T_1)e^{\alpha p} \tag{2.2}$$

An expression that also includes dependence on the temperature is of the form

$$\mu(p,T) = A_3 \exp\left(\frac{A_4 + B_4}{T^n}\right)(1 + \alpha p^m) \tag{2.3}$$

In the foregoing expressions, the A's, B's, T_1, α, n, and m are constants.
    For engineering purposes, reliable viscosity data are usually culled from extensive experimental programs, such as those conducted by the ASTM and the ASME (1). In the area of Newtonian fluids, these are mostly petroleum oils. Typical viscosity and other physical data are given in Table 2.1. Their variation with temperature is portrayed in Figure 2.3. For most common oils, these data agree with temperatures from above the pour point to about 400°F. The plots in Table 2.1 can be represented by the following equation:

$$\ln\ln(v + 0.6) = k_1 \ln(T) + k_2 \tag{2.4}$$

where the viscosity, v, is in centistokes, T is in °R, and the k's are constants corresponding to a particular fluid.

TABLE 2.1
Properties of Petroleum Oils

Oil type	Viscosity (cS) 100°F	Viscosity (cS) 210°F	Density (g/cc at 60°F)	Flash Point (°F)	Pour Point (°F)
Automotive					
SAE 10W	41	6	0.870	410	−15
20W	71	8.5	0.885	440	−10
30	114	11.3	0.891	460	−5
40	173	14.8	0.890	475	0
50	270	19.7	0.902	490	10
Industrial Gear					
SAE 75	47	7.3	0.900	380	−10
80	69	7.9	0.934	365	−25
90	287	20.4	0.930	450	−10
140	725	34	0.937	500	0
250	1220	47	—	490	5
Turbine					
Light	32	5.4	0.872	410	0
Medium	65	8.2	0.877	455	10
Heavy	99	10.8	0.885	470	10
Hydraulic					
Light	32	4.8	0.887	370	−45
Medium	67	7.3	0.895	405	−15
Heavy	196	14	0.901	495	10
Extra low temperature	14	5.2	0.844	230	−80
Wide temperature	56	10.5	0.871	310	−45
General-purpose industrial	22	3.9	0.881	350	25
	44	6	0.898	390	25
	66	7	0.915	365	10
	110	9.9	0.915	390	5
	200	15.5	0.890	455	15
Aviation	5	1.6	0.858	232	−85
	10	2.5	0.864	295	−80
	76	9.3	0.876	420	0
	111	11.3	0.884	435	0

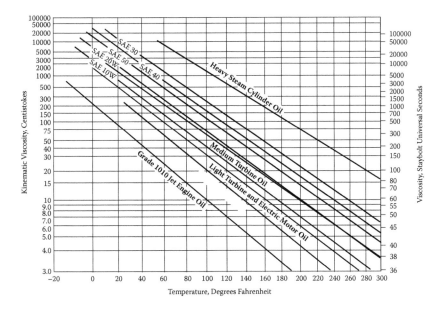

FIGURE 2.3 Temperature–viscosity chart for petroleum oils.

The specific gravity at atmospheric conditions and 60°F ranges from 0.85 to 0.95. At any other temperature, the variation is given by

$$\rho = \frac{\rho_o}{[1+\alpha(T-520)]} \tag{2.5}$$

where ρ_o is the density at 60°F and the coefficient α has the following values (26):

Specific gravity at 60°F	Coefficient $\alpha\left(\dfrac{1}{°F}\right)$
1.03	0.00035
0.92	0.00040
0.81	0.00050

Some nonorganic fluids used in tribology include air, steam, noble gases, water, and some others. Unlike with liquids, the viscosity of air rises with temperature. The dependence of some of these inorganic substances on temperature is given in Figure 2.4.

2.2.2 EXTREME ENVIRONMENTS

While the variation with temperature is most pronounced at standard operating conditions, the dependence of viscosity on pressure does not make itself felt until pressure levels of hundreds of atmospheres are reached. A simple expression for such dependence is given by

$$\mu = \mu_o e^{\beta} \tag{2.6}$$

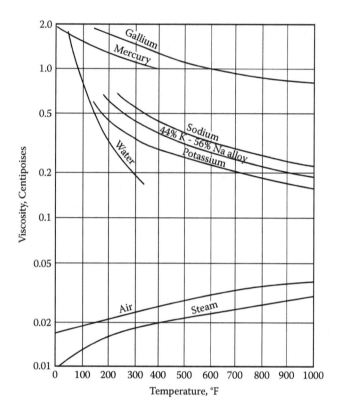

FIGURE 2.4   Temperature–viscosity chart for inorganic substances.

where μ_o is the viscosity at atmospheric conditions and β is a constant that is specific to a given fluid at a particular temperature. Figure 2.5 portrays this relationship for a number of petroleum oils at a temperature range from 32°F to 425°F. As can be seen, higher temperatures have a mollifying effect on the rise of viscosity with pressure.

The density dependence on pressure can be expressed by

$$\rho = \rho_o(1 + mp - np^2) \tag{2.7}$$

where ρ_o is the density at standard conditions, and the constants m and n are given as follows:

Temperature (°F)	m (1/psi)	n (1/psi)
30	4.02×10^{-6}	7.0×10^{-11}
60	4.19	6.4
100	4.38	5.7
140	4.53	5.1
180	4.63	4.7
220	4.68	4.4

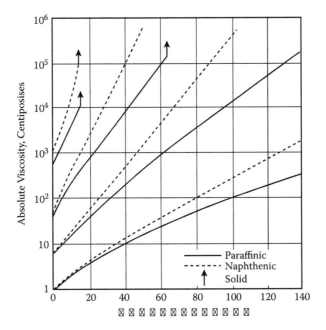

FIGURE 2.5   Pressure–viscosity graph for petroleum oils at constant temperatures.

In many applications it is desirable to have a fluid with a high pressure-viscosity but low temperature-viscosity coefficient. There is, however, a linkage between the two. As shown in Figure 2.6, a rising pressure coefficient implies the same for the temperature coefficient. However, there is considerable scatter in the data, so that a factor of nearly 3 separates the highest pressure coefficient from the lowest. It is, thus, possible, if not to reverse the trend, to at least minimize the adverse relationship.

An attempt is made in Reference 21 to find a mathematical relationship between the two coefficients. It is postulated that, at low shear rates, viscosity is a thermodynamic property; that is, a function of two parameters only. Analogous to the Maxwell equations in thermodynamics, one can write rearranging the terms and defining a parameter:

$$r = \frac{\left(\frac{\partial \ln \mu}{\partial T}\right)_y}{\left(\frac{\partial \ln \mu}{\partial T}\right)_p} \tag{2.8}$$

One obtains from the thermodynamic equivalence:

$$\left(\frac{\partial p}{\partial T}\right)_T = -\left[\frac{\left(\frac{\partial v}{\partial T}\right)_p}{\left(\frac{\partial v}{\partial T}\right)}\right]_T = \left(\frac{Expansivity}{Compressibility}\right)$$

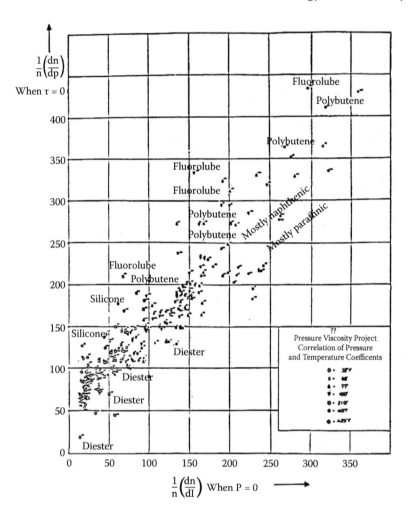

FIGURE 2.6 Relationship between the viscosity pressure and temperature coefficients for Newtonian fluids.

One obtains

$$\left(\frac{\partial \ln\mu}{\partial p}\right)_T = \left(\frac{Compressibility}{Expansivity}\right)(r-1)\left(\frac{\partial \ln\mu}{\partial T}\right)_p \qquad (2.9)$$

The first term in the preceding equation, $\left(\frac{\partial \ln\mu}{\partial p}\right)_T$, is the pressure-viscosity coefficient while $\left(\frac{\partial \ln\mu}{\partial T}\right)_p$ is the temperature-viscosity coefficient. Equation 2.9 provides a relationship between the two in terms of $\left(\frac{\partial T}{\partial p}\right)_v$ and r. The quantity r lies between 0 and 1 because the rate of change of viscosity on heating at constant volume is less than at constant pressure. Typically, r is around 0.5. Because $\left(\frac{\partial \ln\mu}{\partial T}\right)_p$ and $(r-1)$ are both negative, it follows that $\left(\frac{\partial \ln\mu}{\partial p}\right)_T$ must be positive. The value of $\left(\frac{\partial T}{\partial p}\right)_v$ is generally in the

TABLE 2.2

The Temperature-Pressure Gradient at Constant Volume

Fluid	Temperature (°C)	$(\partial T/\partial P)_v \times 10^{-7}$ (°C/Pa)
Hexane	40	11.1
Heptane	40	10.7
2,3,3-Trimethyl butane	40	11.8
Ethyl pentane	40	10.4
Cyclohexane	40	10.5
Squalane	40	8.9
Squalene	40	7.7
Silicone oil 100 cS	40	12.6
Silicone oil 1000 cS	40	11.8
Polyethyl siloxane (Mwt 1900)		12.4
Anisole	40	6.7
Pyridine	60	6.9
Benzene	120	7.4
Toluene	50	7.7
Glycerol		4.2
Triolein		7.4
bis-2-Ethylhexylsebacate		8.4
Santotrac 40	40	8.0
Jojoba oil		8.3
Polybutene (Mwt 450)		7.9
Fluorolube oil		14.7

range of $(6-10)\times10^{-7}$ °C/Pa for most fluids. If both r and $(\frac{\partial T}{\partial p})_v$ are approximately constant, then the pressure-viscosity coefficient will be proportional to the temperature-viscosity coefficient. The potential for reducing r is limited because $(r-1)$ can never be numerically higher than $^{-1}$. This leaves the $(\frac{Compressibility}{Expansivity})$ term as the main factor for improving the ratio of the two coefficients.

The evaluation of $(\frac{\partial T}{\partial p})_v$ requires knowledge of density as a function of p and T. Table 2.2 lists some typical $(\frac{\partial T}{\partial p})_v$ values. These range over $(4-14)\times10^{-7}$ °C/Pa. From Table 2.3, the values of r are seen to be 0.4 to 0.5. Assuming a small range of variation in r and $(\frac{\partial T}{\partial p})_v$, we have

$$\left(\frac{\partial ln\mu}{\partial p}\right)_T = k\left(\frac{\partial ln\mu}{\partial T}\right)_p$$

Using a suitable temperature-viscosity relationship, such as $\mu = Ae^{B(T-C)}$, we can differentiate it to obtain

$$\left(\frac{\partial ln\mu}{\partial T}\right)_p = \frac{-B}{(T-C)^2}$$

TABLE 2.3

Values of the Parameter _r_ for Newtonian Fluids

Fluid	Temperature (°C)	_r_
Santotrac 40	40	0.32
4-Pentyl-(4-cyanophenyl) cyclohexane (a liquid crystal)	80	0.47
Toluene	0	0.34
Pyridine	60	0.23
bis-2-Ethylhexylsebacate	37.8	0.57
	99	0.49
1,1-Di(cycloethylhexyl) nonoane	0	0.54
	37.8	0.50
	95	0.43
Polybutene (Mwt 350)	0	0.46
	37.8	0.42
	99	0.38
Polymethyl siloxane (Mwt 1300)	24	0.28
Polymethyl siloxane (Mwt 6700)	25	0.48

and finally

$$\left(\frac{\partial \ln\mu}{\partial T}\right)_T = kB(T-C)^2 \tag{2.10}$$

For mineral oils, k is approximately 4×10^{-7}, and the constants B and C can be obtained from viscosity-temperature data for a particular fluid.

2.3   VISCOELASTIC SUBSTANCES

It should, of course, be clear that there is a continuous overlap from obviously inviscid fluids to solids devoid of any elastic characteristics. Within our framework of interest, the course we will take is from perfectly viscous fluids, such as water, to oils carrying additives aimed at improving their lubrication properties, and on to synthetic fluids: greases and substances possessing both yield and limiting stress values that impart to them the nature of a Bingham plastic.

To place viscoelastic bodies in their proper perspective, it is helpful to delineate their idealized end states. Work done on a purely viscous fluid subject to shear is immediately dissipated as heat, whereas work done on a purely elastic body is stored and can be recovered by returning the body to its original shape. A viscoelastic fluid exhibits the characteristics of both of the foregoing idealized substances, hence its name. Fluids

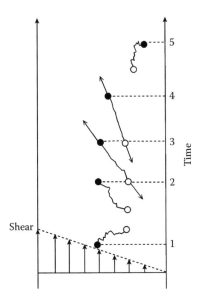

FIGURE 2.7   Model of two polymeric molecules under a shear strain.

exhibiting viscoelastic properties consist of complex long-chain molecules called polymers. These molecular chains are loosely bound or limp and, when stretched, Brownian
motions tend to restore them to their limp state. But this does not happen instantly, and
it takes time before the relaxation is accomplished. Pipkin (17) portrays this notion in
terms of two loosely chained balls, which, as shown in Figure 2.7, are initially free
floating. Under shear, they are stretched and tensed so that they mutually interact until,
with time, they are relaxed again. This kind of mechanism in the intermolecular structure of polymers is the origin of the notion of relaxation time.

While in purely viscous flow, the constitutive equation relating shear stress and
shear strain is the simple $\mu \gamma' = \tau$, the constitutive equation for viscoelastic bodies
must account for its elastic as well as viscous component. Thus, in general, this relationship would appear as follows:

$$\gamma' = \left(\frac{\tau}{G} \right) + \frac{\Phi(\tau)}{u_o} \tag{2.11}$$

where the first right-hand term represents the elastic and the second the viscous element. This expression is generic, and by different investigators, numerous variations
of it have been tried, particularly as to the form of the $\Phi(\tau)$ function.

2.3.1   SHEAR-THINNING FLUIDS

The simplest non-Newtonian fluid is one that undergoes a reduction in viscosity with
a rise in shear strain. Most polymers have this property. In a few, such as concentrated suspensions of very small particles, viscosity rises with shear strain, and these

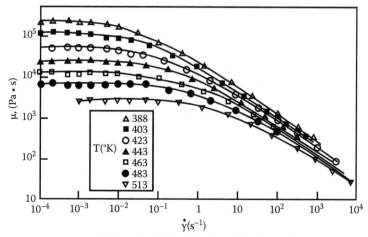

(a) Viscosity of a low density polyethylene melt

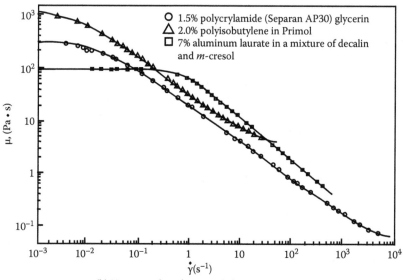

(b) Viscosity of a polymer and aluminum soap solutions

FIGURE 2.8   Viscosity of polymers as a function of shear rate.

are said to be dilatant polymers. Thus, the constitutive equation for shear thinning substances is simply

$$\mu(\gamma')\gamma' = \tau$$

where $\mu(y')$ indicates that the viscosity is a function of the shear strain. Figures 2.8a and 2.8b give the behavior of some shear-thinning substances as a function of the shear strain with temperature as a parameter; the latter is particularly noticeable at low shear rates.

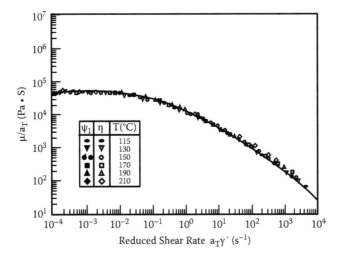

FIGURE 2.9   Normalization of μ-T dependence with parameter $a_T = [\mu_o(T)/\mu_o(T_o)]$.

An adjustment can be made for the effect of temperature by using the ratio $a_T = [\frac{\mu_o(T)}{\mu_o(T_o)}]$, where the numerator denotes the viscosity at very low shear and some reference temperature T_o, while the denominator is the low shear viscosity at any arbitrary temperature. When both abscissa and ordinate are adjusted by that factor, one obtains Figure 2.9, with most points falling on a common curve. When the data are plotted logarithmically, a linear region is obtained at high shear rates with a slope ranging from 0.4 to 0. 9 (4). For a narrow range of molecular weights, the transition from a constant viscosity to linear dependence is fairly narrow; for a wide molecular weight range, the transition region widens. Eventually, at very high rates of shear, the viscosity again becomes independent of shear, forming a second plateau in the $\mu - \gamma'$ relationship. Thus, generically, the behavior of the viscosity curve would look something like that of Figure 2.10.

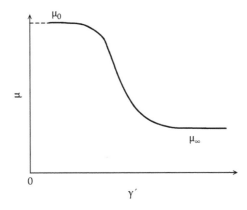

FIGURE 2.10   General μ-γ' curve for shear thinning fluids.

2.3.2 Additives

Organic fluids are usually laced with additives to improve their performance with respect to a number of operational hazards. The most commonly used additives, listed in Table 2.4, are meant to bring about the following protection or improvements:

- Maintain an adequate viscosity at high temperatures—viscosity index improver
- Maintain lubricant fluidity at low temperatures—pour-point depressant
- Improve the load capacity of the lubricant—extreme pressure (EP) additives
- Inhibit oxidation
- Act as anticorrosion agents
- Act as detergents against the formation of sludge, varnish, and other deposits
- Act as defoaming compounds

These additives consist mostly of one sort of polymer or another affecting the nature of the original Newtonian fluid. A brief description of their composition and capacities follows, starting with antioxidation substances, probably the additives' most important function in ordinary lubrication:

1. Oxidation inhibitors. In general, the additives' role is to reduce peroxide formation and break the chain reaction during oxidation. Aromatic amines, phenols, and organic sulfides are typical. These are used in about 0.5% to 2% concentrations. Particularly important is their presence with lead babbitt, zinc, and copper alloys, which are easily corroded by the acids in oxidized oils. For severe service, complex compounds of sulfur and phosphorous are used.
2. Detergents. These are required in heavy-duty machinery, where they are used in concentrations of 2% to 10%. Their role is to keep in suspension or dispersion any insoluble detritus coming either from the outside or

TABLE 2.4
Types of Additives

Function	Typical Chemical Type	Function	Chemical Type
Oxidation inhibitor	Phenolics	Wear preventive	Tricresyl phosphate
	Dithiophosphate	Boundary lubrication	Chlorinated naphthalene
Detergent	Calcium petroleum sulfonate		Sulfurized hydrocarbon
		Viscosity index improver	Polyisobutylene
Rust inhibitor	Organic acids	Pour-point depressant	Polymethacrylate
	Sodium petroleum sulfonate	Defoaming agent	Silicone oil

the mating wear from the mating surfaces. Barium and calcium salts of petroleum sulfomic acids or phenolic compounds are the most common products.

3. Rust inhibitors. These are surface-active compounds, which keep out moisture, such as those that may be present in steam turbines and hydraulic equipment or in the lubricant itself. Long-chain organic acids and their metal salts are used for this purpose. Under more severe conditions, polar organic phosphates, polydric alcohols, and sodium sulfonates are used, which tend to adhere strongly to ferrous substances.

4. Load-carrying additives. For extremely heavy loads and where there is a chance of metal-to-metal contact, sulfur, chlorine, and lead components are introduced. These additives react chemically to form a low-shear strength coating, which prevents welding or excessive wear.

5. Pour-point depressants. When some oils, particularly of the paraffinic family, are taken below the pour point, a waxy material crystallizes, which immobilizes the oil. The cure is to coat the individual wax crystals to prevent their amalgamation. The compounds used to achieve this are high-molecular-weight polymers. With only 1% depressant, the pour point can be lowered by as much as 50°F.

6. Viscosity index improvers. These additives are long-chain polymers with molecular weights in the range of 5,000 to 20,000, containing 350 to 1,400 carbon atoms (versus 20–70 for mineral oils). They are so different from the parent fluid that they are often considered as being a colloidal suspension. Their effectiveness declines at high rates of shear when the large molecules are broken up.

7. Defoaming agents. As an antifoam additive, methyl silicone polymers of viscosities up to 1000 centistokes at 100°F are used but in tiny amounts: a few parts per million. They form isolated droplets of very low surface tension, breaking up the foam bubbles and releasing the trapped air.

2.3.3 POLYMERS

2.3.3.1 The Constitutive Equations

Not only do polymers have complex rheological properties, but under high pressure, they can undergo a change of phase from liquid to solid. Thus, in EHD applications, rolling element bearings, and in gears, the film, though liquid at the inlet, becomes solid in the high-pressure contact zone with a transient regime between the two states. The pressure and temperature relationship at which this transition occurs is shown in Figure 2.11 for a naphthalenic mineral oil. Thus, in a sense, all three modes—viscous, viscoelastic, and plastic—may manifest themselves in the same interface layer. Added to this are the phenomena of yield and limiting shear stresses, which prove to be functions of pressure and temperature. The variation of the elastic shear modulus in a 5P4E polymer fluid is shown in Figure 2.12 and, as can be seen, it rises with temperature. Figure 2.13 shows the variation of viscosity of the same fluid as a function of rate of shear with the value of the limiting shear stress given at 40°C. The limiting shear modulus, $G\infty$, given in Figure 2.12, and the

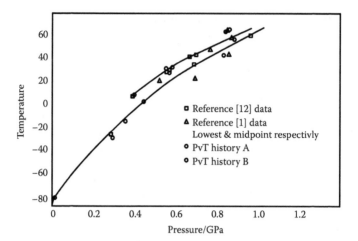

FIGURE 2.11   Transition from liquid to amorphous state in naphthalenic oil.

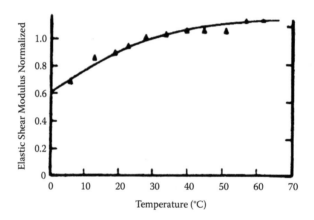

FIGURE 2.12   Elastic shear modulus of 5P4E in amorphous region.

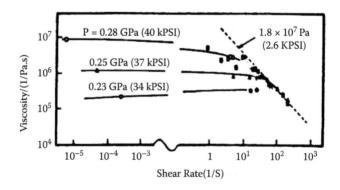

FIGURE 2.13   Viscosity and limiting shear stress of 5P4E at 40°C.

limiting shear stress τ_ℓ are connected via the maximum recoverable elastic shear strain denoted by γ_E or

$$\tau_L = G\infty \gamma_E \tag{2.12}$$

Because, for the fluids shown here, the value of the recoverable elastic strain is small, the limiting elastic shear modulus is one or two orders of magnitude higher than the limiting shear stress.

The standard plots of shear strain versus shear stress are given in the plots of Figure 2.14a and 2.14b for several fluids. These plots could all be fitted on a single

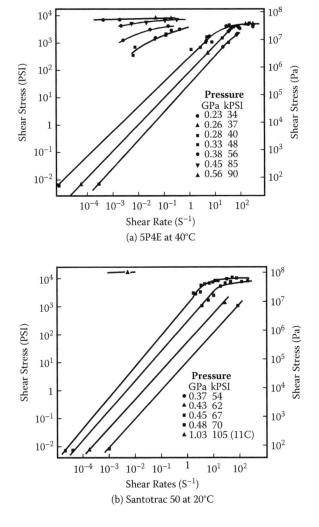

FIGURE 2.14 $\tau - \gamma'$ Curves at various pressures for two polymers.

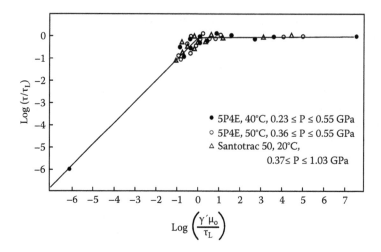

FIGURE 2.15   Normalized relationship of $\tau - \gamma'$.

curve, Figure 2.15, by employing the following normalizations:

$$\tau = \left(\frac{\tau}{\tau_L}\right); \quad \gamma = \left[\frac{\gamma\,\mu_o}{\tau_\ell}\right]$$

Based on the foregoing data, the authors of Reference 5a formulated a constitutive equation for flow. For large strains, the simple form was used:

$$\gamma' = -1n(1-\tau)$$

Introduced as the viscous part in the Maxwell equation, this yielded the expression

$$\gamma' = \left(\frac{1}{G_\infty}\right)\left(\frac{d\tau}{dt}\right) - \left(\frac{\tau_\ell}{\mu_o}\right)1n\left(\frac{1-\tau}{\tau_\ell}\right) \tag{2.13}$$

Three physical quantities are needed to describe the viscoplastic substance: the low shear viscosity μ_o (the first plateau), the limiting shear modulus G_∞, and the limiting shear stress τ_L. As the viscoelastic is distinct from the viscoplastic case, one can let the limiting shear stress become infinitely large, and then Equation 2.13 reduces itself to the familiar form of the Maxwell equation, namely,

$$\gamma' = \left(\frac{1}{G_\infty}\right)\left(\frac{d\tau}{dt}\right) + \left(\frac{\tau}{\mu_o}\right) \tag{2.14}$$

By requiring that the limiting elastic shear modulus become very large, we obtain the nonlinear viscous equation

$$\gamma' = -\left(\frac{\tau_L}{\mu_o}\right)1n\left[1 - \left(\frac{\tau}{\tau_\ell}\right)\right] \tag{2.15}$$

In the foregoing expressions, the viscous term suffers from a lack of symmetry about zero shear stress and is, therefore, unsuitable for analytical treatment. Equation 2.13 can be rewritten in the form of

$$\gamma' = \left(\frac{1}{G_\infty}\right)\left(\frac{d\tau}{dt}\right) + \left(\frac{\tau_\ell}{\mu}\right)\tanh^{-1}\left(\frac{\tau}{\tau_L}\right) \qquad (2.16)$$

and the equivalent of Equation 2.15 then becomes

$$\gamma' = \left(\frac{\tau_\ell}{\mu}\right)\tanh^{-1}\left(\frac{\tau}{\tau_L}\right) \qquad (2.17)$$

2.3.3.2   The Rheological Properties

This section will cite measured rheological properties for a number of polymeric substances listed in Table 2.5, as given in Reference 5a. Data are given for the following relationships:

- Shear stress versus shear strain over a range of pressures and temperatures
- Limiting shear stress at various pressures and temperatures
- Limiting elastic shear moduli at small shear strains
- Viscosity at various temperatures and pressures aimed primarily at obtaining pressure-viscosity coefficients

Except for viscosity, all others are nonequilibrium properties and depend on their history prior to testing. The impact of their history rises the deeper the substances are taken into the amorphous solid state. This can be seen clearly in the diagrams of Figure 2.16. To mitigate this effect, most measurements were taken under conditions of isothermal compression.

The elastic shear modulus for a naphthalenic base oil, shown in Figure 2.17, reached values of 2.5 and 3.2 GPa at the two highest levels of applied pressure. At the lowest pressures, no limiting shear modulus could be noticed. Figure 2.18 carried the experiments into high shear rates, eventually reaching the limiting shear stress. At the low shear rates, Newtonian flow is approached with the curves sloping at about a 45° angle; this slope, therefore, represents the viscosity of the purely viscous base oil.

The next series of plots give the viscosity of a number of polymers. In Figure 2.19, pressure-viscosity isotherms are plotted for polyphenyl ether, while the four diagrams of Figure 2.20 give similar isothermal pressure-viscosity data for several other substances. One interesting feature in Figure 2.20 is that at high pressure at 26°C, the base mineral oils appear to have a higher viscosity than those with a polymer blend. At higher temperatures, no such crossings occur. The viscosity-pressure coefficients resulting from the foregoing series of experiments are given in Table 2.6. The viscosity-temperature coefficients are given in Figure 2.21.

TABLE 2.5
Experimental Fluids Tested in Reference 5a

Symbol:	**R-620-15**
Type	Naphthalenic base oil R-620-15
Properties	Viscosity at 38.7°C, mm²/s, 24.1
	Viscosity at 98.9°C, mm²/s, 3.73
	Viscosity index (ASTM D-2270), −13
	Flash point, °C, 157
	Pour point, °C, −43
	Density at 20°C, kg/m³, 915.7
	Average molecular weight, 305
Symbol:	**R-620-16**
Type	Naphthalenic base oil
Properties	Kinematic viscosity at 38.7°C, mm²/s, 114
	Kinematic viscosity at 98.9°C, mm²/s, 8.1
	Density at 20°C, kg/m³, 930
	Average molecular weight, 357
Symbol:	**PL4520, PL4521, PL4523**
Type	Polyalkylmethacrylate
	(Polymer additive used in solution in R620-15, 4.0% polymer by weight)
	The chemical composition of each is the same. They differ only in molecular weight and are supplied in a carrier oil similar to R620-15.

Properties	PLA4520	PLA4521	PLA4523
Polymer weight	42.6%	36.1%	19.0%
Viscosity	355×10^3	560×10^3	2×10^6

Symbol:	**5P4F**
Type	Five-ring polyphenyl ether
Properties	Viscosity at 38.7°C, mm²/s, 363
	Viscosity at 98.9°C, mm²/s, 13.1
	Flash point, °C, 288
	Pour point, °C, 44
	Density at 22.2°C, kg/m³, 1205
	Density at 37.8°C, kg/m³, 1190
Symbol:	**Santotrac 50**
Type	Synthetic cycloaliphatic hydrocarbon traction fluid
Properties	Viscosity at 38.7°C, mm²/s, 34
	Viscosity at 98.9°C, mm2/s, 5.6
	Density at 37.8°C, kg/m³, 889
	Pour point, °C,−37
	Flash point, °C, 163
	Fire point, °C, 174
	Specific heat at 37.8°C, J/kg*k, 2332
	Additive package includes: antiwear (zinc dialkyl dithiophosphate), oxidation inhibitor, VI improver (polymethacrylate)

(continued)

TABLE 2.5 (CONTINUED)
Experimental Fluids Tested in Reference 5a

Symbol:	**Krytox 143-AB (Lot 10)**
Type	Perfluorinated polyether
Properties	Viscosity at 38.7°C, mm²/s, 96.6
	Viscosity at 98.9°C, mm2/s, 11.5
	Density at 24°C, kg/m³, 1890
	Density at 98.9°C, kg/m³, 1760
	Pour point, °C, –40
	V.I. (ASTM D-2270)116
	Flammability: does not burn
Symbol:	**XRM-177-F**
Type	Synthetic paraffinic hydrocarbon plus antiwear additive
Properties	Viscosity at 38.7°C, mPas, 376
	Viscosity at 98.9°C, mPas, 31.6
	Density at 37.8°C, kg/m³, 838.9
	Pour point, °C < –40

Source: Blair and Winer, *ASME J. Lubrication Technology*, Vol. 101, July 1979; Vol. 104, July 1982.

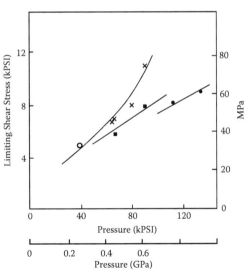

○ High shear viscometer at 38°C [12]
× Isobaric cooling history, measurement at 38°C
■ Isothermal compression history, measurement at 38°C
● Isothermal compression history, measurement at 80°C

FIGURE 2.16 Limiting shear stress for 5P4E with two prior histories.

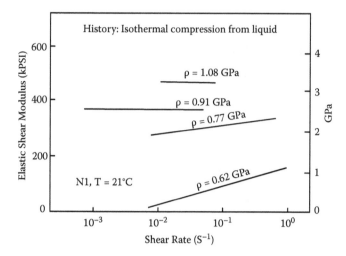

FIGURE 2.17 Elastic shear modulus of naphthalenic oil.

In addition to the polymers discussed earlier, there are a great number of synthetic oils used in tribology. The most common ones are given in Table 2.6, and a brief description of their properties follows:

- Diesters. Most diester oils are formulated from a branched-chain alcohol attached to a dibasic acid. Such fluids have a very high viscosity index, high flash points, and exceptionally low pour points. They, therefore, are found in great use in applications of a temperature range −70°F to 350°F.

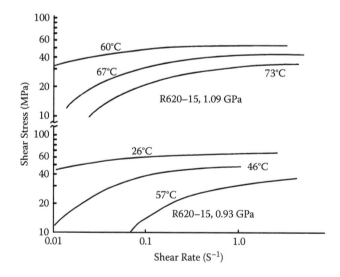

FIGURE 2.18 $\tau - \gamma'$ curves for naphthalenic oil at high shear rates.

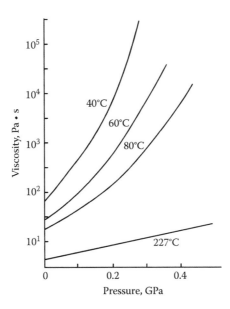

FIGURE 2.19   Pressure-viscosity relationship for 5P4E.

- Polyglycols. These are formed from the polymerization of ethylene and pro-
  pylene oxide to yield polyethylene glycol and polypropylene glycol. Their
  molecular weight runs from 400 with 20 carbon atoms to 3000 with 150
  carbon atoms. They are very volatile and will evaporate after usage instead of
  forming carbonaceous deposits, as happens during the oxidation of mineral
  oils.
- Polybutenes. These are formed by the polymerization of butenes, which
  are the product of petroleum cracking operations. Their utility is also that
  they vaporize, leaving no residue. Their molecules contain 20 to 100 carbon
  atoms and can attain viscosities of 600 centistokes.
- Silicones. These fluids are made up of flexible silicon-oxygen-silicon chains
  from which extend methyl or phenyl hydrocarbon branches. They show little variation in viscosity with temperature, have good thermal stability, and
  are good surface protectors. Many silicones retain their fluidity down to
  temperatures of −100°F.

2.3.4 GREASES

In addition to being widely used as lubricants, greases are of special interest because,
structurally and rheologically, they have much in common with powder films, as
discussed in Section 2.5. Greases cover a variety of materials ranging from some that
can be poured to solids. They all consist of a liquid lubricant base thickened with
solid particles, which may be spherical, elongated, or some other intermediate shape.

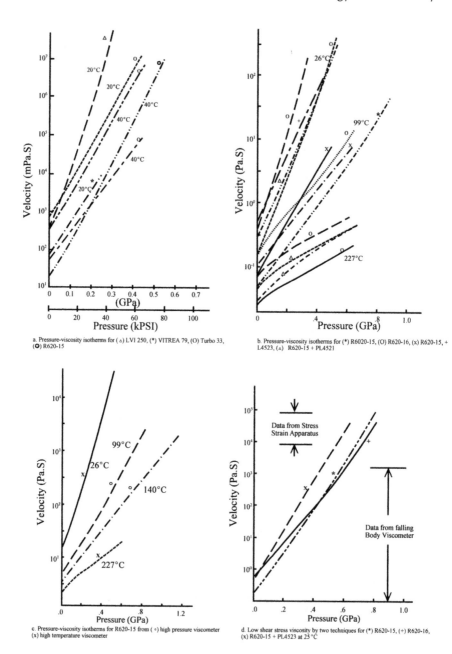

a. Pressure-viscosity isotherms for (△) LVI 250, (*) VITREA 79, (O) Turbo 33, (◉) R620-15

b. Pressure-viscosity isotherms for (*) R6020-15, (O) R620-16, (x) R620-15, + L4523, (△) R620-15 + PL4521

c. Pressure-viscosity isotherms for R620-15 from (○) high pressure viscometer (x) high temperature viscometer

d. Low shear stress viscosity by two techniques for (*) R620-15, (+) R620-16, (x) R620-15 + PL4523 at 25 °C

FIGURE 2.20   Pressure-viscosity plots for various polymers.

TABLE 2.6
Pressure-Viscosity Coefficients

Fluid	T/C	α_{OT}
R620-15	26	27.4
	40	21.9
	99	15.4
	149	10.7
	227	12.0
R620-16	26	35.6
	99	19.8
	227	10.8
R620-15-4523	26	25.5
	99	17.1
	227	16.8
R620-15-4521	26	24.2
	99	15.0
	227	13.8
5P4E	40	40.6
	60	27.6
	80	20.0
	227	6.8
LVI260	30	31.9
VITREA 79	30	22.6
	40	22.6
Turbo 33	30	19.6
	40	16.7

Essentially, the solids serve as a storehouse for the base lubricant, which facilitates their installation and use in machinery.

2.3.4.1 Composition of Greases

Industrial greases can be divided into petroleum base and synthetic greases. The first group consists of the following structures:

Soap Base	Nonsoap Base
Aluminum	Modified clay
Barium	Silica gel
Calcium	Carbon black
Lithium	
Strontium	

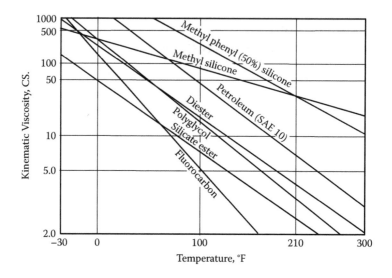

FIGURE 2.21   Viscosity-temperature relation for synthetic oils.

The sodium-type greases (with about 20% calcium) and the lithium soaps are the most commonly used. They can operate in a temperature of up to 250°F with a service life of 10 or more years. The barium strontium and aluminum greases are known for their water-resistant properties and stability under high shear. The use of nonorganic thickeners, such as clay and silica gel, is motivated by their water resistance and non-melting property. The base petroleum oils used are in the range of 45 to 110 centistokes at 100°F. The usual kinds of additives, discussed in Section 2.3.1, are used in generous quantities in grease lubricants. The viscosity of greases at shear rates below 10 (1/s) is close to Newtonian but begins to drop sharply at shear rates of about 1000 (1/s) where, as shown in Figure 2.22, it begins to approach that of the base oil.

A special problem with grease is loss of the liquid carrier: bleeding. Two mechanisms are the cause here. One is the squeezing out of the liquid carrier by the prevailing high pressures in a bearing or gear and the other is the shrinking of the soap structure. A 50%–60% oil loss usually signifies the end of the grease as a lubricant. Another serious problem with greases is their susceptibility to oxidation, yielding gummy materials that solidify upon cooling.

The other family of greases is synthetic products. These are usually designed for temperatures down to −100°F, as well as for elevated temperatures close to 250°F. Some of these synthetics are given in Table 2.7. Silicone greases, in particular, have outstanding bleeding and evaporation characteristics, as well as high thermal stability. Some of these qualities, as well as of the other synthetic greases, as compared to petroleum base products are shown in Figures 2.23 and 2.24. Their lubrication characteristics, however, are not very good.

2.3.4.2   The Rheology of Greases
The rheological complexity of greases is reflected in the difficulties of conducting reliable experiments. In the instrumentation used, there is nonuniformity of flow, edge and

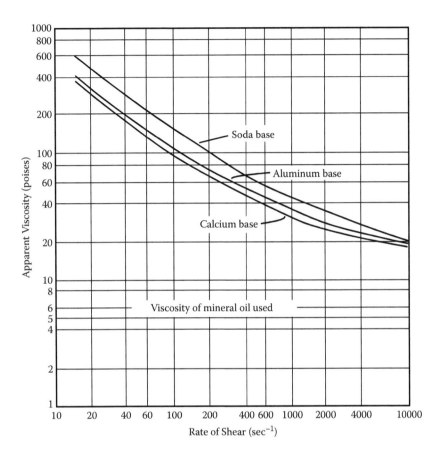

FIGURE 2.22 Viscosity versus rate of shear in greases.

centrifugal effects, and particle migration. Perhaps the biggest problem is the presence of walls. These cause the formation of boundary layers that are more dilute than the bulk of the material or that, in extremis, consist only of the base liquid and the appearance of slip at the walls. In narrow gaps where the boundary layers may extend over their entire width, this is a particularly grave complication in establishing correct bulk properties. Furthermore, the nature of the walls determines the extent of the foregoing phenomena.

In the tests conducted by the authors of Reference 8, grooves were cut in the mating surface, which filled up with the boundary layer, permitting the rest, from the inner diameter of the conduit inward, to be considered the bulk of the grease tested. Compared to an ungrooved surface, Figure 2.25 shows the considerable effect on the frictional torque in the presence of a boundary layer. When it is attempted to establish the value of the yield shear stress with slip and a boundary layer present, the obtained value at the wall may be only a quarter that of the bulk material. These narrow gaps are prone to be more affected by this, as shown in Figure 2.26. In addition, when a base liquid thickens with submicron particles in narrow gaps, the greases have a tendency to crystallize into a substance of excessive consistency.

TABLE 2.7
List of Synthetic Lubricants

Type	Kinematic Viscosity (cS)			Flash Point (°F)	Pour Point (°F)
	210°F	100°F	−65°F		
Diester					
Turbo oil 15	3.6	14.2	12600	430	−90
MIL-0-6085	3.5	13.5	10000	450	−90
MIL-0-6387	4.6	15.8	5000	410	<−80
Phosphate					
Tricresyl phosphate	3.8	30.7	—	465	—
Skydrol	3.85	15.5	>20000	355	−70
Pydraul F-9	5.8	54	—	430	5
Silicone					
SF-96 (40)	16	40	850	600	<−100
SF-96 (300)	122	300	7000	605	<−55
SF-96 (1000)	401	1000	20000	605	<−55
DC-710	40	275	—	575	−10
Silicate					
OS-45	3.95	12.4	2400	—	<−85
Orsil BF-1	2.4	6.8	1400	395	<−100
Polyglycol					
LB-140X	5.7	29.8	—	345	−50
LB-300X	11.0	65.0	—	490	−40
LB-650X	21.9	141	—	490	−20
50-HB-55	2.4	8.9	—	260	−85
50-HB-280X	11.5	60.6	—	500	−35
50-HB-2000	72	433	—	545	−25
Hydrolube 300N	—	66.3	—	None	−55
Chlorinated aromatics					
Aroclor 1248	3.1	48	—	380	20
Aroclor 1254	6.1	470	—	None	50
Polybutenes					
No. 8	7.9	72	—	310	−40
No. 20	106	3600	—	410	10
No. 128	4000	—	—	450	70
Florolubes					
Florolube FS	1.1	3.52	—	None	—
Florolube S	4.6	24.1	—	None	—

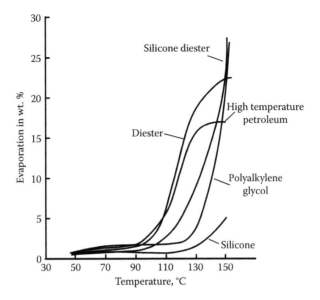

FIGURE 2.23   Evaporation rates for synthetic greases at 1000 h.

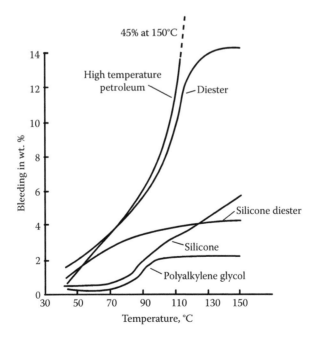

FIGURE 2.24   Bleeding in synthetic greases at 1000 h.

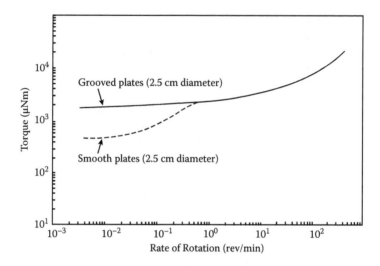

FIGURE 2.25 Effect of grooved and ungrooved surfaces on performance of greases.

Generically, the rheological nature of a grease can be described as follows:

1. The viscosity decreases sharply with increasing shear rate to a point where, at high shears, it equals the viscosity of the base liquid.
2. It exhibits a yield shear stress, below which it will not flow.
3. It is thixotropic.
4. It can be fractured under external mechanical forces.
5. It forms a boundary layer and causes slip at the walls.

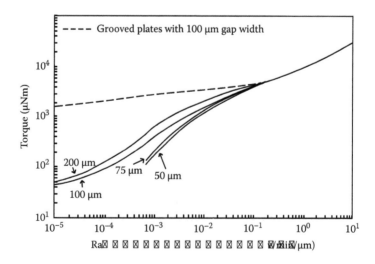

FIGURE 2.26 Gap width effect on extent of boundary layer in greases.

All of these features are also present in powder films, as will be discussed later. As a consequence of the aforementioned features, there are great disparities in the constitutive equations assigned to greases. The most common one is to treat them as a form of Bingham plastic given by

$$\tau = \tau_y + \mu_\infty \gamma' \tag{2.18}$$

This equation ignores the changes in viscosity occurring with a rise in shear rate and, therefore, the following equation has been proposed:

$$\tau = \tau_y + k(\gamma')^n \qquad 0 \le n \le 1 \tag{2.19}$$

where k is the plastic viscosity and a constant, both dependent on the particular grease. Because this form is not symmetrical with respect to a zero shear stress, the authors of Reference 7 propose the following expression:

$$\tau^{\left(\frac{1}{n}\right)} = (\tau_y)^{\left(\frac{1}{n}\right)} + [\mu_\infty \gamma']^{\left(\frac{1}{n}\right)} \tag{2.20}$$

Figure 2.27 shows test results obtained on the calcium grease where the value of n was measured to be close to 2. If it were not for the disturbance at the wall, this line would have intersected the ordinate at a shear value of about 15 (1/s). But because of the presence of a boundary layer and slip, the test points give a yield stress value of only 10. The flow picture at the wall can be represented by the schematic of Figure 2.28. Here, the distribution of the stress can be written as

$$\tau = \tau_y \left[1 - e^{\frac{-(z+d)}{s}} \right]$$

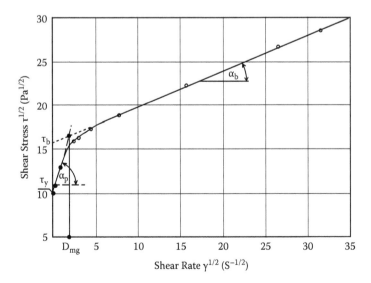

FIGURE 2.27 Apparent and actual wall shear stress due to boundary layer.

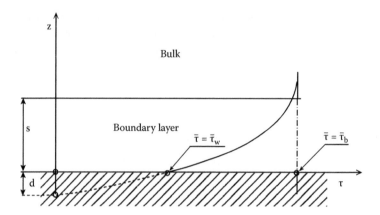

FIGURE 2.28 Geometry of the boundary layer in greases.

with an imaginary line of height d extended into the wall. The yield stress at the wall is given by τ_w, whereas only at the end of the boundary layer thickness s will it reach τ_y: the yield stress of the bulk of the material. Table 2.8 gives results for a number of greases tested, including the effects of the boundary layers and rheological data of interest. Finally, Figure 2.29 shows the effect of temperature and the impact of the boundary layer decreasing at higher temperatures.

2.4   TWO-PHASE SYSTEMS

This section is concerned with nonisotropic substances consisting either of gas–solid or liquid–solid mixtures, the latter commonly referred to as slurries. The progression of the discussion is from very low solid-to-fluid ratios, due either to the extremely small size of the suspensions or low solid concentrations, to ever-larger particles and higher solids fractions. Ultimately, when discussing powder films, the carrier fluid would be so dominated by the suspended solids as to play a negligible role in the dynamics of the system.

2.4.1   COLLOIDS AND SLURRIES

2.4.1.1   Theoretical Expressions

In dilute solutions, the solvent viscosity is the dominant factor. Defining a relative viscosity by the ratio $\left(\frac{\mu}{\mu_L}\right)$, this can be expanded into a Taylor series:

$$\mu_R = \left(\frac{\mu}{\mu_L}\right) = 1 + k_1\phi + k_2[k_1\phi]^2 + \cdots$$

where $\phi = \left(\frac{Volume\,of\,Solids}{Total\,Volume}\right)$ is the solids fraction and the factor k_1 is called the intrinsic viscosity. For dilute solutions, the ϕ^2 term becomes very small and the expression reduces to

$$\mu_R = 1 + k_1\phi \qquad\qquad (2.21)$$

TABLE 2.8

Rheological Properties of a Number of Greases

Grease		Thickener	Value of Parameter n	Lower Limit For Viscosity μ_∞	Yield Stress in the Bulk of the Grease τ_{yb}	Yield stress at the wall τ_{yw}	Parameter For the Boundary Layer Thickness S	Parameter d	Base Oil Viscosity μ_L	$\dfrac{\mu_\infty}{\mu_L}$
					Pa × s	Pa	Pa	μm	μm	Pa × s
R1		Li	2	0.55	260	38	9.5	1.5	0.2	2.75
EP2	Alvania	Li	2.4	1	543	90	5	0.6	0.7	1.43
R3		Li	3	0.2	1367	200	3	0.6	0.2	1
Lt4-S2		Li	4.5	0.04	380	120	5	1.9	0.3	0.13
STP		Ca	2	0.405	216	32	2.7	0.4	0.08	5.06
Lt-12		Ca	1.8	0.82	580	60	1.56	0.16	0.09	9.11
M-2		Ca	2.5	0.36	142	17	1.3	0.2	0.08	4.5
Lt-23		Na-Ca	5	0.008	880	220	0.06	0.016	0.09	0.09

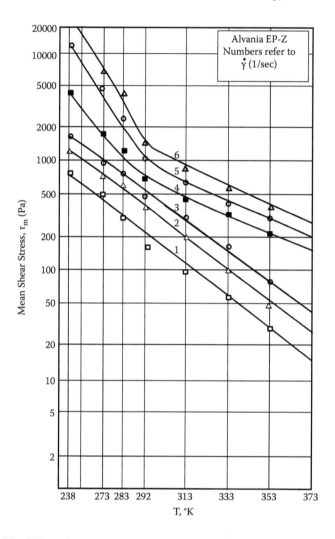

FIGURE 2.29 Effect of temperature on extent of boundary layer.

Albert Einstein, in his study of Brownian motion in 1906 (9), was the first to treat the foregoing expression offering the intrinsic viscosity value $k_1 = 2.5$. The main restriction Einstein put on Equation 2.21 was that it be a dilute solution; that is, $\phi \ll 1$. This was followed by efforts to extend the applicability of this expression, of which the work of Vand in 1948 (25) is perhaps the most comprehensive. The work involves the following steps:

- For solutions of concentrations up to 8%

$$\mu_R = \left(\frac{1 + 0.5\phi - 0.5\phi^2}{1 - 2\phi - 9.6\phi^2} \right)$$

(2.22)

From Einstein's formula, μ_R at 8% is 1.2; from Equation 2.22, it equals 1.32.
- No limitation on concentration—Arrhenius' formula

$$\mu_R = e^{2.5\phi} \tag{2.23}$$

By this formula, the relative viscosity at 8% concentration is 1.22.
- Effect of particle interaction. Due to the collisions between particles, Equation 2.23 is modified to

$$\mu_R = \frac{e^{2.5\phi}}{(1-k^2\phi)} \tag{2.24}$$

When the foregoing factors are combined, the resulting equation for the relative viscosity of a solid suspension in a liquid is as follows:

$$\ln \mu_R = \frac{(2.5\phi + 2.7\phi^2)}{(1-0.61\phi)} \tag{2.25}$$

Vand conducted experiments on a saturated solution of zinc iodine in a carrier of water and glycerol, whose viscosity was 80 cps. Figure 2.30 compares the relative viscosity

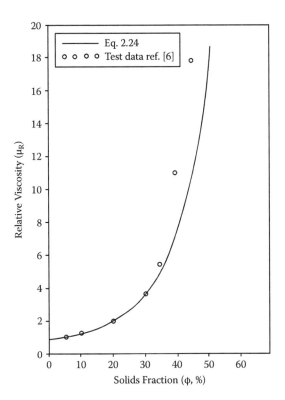

FIGURE 2.30   Viscosity of slurries.

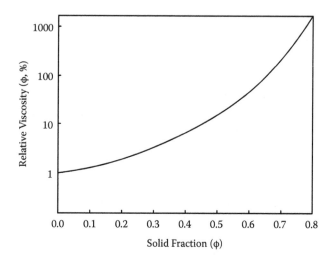

FIGURE 2.31   Relative viscosity of slurry according to Equation 2.24.

obtained from Equation 2.25 and from the experimental data. As can be seen, the theoretical values hold up well up to concentrations of 40%–50%. The relative viscosity, calculated from Equation 2.25, is given in Figure 2.31 up to solid concentrations of 80%. If the carrier fluid is water, whose viscosity at room temperature is about 1 cp, then the ordinate on the figure provides the actual solution viscosity in centipoises.

In many applications, the suspended particles are coated with antiflocculation or dispersant substances to prevent coagulation. When the solids are exceedingly small, the thickness of these coatings constitutes a measurable fraction of particle diameter, and they must be taken into account. The solid fraction then becomes

$$\phi\left(1+\frac{\delta}{a}\right)^{3} \qquad (2.26)$$

where δ is the thickness of the surfactant layer. Thus, the foregoing parentheses contain the increase of particle volume due to the added coating. Equation 2.25 is, thus, still valid with the solid fraction ϕ multiplied by the added volume $(1+\frac{\delta}{a})^{3}$.

All the foregoing expressions are represented as a function of the solids fraction alone. More generally, viscosity is a function of two parameters: the solids fraction and the Peclet number given by

$$Pe = \frac{\gamma'a^{2}}{D_{o}} = \frac{6\pi\mu_{L}a^{3}\gamma'}{k_{B}T}$$

where D_{o} is the Einstein diffusion coefficient and $k_{B}T$ the thermal energy. The Peclet number gives the ratio between the shear and Brownian motion, the latter tending to restore equilibrium to the fluid disturbed by shear. It is, thus, a measure of disequilibrium in the fluid. In general, $10^{-4} \leq Pe \leq 0.2$ can be considered the low-shear region; Peclet numbers in the range $10^{-2} \leq Pe \leq 10$ bring one into the high-shear domain. There are also differences in viscosity arising from different shapes of the suspended solids:

1. Spherical particles. An expression for the viscosity of spherical particles, which takes into account the rate of shear (6), is given by

$$\mu_R = 1 + 2.5\phi + k_2\phi^2 \qquad (2.27)$$

where k_2 is a function of the shear rate. For low shear rates, where Brownian motion dominates, $k_2 = 6.2$ at higher shear rates $k_2 = 5.2$. The foregoing equation, too, is restricted to relatively low concentrations. An empirical expression for high concentration is given by

$$\mu_R = \left[1 - \left(\frac{\phi}{\phi_{max}}\right)\right]^{-k_3\phi_{max}} \qquad (2.28)$$

where ϕ_{max} is the maximum packing or maximum solid fraction, and the value of the exponent is close to 2.
2. Elliptical shapes and slender rods. The viscosity of a suspension of prolate spheroids of semi-axes a, b, and c in the x, y, and z directions in which $a \rangle b = c$ can be written in terms of the solids fraction

$$\mu_R = 1 + k_1\phi \qquad (2.29)$$

where $\phi = [(\frac{4}{3})\pi ab^2\rho]$, with ρ the number density given by $(\frac{N}{v})$. Onsager gives the value of k_1 as

$$k_1 = \frac{\left(\frac{4}{15}\right)\left(\frac{a}{b}\right)^2}{ln\left(\frac{a}{b}\right)} \qquad (2.29a)$$

Another value for k_1, with $e = (\frac{a}{b}) \rangle 1$, was derived in the form of

$$k_1 = e^2\left[\frac{1}{15}\left(ln2e - \frac{3}{2}\right) + \frac{1}{5}\left(\ell n2e - \frac{1}{2}\right)\right] + C \qquad (2.29b)$$

with C given by two different investigators as either $\frac{14}{15}$ or 1.6. The first term in the foregoing expression represents the hydrodynamic contribution and the second, that of Brownian motion, which for large $(\frac{a}{b})$ values is three times as large as the hydrodynamic effect. Generally, when $(\frac{a}{b})$ is large, the precise shape of the particles becomes irrelevant.

A further refinement of the viscosity of slender bodies is given here, which includes second-order effects of the solids fraction, namely

$$\mu_R = 1 + k_1\phi + \left(\frac{2}{5}\right)k_1^2\phi^2 \qquad (2.30)$$

with k_1 as given in Equation 2.29a. The inclusion of the second order in ϕ extends its applicability to higher concentrations but, still, all of the foregoing relationships are restricted to solutions in which the distance between the particles is much larger than their length. When this distance is reduced, as happens in high concentrations, interaction between the particles becomes important. For this case, one of the expressions given is

$$\mu_R = \left(\frac{32}{15\pi^2} \right) \left[\left(\frac{e^6}{\beta \ln e} \right) \phi^3 (1 - ae\phi)^{-2} \right]$$

where β is of the order $O(10^3)$ while a is of the order $O(10^{-1})$.

3. Disks. Similar to rodlike particles, which can be modeled by prolate ellipsoids, disks can be represented by oblate ellipsoids of revolution. For these shapes, the two values offered for k_1 are

$$k_1 = \frac{4}{9} + \frac{32}{15\pi e} \tag{2.31a}$$

$$k_1 = 2.5 + \left(\frac{32}{15\pi} \right) \left[\frac{(1-e)}{e} \right] - 0.625 \left[\frac{(1-e)}{(1-0.075e)} \right] \tag{2.31b}$$

where $e = \left(\frac{a}{b} \right)$ is less than one. Rodlike particles with a high aspect ratio $\left(\frac{a}{b} \right)$ have a much higher intrinsic viscosity (k_1) than disklike particles with a high $\left(\frac{a}{b} \right)$ ratio, the difference sometimes being close to an order of magnitude.

2.4.1.2   Experimental Results

The authors of Reference 20 conducted tests on the magnetic fluid to obtain a relationship between the parameter

$$\theta = \frac{\delta_s [\delta - \delta_L]}{(\delta_s - \delta_L)}$$

and the viscosity temperature variation. The results, given in Figure 2.32, were compared to the Andrade equation (2)

$$\mu_R = Ae^{\frac{E}{RT}}$$

where the quantity E is the activation energy of the flow. Using the experimental data of Figure 2.31, the obtained value of E was 3.6 kcal/mol for solid concentrations below 45%. At higher concentrations, the value of E starts rising, indicating that the solution has lost its Newtonian character. Another series of tests, shown in Figures 2.33 and 2.34, succeeded in demonstrating two crucial aspects of slurries. The first shows at what solids fraction the strain-stress dependence becomes nonlinear, which happens here at some 53%. The second figure shows that the non-Newtonian fluid contains both static and dynamic viscosities. This latter viscosity is a function of the shear rate and shows the elastic component G inherent in its constitutive equation.

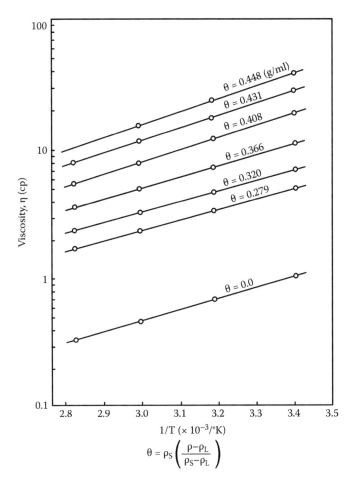

FIGURE 2.32 Viscosity of magnetic fluid as a function of $(1/T)$.

A comprehensive set of experiments, relating to the previously formulated theoretical expressions, was conducted by Bergstron (5b) on a silicon nitride powder stabilized with a surfactant 0.01 microns thick. The particle sizes ranged from 0.2 to 2.2 microns. The carrier fluid was decahydro-naphthalene, known as Decalin, whose viscosity at 25°C was 3 MPa-s. The results were found to depend on particle size and shape, the solids fraction, and the generated interparticle forces. While it was realized that, in its dependence on shear, the slurry would have plateaus at both the low- and high-shear regions, it was not possible to obtain low enough shears to generate the first plateau. Therefore, the equation against which the results were checked was a modified Cross equation, which, in its complete form, is as follows:

$$\frac{(\mu - \mu_\infty)}{(\mu_o - \mu_\infty)} = \frac{1}{[1 + b(\gamma')^c]} \tag{2.32}$$

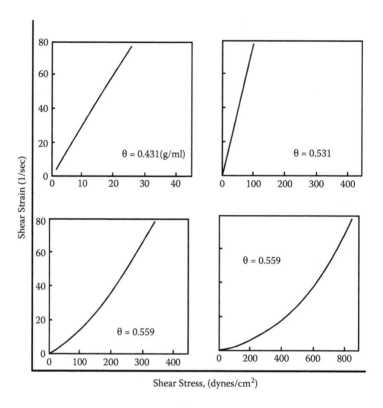

FIGURE 2.33   Stress-strain curves in a 50 Å magnetic slurry.

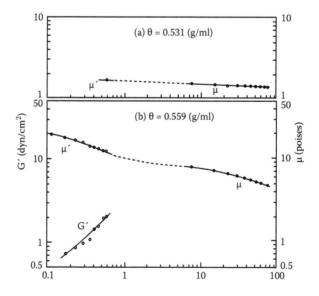

FIGURE 2.34   Static and dynamic rheodynamic components in a magnetic slurry.

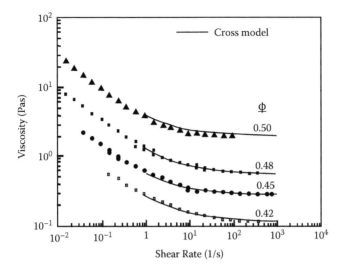

FIGURE 2.35 Viscosity of LC 10 silicone nitrite–Decalin slurry.

where b and c are constants. For high shear rates, this equation can be reduced to

$$\mu = \mu_\infty - \left[\frac{(\mu_0 - \mu_\infty)}{b\left(\frac{du}{dy}\right)^p} \right] \tag{2.33}$$

Figure 2.35 shows the experimental data for the $\mu - \gamma'$ relationship fitted against the preceding equation with a value of $c = 0.6$, as obtained from all the solids fractions tested. The relationship between the viscosity and the solids fraction is shown in Figure 2.36. Here, an empirical relationship was formulated, namely

$$\mu_R = \left[1 - \left(\frac{\phi}{\phi_m} \right) \right]^{-n} \tag{2.34}$$

where $\phi_m = 0.54$ and $n = 2.4$. The factor m denotes the solids fraction at which the viscosity tends to rise to infinity, and the exponent n is an index of particle concentration. The behavior of the test results plotted against Equation 2.34 is shown in Figure 2.36. No viscoelastic effects were observed at concentrations below 45%.

In considering the viscoelastic nature of slurries, it is often postulated that the two interacting forces are the van de Waals attraction forces, which make the particles aggregate in clumps, opposed by the repulsion forces of the electrostatic charges, imparting to the slurry a more Newtonian flow. One set of experiments with coal and sand slurries, reported in Reference 24, shows the rheological dependence on both the Peclet number and shear rate. The characteristics of the slurries are given in Table 2.9, while the test results are shown in Figures 2.37 and 2.38. The dependence

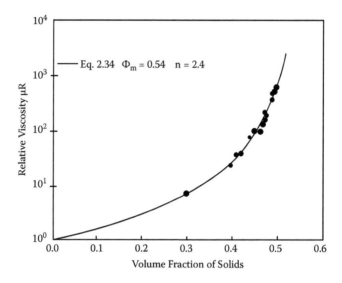

FIGURE 2.36 Viscosity of the silicon nitrite–Decalin slurry versus solid fraction.

of viscosity on the shear rate is here expressed in the form of

$$\mu = K(\gamma')^{n-1} \qquad (2.35)$$

with the values of n listed in Table 2.9.

A number of experiments on relatively high concentrations are reported by Frith and Mewis (10). The particle sizes on the low range were of the order of 0.1 microns and on the high side were about 10 microns. The two-phase system consisted of poly-methyl-methacrylate particles dispersed in a hydrocarbon carrier. A surfactant

TABLE 2.9
Properties of a Water-Based Coal Slurry

	Slurry Index n	Coal vol. Fraction ϕ	Suspending Liquid	
			Concentration Wt%	Viscosity cP
A	0.54	0.47	40% glycerol	3.6
B	0.52	0.43	42% glycerol	3.0
C	0.82	0.40	40% glycerol	3.6
D	0.70	0.45	water/disp[a]	1.0
E	0.94	0.46	water/disp[a]	1.0

[a] Dispersant (wt%) concentrations of 0.32% and 0.49% anionic polymer by wt. of coal for slurries D and E, respectively.

Source: Tsai, 1991.

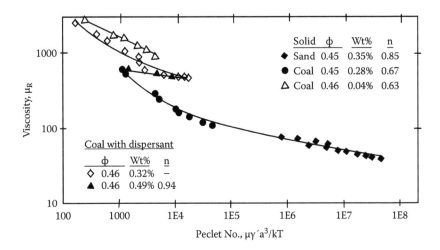

FIGURE 2.37 Viscosities of coal and sand slurries as a function of Peclet number.

was used to reduce the tendency to flocculate, whose thickness, relative to the particle, varied from 0.0016 to 0.2. The shearing of the system was done in a 1 mm (40 mil) wide gap. An attempt was made to obtain a reduced expression in which the variation of viscosity would embrace the effects of particle size, temperature, and concentration—at least in the high range. In general, as has been pointed out before, the viscosity curve showed two plateaus with a viscosity decrease at intermediate shear rates. This reduced expression is as follows:

$$\mu_R = \mu_\infty + \frac{(\mu_o - \mu_\infty)}{[1 + (b\sigma)^m]} \tag{2.36}$$

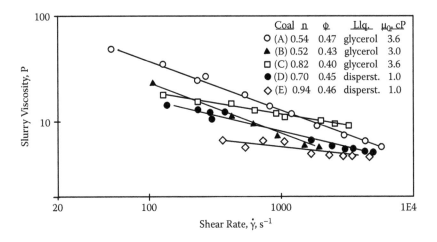

FIGURE 2.38 Viscosity of coal–water slurry as a function of rate of shear.

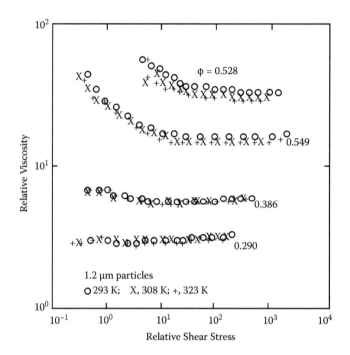

FIGURE 2.39 Dilatancy effects at low solid fractions.

where $\sigma = (\frac{\mu\gamma'a^3}{kT})$, the exponent m is close to unity, and b is some function of the concentration.

Test results, plotted against the relative viscosity given in Equation 2.36, are given in Figures 2.39 and 2.40. As can be seen in Figure 2.39, at low concentrations there is a dilatancy effect, which deviates strongly from the generic pattern suggested by Equation 2.36. In Figure 2.40 there is evidence of a slight hysteresis effect as the shear is first increased and then decreased.

The previous experiments were all conducted for a nominally uniform particle size. The data taken from Reference 22 are from tests on a wide size distribution, which is the more common practical case. The size distribution—given in Figure 2.41—ranges from 1 to nearly 1000 microns with a 50% accumulation below 10 microns. The liquid used was a mixture of perchloroethylene and aliphatic naphthas of a specific gravity—2.64—for quartz particles and 1.36 for the coal, which gave the fluid the same density as the solids, thus preventing sedimentation. The solids fraction ranged from 30% to 60%.

The first simple tests on torque versus speed, which relates to shear rates, immediately distinguished between the pseudoplastic and the Bingham plastic regimes. The data in Figure 2.42 corroborate the results of Figure 2.43 as to the boundaries between the two regimes, which here occurred at shear stresses of 20 to 40 (1/s). The dependence of yield stress on concentration is shown in Figures 2.43a and 2.43b, and levels of slurry viscosity are given in Figure 2.44.

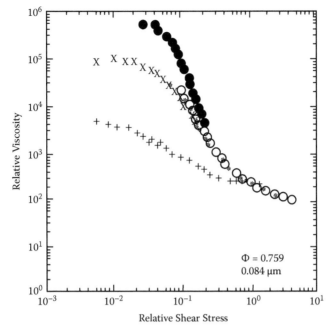

●, 293 K; ○, 293 K after cooling from 323 K; X, 308 K; +, equilibrium at 323 K; *, after high shear.

FIGURE 2.40   Shear hysteresis and temperature effects in an Exxsol slurry.

A constitutive equation offered for such a wide particle size distribution is given as

$$\frac{\tau}{\mu} = \frac{\tau_y}{\mu} + \left[\left(\frac{\tau_B}{\mu}\right) + \left(\frac{\mu_p}{\mu}\right)\gamma' - \left(\frac{\tau_y}{\mu}\right)\right]Q \qquad (2.37)$$

In the preceding equation, the subscripts B and P refer to the Bingham and plastic values of the yield shear stress, and Q, shown in Figure 2.45, accounts for behavior at low shear rates. At high shear rates, Q becomes 1 and, as shown in Equation 2.37, reduces to that of a Bingham plastic.

2.4.2 BIO-RHEODYNAMICS

The human body contains a number of humors—various fluidlike substances—in the organs and tissues. These include blood, the composition of which is shown in the diagram of Figure 2.46; synovial fluids of the joints, forming the ostheoarthrosis: the linkage between the adjacent bones; and fluids in the lymphatic system, which contain a small number of red cells and of tissue fluids, free of any cellular suspensions. Whole blood, a total of 5.2 L, constitutes about 8% of the body weight and has a specific gravity of 1.057. It is about 3 to 6 times as viscous as water. When all the

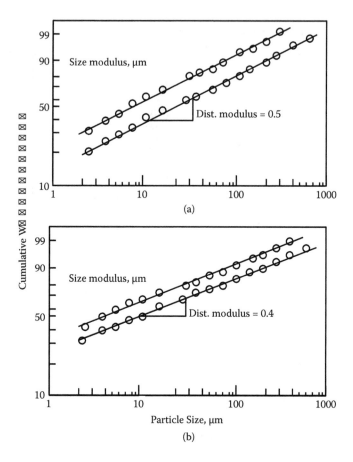

FIGURE 2.41   Particle size distribution for experiments in Reference 22. (From Tangsathitkuchai and Austin, 1988.)

hematocytes are filtered out, it leaves an aqueous saline solution with a viscosity of 1.5 to 2 times that of water and a specific gravity of 1.035. Some 6.5% of its weight consists of proteins called fibrogens. Removal of these leaves a serum, which is a pure fluid with a density of 1.016 and a viscosity 50% higher than water.

The tribological role of blood appears in the course of its flow in narrow passages or more prominently in various prosthetic organs, such as synthetic valves, joint caps, vascular assist devices (pumps), and other artificial parts. In complex surgical procedures, as when installing bypass arteries or in open heart surgery, the blood is passed externally through a lung-heart machine, where it is exposed to shears and stresses inherent to mechanical devices. In the case of synovial fluid, its function as a lubricant emerges in its most classical and sophisticated role.

It is of relevance to note that Poiseuille (18) took an interest in the fluid dynamics of human arteries by testing, in vitro, the flow of blood in capillaries. It was, in fact, he who, in 1847, first noted that blood does not obey his formula of viscous flow in tubes. It is only blood plasma or serum (see Figure 2.46) that comes close to Newtonian

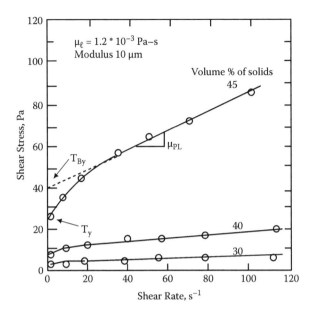

FIGURE 2.42 $\tau - \gamma'$ Relationship for coal slurries as a function of solids fraction.

behavior but not whole blood, which contains a complex suspensions of red and white cells, platelets, proteins, and fibrogens. In addition, blood is unstable because it is prone to hemolysis; that is, separation of the hematocytes from the solution, degeneration of the cells due to shear, and the effects of time exposure when it tends to clot and gel.

As was discovered by Poiseuille, rheologically, blood is a non-Newtonian substance. It is not feasible to run experiments on blood the way it is possible with other fluids. In the first place, human blood is not available in sufficient quantities. In addition, whole blood is opaque and unstable; it is also prone to hemolysis and tends to clot. It is, therefore, necessary to find analogs having similar viscoelastic properties. In Reference 23, three substances were examined as possible blood analogs: an aqueous solution of hydrolyzed polyacrylamide, a polysaccharide Xanthan gum, and a suspension of fragments of a Xantham gum gel. For reference purposes, a sample of human blood was first tested under steady and unsteady conditions with the results shown in Figure 2.47. Under nonsteady conditions, the expected results are obtained: a high elastic component at high shear rates. The results for the three analog substances are shown in Figure 2.48a, 2.48b, and 2.48c. On a purely comparative basis, the fragmented Xantham gum gels give the closest approximation to whole blood, but equivalence could not be obtained.

Similar tests were performed by the authors of Reference 16 using two of the same analog substances, as well as bovine blood. First, Newtonian flow comparisons were made and the results correlated with the familiar power law expression

$$\mu = m(\gamma')^{n-1}$$

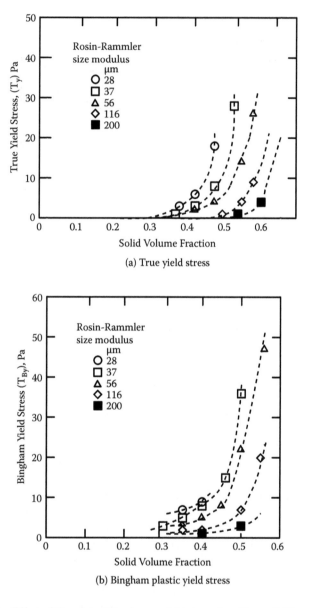

FIGURE 2.43 Effect of Φ and particle size on yield stress (see Figure 2.42).

The values of the constants *m* and *n* are given in Table 2.10, where the two analog fluids are seen to have values close to human blood and closer than bovine blood. This, however, was obtained in steady-state tests. In the oscillatory tests, shown in Figure 2.49, when both viscous and elastic components were present, the results corroborated the previous conclusions. While the viscous components for blood and the two analog substances were close enough, the elastic components were quite

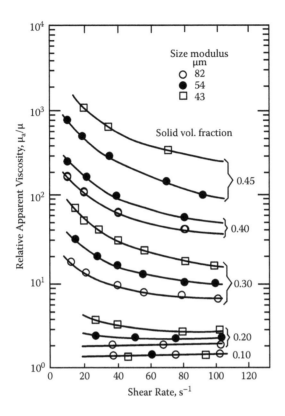

FIGURE 2.44   Viscosity versus solids fraction and particle size in quartz slurry.

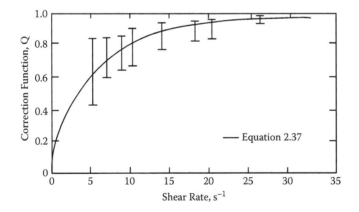

FIGURE 2.45   Value of Q at low shear rates.

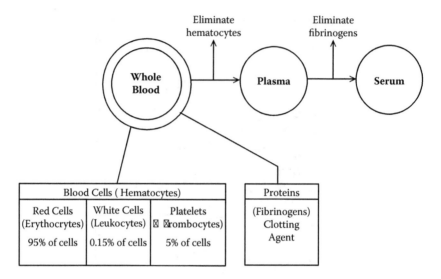

FIGURE 2.46   Composition of human blood.

different. For example, at a shear rate of 100 (1/s), the polyacrylamide solution was 10 times as high as human blood and 3 times that of the Xantham fluid. In all cases, the analog fluids are much too elastic to simulate human blood, though it is not clear whether this is the sole reason for the discrepancy.

Synovial fluids residing at the ostheoarthrosis of mammalian joints are probably the most exotic of lubricants, whether produced by nature or synthesized by humans.

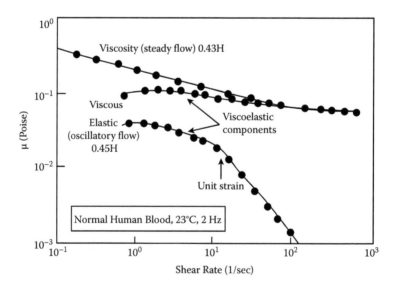

FIGURE 2.47   Rheological characteristics of human blood.

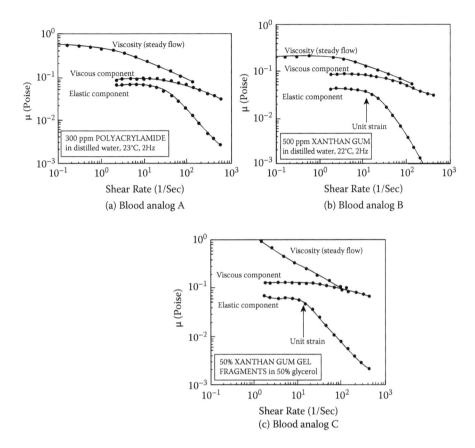

FIGURE 2.48 Rheological properties of blood analogs.

TABLE 2.10

Viscosity Power Law Constants for Blood Analog Fluids at Room Temperature

Fluid	m (dynes s/cm²)	n
0.0375% xanthan gum + 0.02% NaCl	0.118	0.732
0.020% Separan AP30 + 0.02% NaCl	0.123	0.753
Bovine blood (H = 30%)	0.04–0.07	0.86–0.92
Human blood (H = 43.2%, T = 37°C)	0.135	0.784

Note: (m,n) are the constraints in the correlation $\mu = mS^{n-1}$.

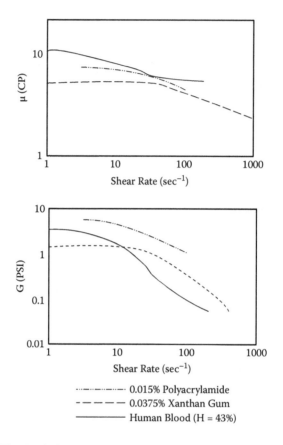

FIGURE 2.49 The rheological characteristics of human blood and of two analogs.

Many of its functions are still not fully known, much less understood. A synovial fluid acts not only as a lubricant but also as a provider of still unknown biological, nutritional, and therapeutic services. It plays a vital role in alleviating shocks in human joints by its thixotropic characteristics when it acts to cushion and protect the cartilage from damage. Among its outstanding characteristics is a low friction coefficient of the order of 0.006. Its chemical composition is described by the authors of Reference 11 thus: "a dialysate of plasma plus a long chain glycosaminoglycan hyaluronical acid of molecular weight 6×10^6. Its rheology is dominated by the complex formed between hyaluronic acid and protein. This results in gel formation responsible for its elasticity, thixotropy, etc."

As would be expected, synovial fluid is quite non-Newtonian in character. It is highly viscous in low shear and drops rapidly at high shears. Figure 2.50 shows the variation of viscosity with shear taken from a postmortem joint, while Figure 2.51 shows the rheology of a synovial fluid taken from a patient with rheumatoid arthritis, which is of low viscosity and Newtonian in character. Some of the substances experimented with as substitutes for synovial fluids were 200 centistokes silicone oil

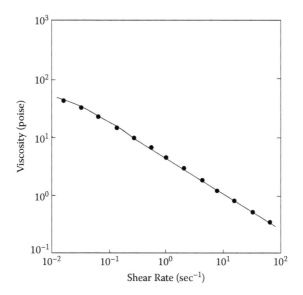

FIGURE 2.50 $\mu - \gamma'$ Relation for a postmortem synovial fluid.

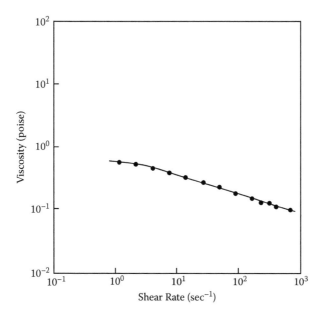

FIGURE 2.51 $\mu - \gamma'$ Relation in synovial fluid from a patient with rheumatoid arthritis.

and amniotic fluid. Although the latter—being a biological substance—was more congenial than the first, it was eliminated from the body after a week.

2.5   POWDER FILMS

As stated in the introduction, the book's theme is the contiguous tribological processes. The impetus to this approach is the realization that particulate films, present either as a result of normal wear or introduced deliberately, perform as legitimate lubricants in that they exhibit features of a quasi-hydrodynamic nature. Powder films constitute the link between fluid film lubrication and the action of solid lubricants and dry surfaces, and occupy a central place in the tribological continuum. On their own, they represent a new branch of tribology with great potential for application in bearings, seals, structural dampers, and other technological areas. For these reasons, they will be treated in greater detail than some of the more familiar modes of operation. Before delving into the rheological aspects of powder films, we shall first take a look at their morphology, both as single particles and as compressed layers. This is a necessary step in order to comprehend their response when compacted into lubricant films.

2.5.1   Characteristics of Triboparticulates

While single particles behave like an elastic body (that, when strongly compacted, resemble a solid), there is a range in between when the assembled particles exhibit a layer-like flow reminiscent of fluids. The state of these particles in this regime will be referred to as triboparticulates (12,13,27–30). While the most common designation of particles is probably size, this is a relative term and, in the present context, particles will be deemed to be small when they are so, relative to the clearance space in which they act as an intermediate layer. As shown in Table 2.11 and Figure 2.52, in addition to size, mass, volume, shape, material properties, and spatial distribution, other factors have an effect on the nature of the flow. Some of these factors are discussed in the following text:

- Particle shape. It is impossible to assign to triboparticulates a specific shape. Furthermore, even if this were possible, one must keep in mind that the particles undergo tricbocomminution during application, when both size and shape undergo substantial changes. Figure 2.53 shows standardized shapes to which some of the triboparticulates could be compared. Figures 2.54 to 2.59 show optical micrographs of a number of common tribological powders: spheroidal nickel powder; a packed nickel oxide powder; a rutile form of titanium dioxide powder isostatically compressed at 10,000 psi; the boron nitride powder, shown in Figure 2.58, exhibits polyhedric, isometric, nonisometric, as well as platelike shapes 10 to 20 microns in diameter, 1 to 2 microns thick, and Figure 2.59 shows a mixed powder of rhenium and copper oxides in which it is difficult to recognize any particular pattern. One should also remember that even a high-resolution micrograph can, at best, provide only two-dimensional information on particle shape.

TABLE 2.11
Characteristics of Some Powder Films

Dimensional Groups		Bulk Powder Properties
Volume: Volume (solid fraction)	Packing	Densification
Mass: Volume (mass density)	• Single particle	• Porous
		• Nonporous
	• Ensemble of particles	• Compressed
		• Sheared
		• Aerated
		• Vibrated
Mass: Mass (interaction)	• Particle-particle	• Cohesiveness/tensile strength
		• Densification
		• Expansion
		• Fluidization
	• Particle-fluid	• Pneumatic transport
		• Fluidization
		• Catalytic
		• Agglomeration
Mass: Time (flow rate)	• Particle-particle	• Flowability
		• Floodability
Volume: Time (transportation)	• Particle-fluid	• Gaseous
		• Liquid

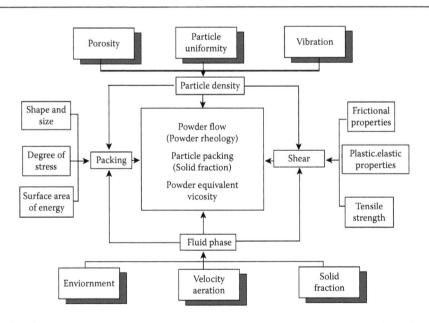

FIGURE 2.52 Variables affecting powder flow behavior and powder properties. (From Heshmat, 1991.)

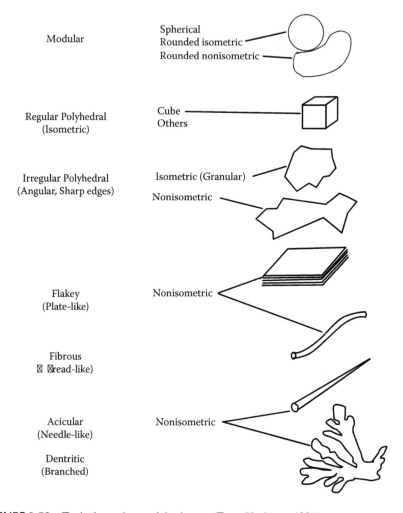

Modular — Spherical / Rounded isometric / Rounded nonisometric

Regular Polyhedral (Isometric) — Cube / Others

Irregular Polyhedral (Angular, Sharp edges) — Isometric (Granular) / Nonisometric

Flakey (Plate-like) — Nonisometric

Fibrous (Thread-like)

Acicular (Needle-like) — Nonisometric

Dentritic (Branched)

FIGURE 2.53 Typical powder particle shapes. (From Heshmat, 1991.)

- Particle size. In view of the foregoing, it should not be surprising that there is ambivalence in assigning size to the particles. Usually, attempts are made to assign an equivalent spherical diameter. The following are commonly used definitions for this quantity:

  1. Volume diameter: $d_v = (\frac{6\pi}{V})^{\frac{1}{3}}$ with V = particle volume
  2. Surface diameter: $d_s = (\frac{S}{\pi})^{\frac{1}{2}}$ where S = surface area
  3. Surface-volume diameter: $d_{sv} = \frac{6V}{S}$
  4. Projected area diameter: $d_a = (\frac{4A}{\pi})^{\frac{1}{2}}$
  5. Perimeter diameter: $d_p = \frac{P}{\pi}$ where P is particle perimeter
  6. Martin diameter: $d_m = \frac{(d_{max}+d_{min})}{2}$
  7. Sieve diameter: d_{si}

FIGURE 2.54   Appearance of nickel particles prior to exposure.

- Particle shape index. Utilizing the foregoing definitions for particle size, attempts are often made to generate a shape index. One widely used shape index is the Wadell index, which is given by $\Phi = [\frac{d_V}{d_p}]^2$; it represents the ratio of the diameter of a sphere having the same volume as the particle and the diameter of a sphere with the same surface area as the particle. Because a

FIGURE 2.55   Appearance of nickel particle surface after exposure at 1500°F (SEM Photograph).

FIGURE 2.56   Compressed NiO particles: fractured face diagonal to compressed face (light micrograph; 300 psi pressure).

sphere has the smallest surface area for a given volume, the maximum value of Φ is 1; for a cube, $\Phi = 0.806$; round sand particles have a $\Phi = 0.8–0.9$. One of the shortcomings of the Wadell index is that it is not unique; a wide range of shapes can have the same Φ. There are other, more sophisticated methods of measuring and describing particle shape but none of them is ideal.

FIGURE 2.57   Compressed TiO particles (rutile form; light micrograph; 10000 psi iso-static pressure).

FIGURE 2.58   Boron nitride powder.

- Properties of selected powders. Among the properties of interest to tribological application are the 50% size range, thermal properties, and micrographic appearance (SEP). Additional properties will be cited later, in connection with powder rheology. Three of the more common tribological powders have the following characteristics:

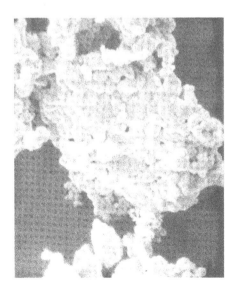

FIGURE 2.59 Re_2O_7 and CuO powder mixture after exposure at 1500°F.

- Rutile form of titanium dioxide (TiO_2). On a 50% accumulative size distribution, sample A reads 2 microns, and sample B reads 6 microns. SEP pressed A and B samples are presented in Figures 2.60 and 2.61. A common feature in both samples is the appearance of the layer adjacent to the boundaries, the sheared faces. These regions are more densely packed and the particles smaller than in the other areas. Fracture marks, where they do appear, seem to be more pronounced on the larger particles.
- Molybdenum disulfite (MoS_2). This substance is commonly used as a dry lubricant or as an additive to other lubricants to reduce friction and wear. Two samples differing only in size were used. The 50% accumulative size distribution was 1 to 2 microns for sample D and 12 microns for sample E. While powder E readily formed aggregates, sample D retained its discrete particle distribution. The SEP pressed samples are shown in Figures 2.62 and 2.63. Noteworthy in these micrographs is the distinct appearance of the sheared faces of a cauliflower-like pattern.
- Zinc dithiomolybdate ($ZnMoO_2S_2$). The interest in this powder is due to its thermal resistance. Its melting temperature is reported to be 1049°C in air, its specific gravity is 3.17, and molecular weight is 474. The average particle size was about 5 microns. Figure 2.64 presents its micrograph appearance, including the sheared site at the boundary.

Features common to all these powders under compression are that, even under minute stress, triboparticulates undergo comminution, and there is always an adhered layer to the tribological surfaces.

2.5.2 RHEOLOGICAL PROPERTIES

The three features of powder films considered here are the bulk density and, related to it, the solids fraction; the viscosity of the compacted powder; and the stress–strain relationship, including a constitutive equation expressing this relationship. In addition, experimental methods of determining these features quantitatively will be described. At the end, rheological data for a number of powders will be provided.

2.5.2.1 Bulk Density and Solids Fraction

While a relatively simple notion in isotropic substances, density and solids fraction in compacted powders are somewhat problematic; however, they are crucial because they affect viscosity, compressibility, and heat transfer properties of triboparticulate films. In some sense, density does not have a unique value, being a function of the size and shape of the particles, the degree of compaction, and even the method of utilization of the particle ensemble. The apparatus used for determining density is shown in Figure 2.65. It consists of a porous cylinder–piston assembly, which enables the carrier fluid to escape upon compression. The powder sample tested is first weighed and measured, then compressed in increments. At each increment, piston displacement is measured, which provides bulk density as a function of applied

TiO$_2$ (Rutile form); P$_d$ = 2 μm; [A]
Compressed Face

Fractured Face

Boundary Layer Face

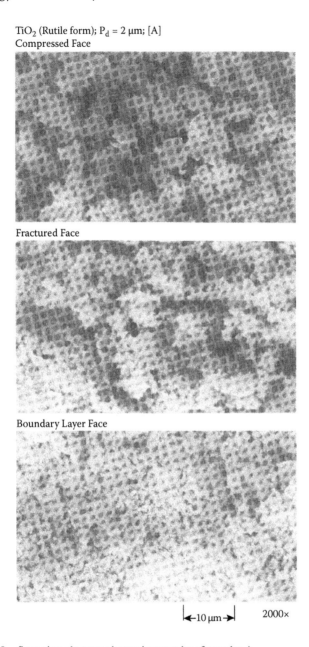

|←10 μm→| 2000×

FIGURE 2.60   Scanning electron photomicrographs of powder A.

TiO$_2$ (Rutile form); P$_d$ = 6 µm; [B]
Compressed Face

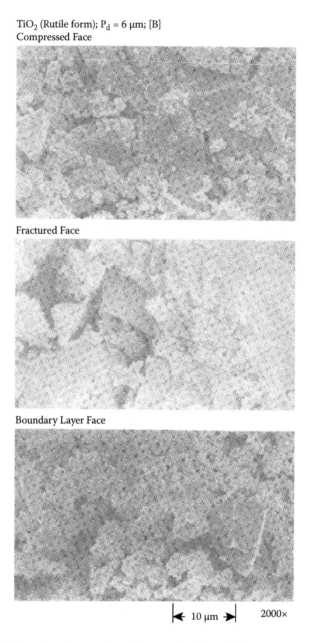

Fractured Face

Boundary Layer Face

|← 10 µm →| 2000×

FIGURE 2.61   Scanning electron photomicrographs of powder B.

MoS$_2$ (Finer size particles); $\overline{P}_d \sim 2$ μm; [D]
Compressed Face

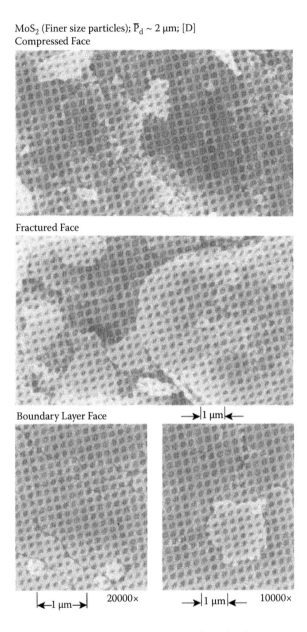

FIGURE 2.62   Scanning electron photomicrographs of powder D.

MoS$_2$ (moderate size); P$_d$ ~ 12 µm; [E]
Compressed Face

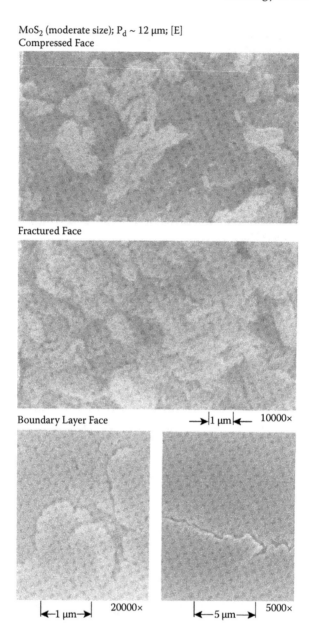

Fractured Face

Boundary Layer Face   →|1 µm|←   10000×

|←1 µm→| 20000× |←—5 µm—→| 5000×

FIGURE 2.63 Scanning electron photomicrographs of powder E.

$ZnMoO_2S_2$, $\bar{P}_d$ = 5 μm; [G]
Compressed Face

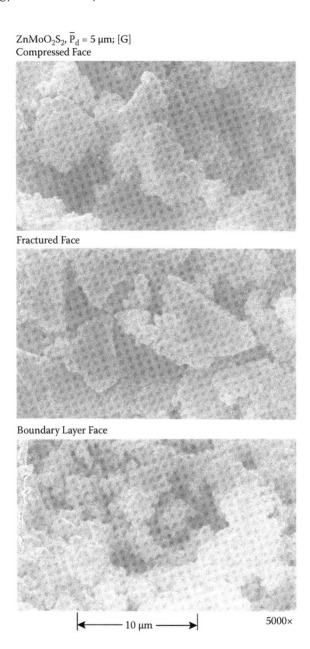

Fractured Face

Boundary Layer Face

←——— 10 μm ———→   5000×

FIGURE 2.64 Scanning electron photomicrograph of powder G.

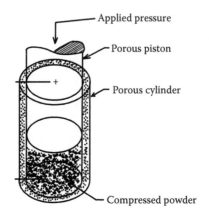

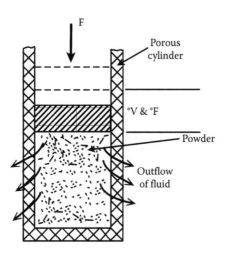

FIGURE 2.65   Container for measuring powder density.

pressure. The solids fraction is the ratio of solid powder density to the powder assembly density as obtained from the experimental measurement or $\phi = (\frac{\rho_s}{\rho})$. This is given in Figure 2.66, where the data are extrapolated to $\Phi = 1$, at which value the assembly is close to being a solid body. The experimental data for various powders show that there are two regimes of compacted density: primary and secondary. Below 35 MPa (350 atm), the density and solid fraction are strongly responsive to the applied pressure, followed by a transition regime between 26 and 50 MPa. Above 50 MPa (500 atm), the packed powder stiffness increases linearly with the load. Thus, above this region one can write for the pressure-density variation

$$\rho = \rho(0)[1 + K_p p] \qquad (2.38)$$

where K_ρ is an appropriate pressure-density coefficient (27).

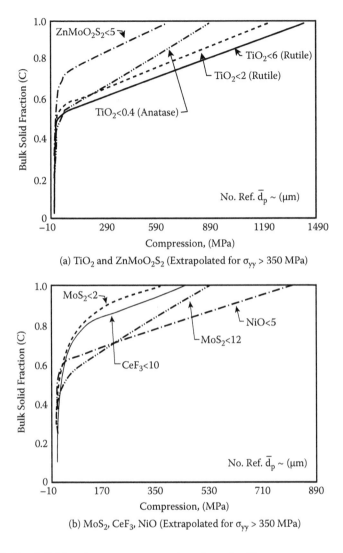

(a) TiO_2 and $ZnMoO_2S_2$ (Extrapolated for $\sigma_{yy} > 350$ MPa)

(b) MoS_2, CeF_3, NiO (Extrapolated for $\sigma_{yy} > 350$ MPa)

FIGURE 2.66   Solid fraction versus compression of powder films.

2.5.2.2   Viscosity

The apparatus designed to measure the equivalent viscosity of powder films is shown in Figure 2.67. About 50 cm^3 of powder is poured into the top of the viscometer. Inert nitrogen is introduced to flood the clearance gap. The shear force transmitted from the lubricant film to the mating collar is measured. Because the values of ΔR and e are small, relative to the mean radius of the rotating collar, the viscous shear force F_τ developed at the mating ring may be obtained from the

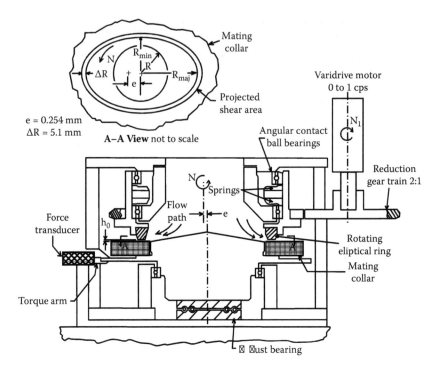

FIGURE 2.67   Schematic of powder viscometer.

following equalities:

$$F_T R_T = F_\tau R$$

$$F_T = \frac{(2\pi\mu\omega)}{(R_T h_o) r^3 dr}$$

$$F_T = \frac{(\mu N \pi^2)}{60 R_T h_o}\left(R_2^4 - R_1^4\right)$$

where
    F_T = Measured force at torque arm
    R_T = Torque arm radius
    F_T = Viscous shear force
    R = Average radius of ring trajectory
    ω = Anuglar velocity (rad/s)
    N = Rotational speed (rpm)
    h_o = Film thickness
    μ = Viscosity (lb-s/in.^{2})

With all the foregoing quantities known from the dimensions of the apparatus or from measurements, the last equation can be solved for the value of the viscosity μ_o.

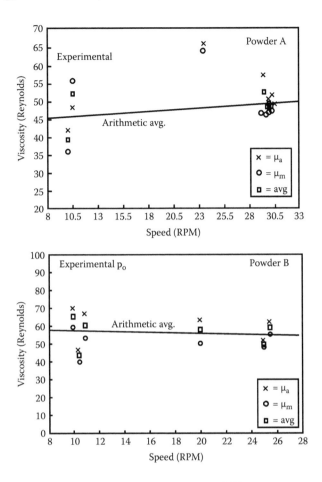

FIGURE 2.68 Powder film viscosity versus shear rate at 70°F.

Before powder viscosities were measured, several fluid lubricants of known viscosities were tested to verify the reliability of the device.

Figure 2.68 shows viscosity measurements on a rutile TiO_2 powder for two size groups: 1–2 microns (sample A) and 5–6 microns (sample B). Their viscosities are plotted as a function of viscometer speed, of which they should be independent. The obtained average viscosities for samples A and B were 55 and 47 micro-Reyn, respectively. These values fall between SAE 30 and SAE 40 weight oils, plotted in Figure 2.3.

2.5.2.3 Stress–Strain Relationship

The apparatus designed to establish the basic rheological quantity of the shear response to applied stress force is shown in Figure 2.69. Of the three short cylinders, the center cylinder is separated from the others by a clearance gap and the powder specimen occupies this space. The desired load is applied via a hydraulic pump and a load cell to a loading piston. The central cylinder, which provides the shearing force, is attached to a pneumatic load cylinder via a connecting rod and force transducer.

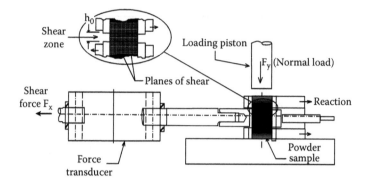

FIGURE 2.69   Apparatus for measuring stress and viscosity as a function of shear rate.

The shear force F_x is applied to the central cylinder by a differential pressure using nitrogen gas. Stopper blocks are applied to the base to prevent horizontal displacement of the shear plate, limiting total motion to 2 mm from its initial position. The test procedure was as follows:

1. Shims 203 μm (8 mil) thick are inserted between the rings to ensure a constant gap h_o in the clearance space.
2. A powder sample is poured into the clearance space and loading cylinder and packed until the powder level is flush with the top cylinder.
3. A constant force is applied via the hydraulic pump on the loading piston.
4. The shims are removed.
5. The loading device is activated, and measurements of shear plate displacement and shear force are recorded.

Shear force and displacement as functions of test time are plotted in Figure 2.70. Force continuously increases up to a limiting value followed by a drop and a further

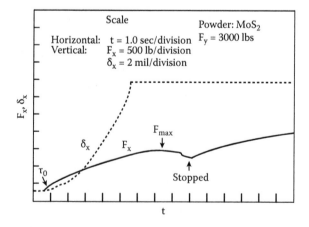

FIGURE 2.70   Sample test data from the apparatus shown in Figure 2.69.

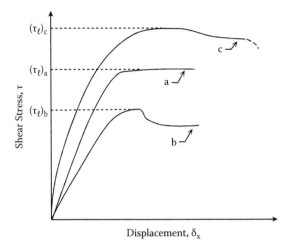

FIGURE 2.71   Results of $\tau - \delta_x$ tests for three different powders.

increase, the latter due to the shear plate contacting the stopper blocks. Loads from 22.73 to 2273 kg (50 to 5,000 lb) were employed in the tests, and typical plots are shown in Figure 2.71. In general, the material deformed elastically until a certain level, after which it either deformed plastically or began to flow. The porosity of the powder films was seen to vary both along and across the film. These differences produced the several resulting strain–stress curves. The behavior of an MoS_2 powder under a stress force is shown in Figure 2.72. The curves exhibit the characteristics of a Bingham plastic in that they possess both a yield stress and a limiting stress. Figure 2.73 is a plot of the yield shear stress versus the applied stress. The relationship appears to be linear until a critical value of applied stress is reached, after which

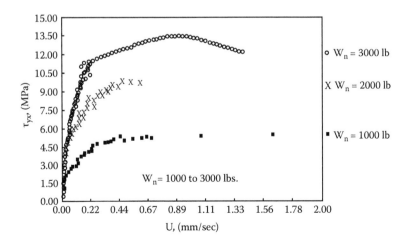

FIGURE 2.72   Shear stress versus strain rate for MoS_2.

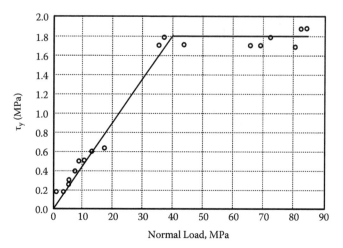

Yield Stress Versus Normal Load for TiO$_2$

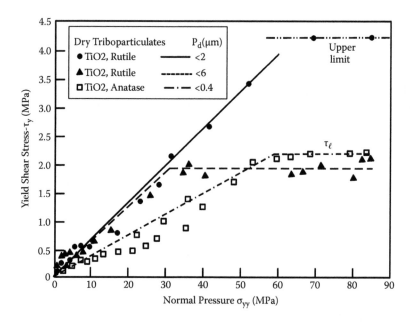

FIGURE 2.73 Yield stress in a powder film as a function of pressure.

the yield point remains constant. Likewise, the limiting shear stress is linear with the applied force, as shown in Figure 2.74.

The general constitutive equation (31) for a viscoelastic substance has the form of

$$\gamma' = \left(\frac{1}{G}\right)\tau + \left(\frac{1}{\mu_o}\right)\phi(\tau)$$

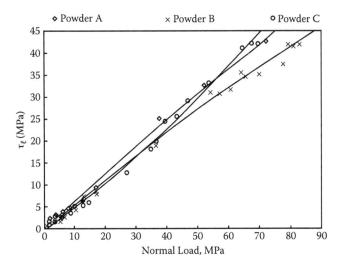

FIGURE 2.74   Limiting shear stress as a function of applied load for three powders.

with G representing the elastic and μ_o the viscous component of the substance. In order to ascertain the relative weight of these two elements, the shear stress rate was plotted against the strain rate, as shown in Figure 2.75. As can be seen, the impact of the elastic term makes itself felt briefly over a small initial interval and reaches an asymptotically constant value at a shear strain of 1.5 s^{-1}. It is, therefore, concluded that in dealing with powder films, the elastic term can be ignored.

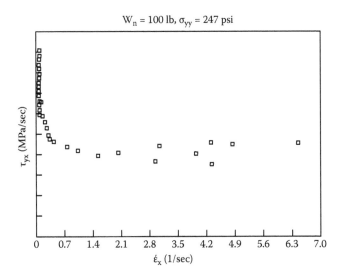

FIGURE 2.75   Shear stress rate versus strain rate for TiO_2.

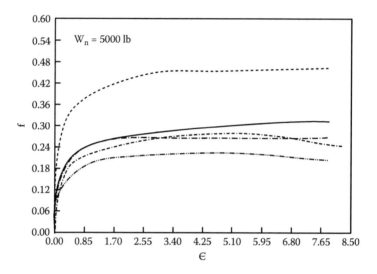

FIGURE 2.76   Coefficient of friction versus normalized strain ε, for MoS_2.

Figure 2.76 shows the friction (f) coefficient of powder films for a range of forces from 500 to 4500 lb. Here, too, the value of f is strongly variable at low values of shear stress (100–200 psi), after which it remains approximately constant and even tends to decrease.

Given that the elastic term can be ignored and that powder films show the presence of both yield and limiting shear stresses, the constitutive equation for powder films can be written in the form of a quintic:

$$\gamma' = \left(\frac{1}{\mu_o}\right)\tau + \alpha\tau^3 + \gamma\tau^5 \qquad (2.39)$$

where a and γ are constants specific to a given powder. The behavior of this function is sketched in Figure 2.77. The test results also showed that the dependence of the yield and limiting shear stresses are linear functions of the pressure and, therefore, parallel to the density expression. These can be written as

$$\tau_y(p) = \tau_y(0)[1 + K_y p] \qquad (2.40)$$

$$\tau_\ell(p) = \tau_\ell(0)[1 + K_\ell p] \qquad (2.41)$$

2.5.2.4   Candidate Powder Materials

In considering certain materials for use as interface layers, some characteristics are clearly preferable over others. The usual tribological requirements of being able to operate under high pressures, high temperatures, and, perhaps, hostile environments

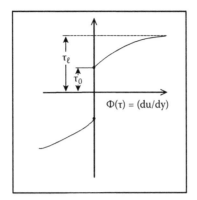

(a) Notions of yield and limiting shear stresses

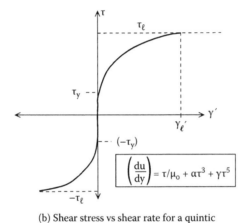

(b) Shear stress vs shear rate for a quintic

FIGURE 2.77   Simulation of yield and limiting shear stresses by a quintic equation.

must be met. They must also be able to withstand the tendency to corrosion and oxidation, as well as any chemical reaction with the mating surfaces. One handicap, as compared with more conventional lubricants, is the lack of prior experience in actual application. Until direct experience with powders is available, one has to rely, to a large extent, on comparisons with known substances. Thus, of all the powders listed in the following tables and figures, two are known for their good tribological characteristics: molybdenum disulfite (MoS_2), which is being used extensively as a dry lubricant and additive, and the rutile form of titanium dioxide (TiO_2), which has a naturally lubricious quality. With this said, Tables 2.12, 2.13, and 2.14, and Figures 2.78a, 2.78b, 2.79a, 2.79b, and 2.79c provide a wide spectrum of candidate powders along with some of their rheological and physical properties relevant to tribological applications.

TABLE 2.12
Listing of Some Available Triboparticulates

	Formula	Name	d (μm)	ρ_s (gm/cc)
A.	TiO_2	Titanium dioxide (rutile form)	1–2	4.26
B.	TiO_2	Titanium dioxide (rutile form)	5–6	4.26
C.	TiO_2	Titanium dioxide (rutile form)	0.4	3.84
D.	MoS_2	Molybdenum disulfide (moderate-size particles)	1–2	4.80
E.	MoS_2	Molybdenum disulfide (moderate-size particles)	12	4.80
F.	CeF_3	Cerium trifluoride (cerous fluoride)	10	6.16
G.	$ZnMoO_2S_2$	Zinc oxythiomolybdate	5	3.17
H.	Carbon	Carbon graphite (finer-size particles)	2	2.25
I.	SiO_2	Fire-dry fumed silica	0.012	2.20
J.	NiO	Nickel oxide, green powder	5	6.67

TABLE 2.13
Experimental Viscosities of Powder A

	$^*F_a$	F_m	F_{avg}	μ_a	μ_m	μ_{avg}
N (rpm)		(g)			(μReyn)	
10	54.1	44.5	49.3	71.2	58.6	65.4
10	35.0	30.0	33.0	47.1	39.5	43.4
10	51.5	39.8	45.7	67.8	52.3	60.1
20	50.0	41.0	46.0	—	—	—
20	95.2	73.3	85.2	62.7	48.2	56.1
25	63.6	51.0	57.3	—	—	—
25	92.2	86.0	89.0	48.6	45.3	46.9
25	115.0	99.0	107.0	60.6	52.2	56.4

Note: Clearance gap 0.0035 in., $T_a = 70°F$.
* $F \sim$ measured force at the torque arm.

TABLE 2.14
Experimental Viscosities of Powder B

		F_a	F_m	F_{avg}	μ_a	μ_m	μ_{avg}
Test No.	N rpm		(g)			(μReyn)	
1	10	15.0	13.0	14.0	42.3	36.5	39.4
2	10	37.0	42.4	37.0	48.7	55.8	52.3
3	29.7	128.0	104.0	116.0	56.7	45.6	51.4
4	30	113.0	103.0	108.0	49.6	45.2	47.4
5	30	116.0	104.0	110.0	50.9	45.6	48.3
6	30	109.0	105.0	107.0	47.8	46.1	46.9
7	30	112.6	104.1	108.3	49.0	46.0	47.5

Note: Clearance gap 0.0035 in. except for Test 1 ($h_0 = 0.0075$ in.), $T_a = 7°F$.

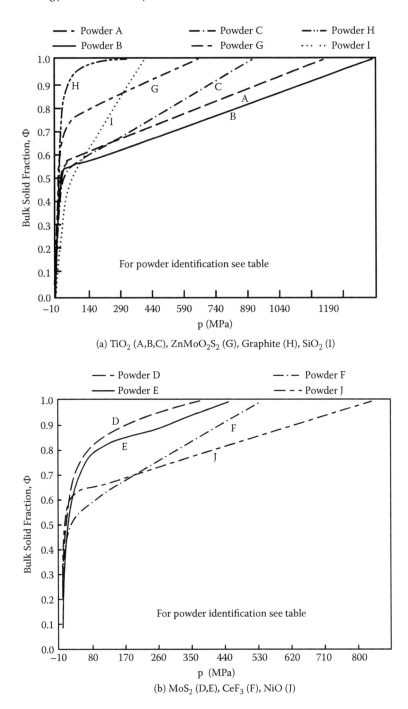

FIGURE 2.78 Bulk solid fraction versus compaction for 10 different powders.

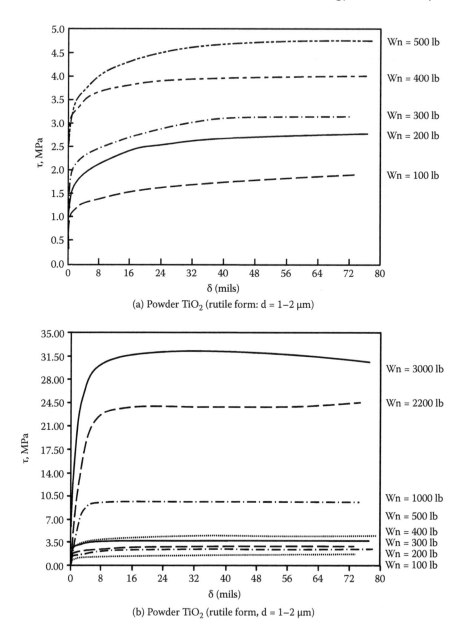

(a) Powder TiO_2 (rutile form: d = 1–2 μm)

(b) Powder TiO_2 (rutile form, d = 1–2 μm)

FIGURE 2.79a,b   Shear stress versus strain for two particle sizes.

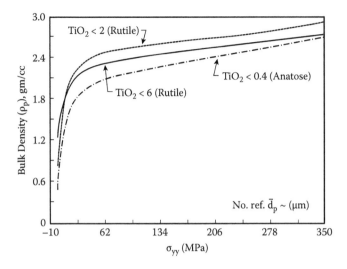

FIGURE 2.79c   Bulk density versus compression: TiO_2.

REFERENCES

1. ASME Pressure-Viscosity Report, ASME, New York, 1953.
2. Andrade, E.N. da C., *Philosophical Magazine*, Vol. 17 (1934), p. 698.
3. Bird, R,B. Material functions for polymeric liquids, *Dynamics of Polymeric Liquids*, Vol. I, Fluid dynamics, R.B. Bird et al., Wiley, New York (1977).
4a. Blair, S. and Winer, W.O., A rheological model for elastohydrodynamic contacts based on primary laboratory data, *ASME J. Lubrication Technology*, Vol. 101, July 1979; also, some observations in high pressure rheology of lubricants, *ASME J. Lubrication Technology*, Vol. 104, July 1982.
4b. Bergstron, L., Rheological properties of concentrated, non-aqueous silicon nitrite suspensions, *J. Am. Ceramic Soc.*, Vol. 79, No 12, pp. 3033–3040.
5. Czarny, R. and Moes, H., Some Aspects of Lubricating Grease Flow, Trib. 1981 (Poland), pp. 96–107.
6. Dobson, G.R. and Tompse, H A.C., The rheometry of lubricating greases, *The Rheology of Lubricants*, ed. Davenport, T.C., Applied Science Publ., pp. 96–107.
7. Einstein, A., On a new determination of molecular dimensions, *Annalen der Physik* (4), 19, 1906, pp. 289–306.
8. Frith, W.J., Mewis, J., and Strivens, T.A., Rheology of concentrated suspensions: Experimental investigation, *Powder Technology*, Vol. 51, 1987, Elsevier Science Publ., pp. 27–34.
9. Ferguson, J. and Nuki, G., The rheology of a synthetic joint lubricant, *The Rheology of Lubricants*, ed. Davenport, T.C., Applied Science Publ., pp. 89–107.
10. Heshmat, H., The rheology and hydrodynamics of dry powder lubrication, *STLE Trans.* Vol. 34, No 3, 1991 pp. 433–39.
11. Heshmat, H. and Brewe, D.E., On some experimental rheological aspects of triboparticulates, *18th Leeds-Lyons Symp. on Wear Particles*, Trib. Series 18, 1992, Elsevier Science Publ.
12. Houwink, B., Rheology in industry, *Proc. Int. Conf. on Rheology*, N. Holland Publ. Co., Amsterdam, 1949.

13. Hutton, J.F., The rheology of petroleum-based lubricating oils and greases: A review, *The Rheology of Lubricants*, ed. Davenport, T.C., Applied Science Publ., pp. 16–29.

14. Mann, D.E. and Tarbell, J.M., Flow of non-Newtonian blood analog fluids in rigid curves and straight artery models, *Biotribology* 27, Pergamon Press, pp. 711–20.

15. Pipkin, A.C., Approximate methods in viscoelastic fluid dynamics, *The Mechanics of Viscoelastic Fluids*, ed. R.S. Rivlin, AMD Vol. 22, ASME, 1977, pp. 1–17.

16. Poiseuille, J., Recherches Experimentales sur le Mouvement des Liquides dans les Tubes de tres Petits Diametres, *Compt Rend*, 1841, 112.

17. Schneck, D.J., On the development of a rheological constitutive equation for whole blood, Tech. Report VPI -E-88-14, VPI, Blacksburg, VA, 1988.

18. Junzo Shimoiizaka et al. Rheological characteristics of water-based magnetic fluids, *Thermomechanics of Magnetic Fluids*, Hemisphere Publ. 1978, pp. 67–75.

19. Spikes, H.A., A Thermodynamic Approach to Viscosity, STLE Preprint 88-TC-6C-2, 1988.

20. Tangsathitkulchai, C. and Austin, L.G., Rheology of concentrated slurries of particles of natural size distribution produced by grinding, *Powder Technology*, Vol. 56, (4) 1988, Elsevier Science Publ., pp. 293–299.

21. Thurston, G.B., Rheological analogs for human blood in large vessels, *Proc. 2nd Int. Symp. On Biofluid Mechanics etc.*, Munich, Germany, 1988.

22. Tsai, S.C. et al., Rheology and atomization of micronized coal—water slurries, FED Vol. 118, ASME, 1991.

23. Vand, V., Viscosity of solutions and suspensions, *J. Physical Chem.*, Vol. 52, pp. 227–314, 1948.

24. Wilcock, D.F. and Booser, E.R., *Bearing Design and Application*, McGraw-Hill, New York, 1957.

25. Heshmat, H., Development of rheological model for powder lubrication, Final Technical Report, NASA Contract Report CR-189043, MTI91TR31, prepared for U.S. Army Aviation Research & Technology Activity (AVSCOM), October 1991.

26a. Heshmat, H., The quasi-hydrodynamic mechanism of powder lubrication—Part I: Lubricant Flow Visualization, presented at the *STLE 46th Annual Meeting*, April–May 1991, Paper No. 91-AM-4D-1, *STLE Lubrication Eng.*, Vol. 48, No. 2, February 1992, pp. 96–104.

26b. Heshmat, H., The quasi-hydrodynamic mechanism of powder lubrication: Part II: Lubricant film pressure profile, presented at the *STLE/ASME Trib. Conf. in St. Louis, MO,* October 14–16, 1991, STLE Preprint No. 91-TC-3D-3, *J. STLE*, Vol. 48, No. 5 (1992), pp. 373–383.

27. Heshmat, H., On the theory of quasi-hydrodynamic lubrication with dry powder: Application to development of high speed journal bearing for hostile environment, presented at the *20th Leeds-Lyon Symposium on Dissipative Processes in Tribology*, Lyon, France, September 7–10, 1993, Dowson et al., eds., published by Elsevier Science B.V., Tribology Series 27, 1994, pp. 45–64.

28. Heshmat, H., High-temperature, solid-lubricated bearing development—dry-powder-lubricated traction testing, *AIAA/SAE/ASME 26th Joint Propulsion Conference Proceedings*, Paper No. 90-2047 (July 1990), *AIAA J. Propulsion and Power,* Vol. 7, No. 5, September–October 1991, pp. 814–820.

29. Heshmat, H., The quasi-hydrodynamic mechanism of powder lubrication: Part III: On theory and rheology of triboparticulates, presented at the STLE Annual Meeting, Calgary, Canada, May 17–20, 1993, *STLE Tribology Trans.*, Vol. 38, No. 2, 1995, pp. 269–276.

3 The Phenomenology of Lubrication

The hydrodynamic theory of lubrication, as launched by Osborne Reynolds in 1886 and subsequently expanded upon by a succession of analysts, provides a fairly solid foundation for a description of a wide variety of tribological devices and processes. These range from giant thrust bearings in hydroelectric power plants to piston rings in internal combustion engines, tiny sliders in electronic computers, and air bearings in medical devices running close to half a million rpm. On the whole, they confirm the predictions of theoretical analysis based on the following postulates:

1. For steady-state operation, there must be relative motion between the interacting surfaces in the tangential direction.
2. While separation between the surfaces must be small—some three orders of magnitude smaller than the dimensions of the surfaces—there must, at the same time, be no contact between them.
3. The configuration of the hydrodynamic fluid film must form a convergent region in the direction of motion.

The foregoing requirements are not simply desirable characteristics but constitute a sine qua non of hydrodynamic action. This can be gleaned from the basic differential equation of hydrodynamic lubrication:

$$\frac{\partial}{\partial x}\left[\frac{h^3}{\mu}\left(\frac{\partial p}{\partial x}\right)\right] + \frac{\partial}{\partial z}\left[\frac{h^3}{\mu}\left(\frac{\partial p}{\partial z}\right)\right] = 6U\left(\frac{\partial h}{\partial x}\right) \tag{3.1}$$

The right side of the equation states that both U and $\left(\frac{\partial h}{\partial x}\right)$ must be different from zero, which corresponds to postulates (1) and (3). When $\left(\frac{h}{x}\right)$ or $\left(\frac{h}{z}\right)$ is much larger than 10^{-3}, the preceding equation yields negligible hydrodynamic pressures and a near-zero load capacity.

Yet, experience with hydrodynamic bearings and seals, whether in the laboratory or in industrial applications, refuses to conform to the dictates of theoretical predictions. Evidence has accumulated that the foregoing postulates can be violated and yet a bearing or seal can survive and continue operating in a manner no worse than with an optimum design. These contradictions occur in the area of parallel surface bearings; in flat face seals; in centrally pivoted pads; in boundary lubrication where, over large portions, there is contact between the surfaces; and in a number of other tribological processes. This chapter provides phenomenological evidence of these departures from basic theory.

3.1 CONTRADICTIONS TO HYDRODYNAMIC THEORY

The essence of hydrodynamic bearings is that they generate forces in the fluid film by virtue of the relative motion only. Thus, contrary to hydrostatic bearings, which depend on an external supply of pressurized fluid, these generate load-supporting pressures by shearing of the viscous fluid between the surfaces. Consequently, when $U = 0$ and all boundaries remain at fixed ambient conditions, we have the resulting Poisson equation:

$$\frac{\partial}{\partial x}\left[\frac{h^3}{\mu}\left(\frac{\partial p}{\partial x}\right)\right]+\frac{\partial}{\partial z}\left[\frac{h^3}{\mu}\left(\frac{\partial p}{\partial z}\right)\right]=0 \qquad (3.2)$$

yielding a zero solution for $p(x,z)$. Likewise, should there be no converging wedge, implying h = constant, we obtain the Laplacian

$$\nabla^2 = 0 \qquad (3.3)$$

likewise yielding an identically zero solution for $p(x,z)$.

It is in the latter instance that we find a glaring departure from theoretical predictions, for many parallel surface devices perform exceedingly well over prolonged periods.

3.1.1 PARALLEL PLATE SLIDERS

A great number of experimental programs on parallel surface devices are reported in the literature, yet only a few of these can be cited here. Extensive work, both theoretical and experimental, was conducted in 1965 by Ettles and Cameron (3) for the specific purpose of elucidating the modus operandi of parallel plate bearings. The first series of tests used the bearing shown in Figure 3.1, which had circumferential grooves adjacent to three of the pads to ensure equal boundary pressures. For the second series of tests, alternate pads were removed to obtain heavier unit loadings and, thus, smaller film thicknesses. As shown in Figures 3.2 and 3.3, the pressure profiles changed radically in going from larger to smaller values of h. While with the larger films there were regions of negative pressure, a decrease in film thickness eventually produced completely positive profiles and substantial load capacity. The results for the second series of tests are shown in Figure 3.4. In these tests, the pressures given are relative to an ambient pressure of 40 psi and, as can be seen, at no instance were negative pressures developed as with the large films in the first series.

A more elaborate series of tests conducted in 1987 by Heshmat et al. (6) aimed at eliciting the modalities of parallel plate operation. Here, too, two configurations, bearing A with 12 pads and bearing B with 3 pads, were tested. The test bearings are shown in Figures 3.5 and 3.6, while the experimental rig is shown in Figure 3.7. Three sets of tests were conducted with bearing A:

- Perfectly flat surfaces extending over the entire 27° of pad arc
- Tapered land bearing, as shown in Figure 3.5b, with reverse rotation
- Perfectly flat surfaces with a tapered inlet at the pad's leading edge

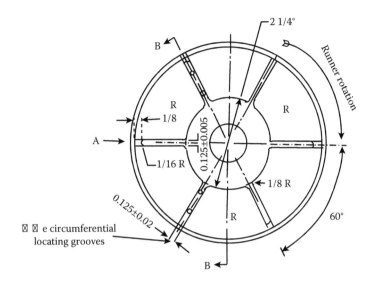

FIGURE 3.1 Configuration of parallel surface bearing tested in Reference 3. (Ettles, C.M. and Cameron, A. 1966.)

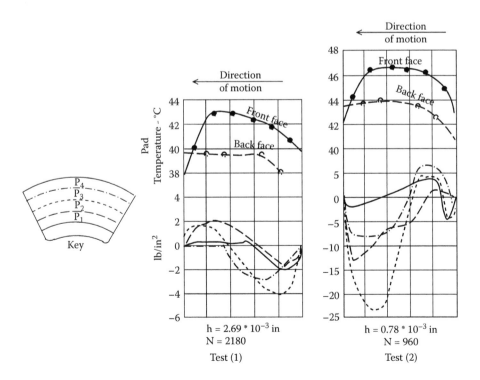

FIGURE 3.2 Pressure profiles in six-pad bearing. (Ettles, C.M. and Cameron, A. 1966.)

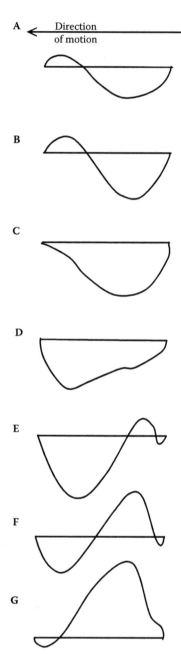

FIGURE 3.3 Changes of pressure profile in six-pad bearing with increase (A to G) in film thickness. (Ettles, C.M. and Cameron, A. 1966.)

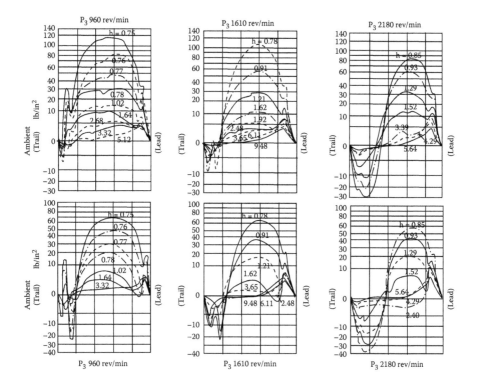

FIGURE 3.4 Pressure profiles for three-pad bearing at various film thicknesses. (Ettles, C.M. and Cameron, A. 1966.)

In manufacturing bearing B, extreme care was taken to produce a perfectly flat surface with a surface finish of 4 μ in. and a flatness of 2 light bands. Likewise, the test rig was provided with a hydrostatically loaded spherical seat for ensuring parallelism between runner and bearing. Oil of a very low viscosity was used to avoid any possible thermal distortion of pads.

For the fully flat bearing A, a mapping of the pad temperatures yielded the distribution given in Figure 3.8. The temperature rise was small everywhere, indicating a uniform heat conduction from the bearing along its path; in the trailing half of the pad, the temperature was mostly constant. An inspection of the bearing after testing showed no signs of distress over any portion of the pads.

For the tapered land bearing A with reverse rotation, which essentially presented a flat land bearing of 6.75° angular extent (the rest was diverging), the following interesting results were obtained:

- Figure 3.9 provides maximum pressure values as a function of the applied load. In terms of the small parallel area, which carried the total load, a maximum unit loading of 800 psi was obtained, which is on the order of maximum load capacity of optimally design thrust bearings. The reason that Figure 3.9 shows a leveling off of p_{max} past a unit loading of

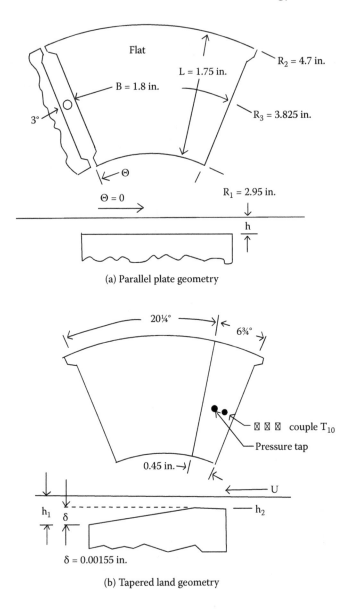

(a) Parallel plate geometry

(b) Tapered land geometry

FIGURE 3.5 Twelve-pad bearing "A" tested in Reference 6. (Heshmat, H., Pinkus, O., and Godet, M. 1989.)

400–500 psi is most likely a shift in the location of p_{max} with a rise in unit loading.

- An indication that the narrow strip of parallel surface carried all the load is provided by the temperature plot of Figure 3.10, where it is seen that, over the diverging portion, the temperatures actually decreased, with cavitation present over most of that region.

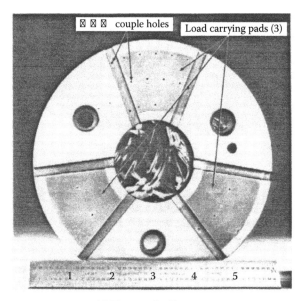

(a) Photograph of bearing

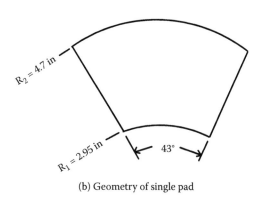

(b) Geometry of single pad

FIGURE 3.6 Geometry of bearing "B" tested in Reference 6. (Heshmat, H., Pinkus, O., and Godet, M. 1989.)

Figure 3.11 shows a comparison of the temperature readings of the same bearing with the correct direction of rotation. With a unit loading 4 times the reverse rotation, T_{max} was 202°F versus 183°F in the conventional bearing—a small difference of 19°F.

Figure 3.12 shows the different T_{max} values obtained for the parallel surface pads with and without an inlet taper. As can be seen, the differences are minimal. The reason this experiment was considered worth trying is that, quite often, the hypothesis is advanced that the success of flat land bearings is due to the presence of an unintended initial taper at the inlet to the pads.

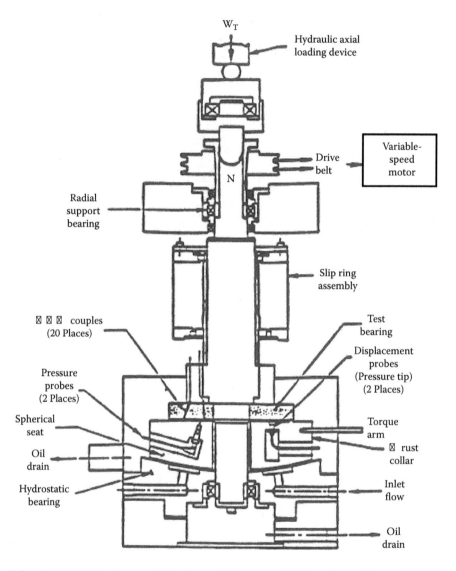

FIGURE 3.7   Test bearing used in Reference 6. (Heshmat, H., Pinkus, O., and Godet, M. 1989.)

Test results with bearing B are as follows:

*   Figure 3.13 shows the recorded pad pressures indicating that it was primar-
    ily the upstream part of the pad that carried the imposed load. More impor-
    tantly, the shape and level of the generated pressure profile are very close to
    those produced in a hydrodynamic film.
*   As happened with bearing A, the temperatures tended to level off in the
    downstream portion of the pad. Often, this is explained by the inflow of
    cold oil from the downstream groove; however, in the present case the film

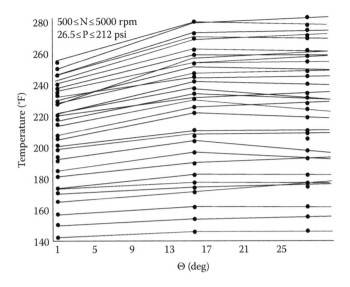

FIGURE 3.8 Temperature profiles in parallel surface bearing "A". (Heshmat, H., Pinkus, O., and Godet, M. 1989.)

thickness was sufficiently small to make this unlikely. Next, it was found that power loss was independent of oil viscosity but could be related to speed and unit loading. This would indicate that the main variable is film thickness: a function of these two parameters.

- Upon examination of the pads at the end of testing, it was noted that whenever burnishing did occur, it took place at the center of the pad, unlike in tapered land bearings, where this would most likely occur at the trailing edge. This ties in with the comments made under the first point earlier.

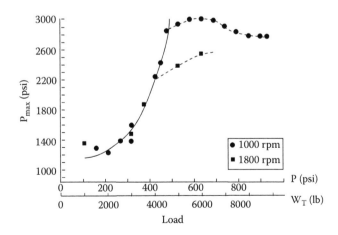

FIGURE 3.9 Maximum pressures in tapered land bearing "A" with reverse rotation. (Heshmat, H., Pinkus, O., and Godet, M. 1989.)

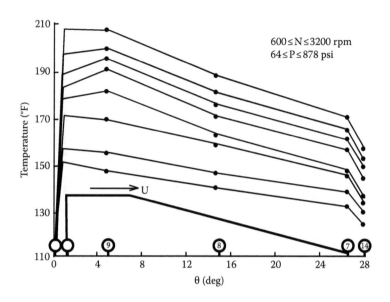

FIGURE 3.10 Temperature profile in tapered land bearing "A" with reverse rotation. (Heshmat, H., Pinkus, O., and Godet, M. 1989.)

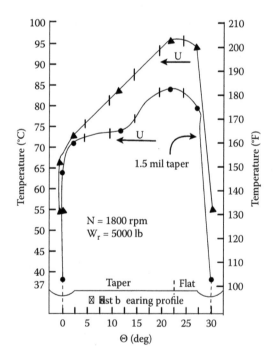

FIGURE 3.11 Comparison of tapered land bearing "A" with two opposite directions of rotation. (Heshmat, H., Pinkus, O., and Godet, M. 1989.)

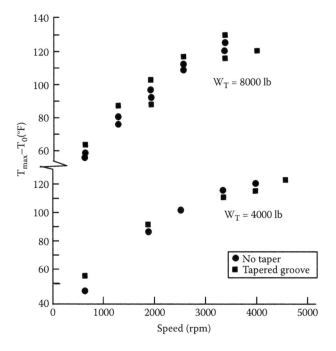

FIGURE 3.12 Comparison of flat surface bearing "A" with and without inlet edge taper. (Heshmat, H., Pinkus, O., and Godet, M. 1989.)

A comprehensive survey of a wide range of experiments on flat land bearings and seals, dating from 1946 to the present, was conducted by Lebeck (11). It showed that a strong load support mechanism is present in all of these cases. Table 3.1 and Figure 3.14 summarize the findings obtained by a number of investigators. The curves plot the coefficient of friction f versus the basic hydrodynamic parameter:

$$S' = \left(\frac{\mu U B}{W} \right) = \left(\frac{\mu U}{W'} \right) \tag{3.4}$$

where W' is the load per unit bearing width. The earliest experiments hark back to Fogg (2), who used a 4-pad thrust collar. In his data, friction decreases with an increase in S', which attests to the appearance of a fluid film. Like the previous authors, Fogg rounded off the entrances to the pads on the leading side and then ran the bearing forward and backward, finding no difference in bearing response. The explanation given by Fogg for the success of his flat land collar is that a "thermal wedge" made this operation possible. On the same figure, we have results obtained by Lenning in 1960 (13) for a range of viscosities, thus varying the value of S'. In all cases, an increase in speed yielded a reduction in friction associated with an increase in film thickness. In 1984, Kanas (10) ran a parallel plate bearing using water having a viscosity two to three orders of magnitude lower than the oil. In this case, it is most

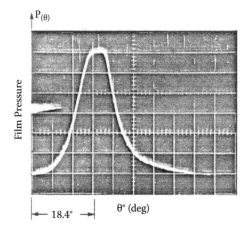

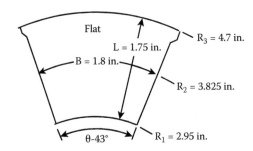

FIGURE 3.13 Measured pressure profiles in parallel surface bearing "B". (Heshmat, H., Pinkus, O., and Godet, M. 1989.)

unlikely that there was any effect of a "thermal wedge." In 1961, Summers-Smith (14) tested a flat face seal and here, too, separation was obtained between the mating surfaces. The fact that the inside sealing pressure caused a film convergence in the radial direction would have no part in producing a hydrodynamic load component, which needs a convergence in the tangential direction for producing a load capacity. Further data on parallel face seals using both oil and water were obtained in 1961 by Ishiwata and Hirabayashi (9), which, likewise, showed a decrease of friction with a rise in speed. Finally, Lebeck (12) supplied data on parallel face seals in water at very low values of S'. Although, as expected, these produced higher levels of friction than in the Kanas experiments, a reduction of friction with a rise in speed occurred here, too.

A comparison of parallel plate bearings with a similar tapered bearing shows that the flat surface bearing had only a slightly higher level of friction than the tapered land bearing, as shown in Table 3.2 and Figure 3.15. This simply means that the film thickness in the parallel plate device was somewhat lower than in the tapered configuration.

TABLE 3.1
Parallel Sliding Data

Lubricant	Configuration	Materials	Roughness (μm)	U (m/s)	μ (Pa·s)	W_n/pad	L (mm)	B (mm)	Source
Oil	Four 13 mm dia. pads/disk	Phosphor bronze/Mi-Cr steel	*/0.15	1–5	$70 \cdot 10^{-3}$	0–2000	13	13	Brix, 62
Oil	Four rectangular pads/disk	White metal/*	*	*	Varies	0–2500	60	14	Fogg, 46
Oil	Six rectangular pads/disk	Cast iron/51 100 steel	0.05/0.13	0–0.2	$17 \cdot 10^{-3}$ $67 \cdot 10^{-3}$	220	51	6	Lenning, 60
Water	Three circular pads/disk	Carbon/tungsten–carbide	0.2/0.05	0.01–11	$0.7 \cdot 10^{-3}$	65	6	6	Kanas, 84
Water + glycerol	Face seal	Various carbons/various hard faces	-.7/*	1–7	$0.9 \cdot 10^{-3}$ 1.4	*	*	*	Summers-Smith, 61
Water and oil	Face seal	Carbon/stainless	*	Constant	*	*	*	*	Ishiwata-Hirabayashi, 61
Water	Face seal	Carbon/tungsten–carbide	0.2/0.05	0.01–10	$0.7 \cdot 10^{-3}$	5200 Constant	319	5	Lebeck, 81

Source: Lebeck, A.O. January 1987.

U = Velocity
μ = Viscosity
W_n = Normal Load
L = Length
B = Bredth
* = Unknown

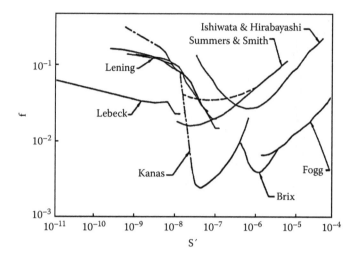

FIGURE 3.14 Survey of various experimental results on parallel surface devices. (Lebeck, A.O. January 1987.)

3.1.2 CENTRALLY PIVOTED PADS

Tilting pad journal or thrust bearings consist of a tandem arrangement of pads, each supported by a pivot around which the pad is free to tilt in accordance with the requirements of operational equilibrium. While, theoretically, the location of the support pivot can be chosen at will, nearly all applications have the pivots located midway of the pad. This is done for a number of reasons, the most obvious one being that it permits rotation in either direction. A schematic of a single pad is shown in Figure 3.16a.

It is immediately clear from the sketch that such a centrally pivoted pad should, from a hydrodynamic standpoint, be an impossible arrangement. To support a load, a hydrodynamic pressure profile of the sort sketched in Figure 3.16a must be generated. Because its resultant would be to the right of the pivot, this would force the pad to assume an ever more horizontal position until, for a balance of forces, the resultant passed through the pivot. This, however, would lead to the formation of a parallel film and, according to theory, zero load capacity.

Yet, centrally pivoted bearings are a common device used throughout industry. They are, in fact, the most stable bearing with an exceptionally high load capacity. As with parallel plates in general, there are various suppositions regarding why they perform so well. The most common is the bending theory shown in Figure 3.16b. Such a situation would indeed permit a centrally pivoted pad to generate hydrodynamic pressures. Given the ineffectiveness of a large downstream part of the pad under bending conditions, the design is equivalent to having the pivot displaced off-center, relative to the effective surface of the pad. However, most tilting pad bearings are made of hardened, thick-walled steel in which pure elastic bending is most unlikely. Furthermore, they work very well under low speed and with low-viscosity lubricants

TABLE 3.2
Parallel Surface to Tilted Surface Comparison

Lubricant	Configuration	Materials	Roughness (μm)	Velocity U (m/s)	Viscosity μ (Pa·s)	Normal Load W N/pad	Length L (mm)	Breadth B (mm)	Source
Oil	Four flat rectangular pads/disks	White metal/?	*	*	*	0–2500	60	14	Fogg
Oil	Four Michell pads	White metal/?	*	*	*		14	14	Fogg
Water	Flat face seal	Carbon/tungsten–carbide	0.2/0.05	0.01–19	$0.7 \cdot 10^{-3}$	5200 Constant	319	5	Lebeck
Water	Three wave face seal	Carbon/tungsten–carbide	0.2/0.05	0.01–19	$0.7 \cdot 10^{-3}$	1700 Constant	106	5	Lebeck

Source: Lebeck, A.O. January 1987.
* = Unknown

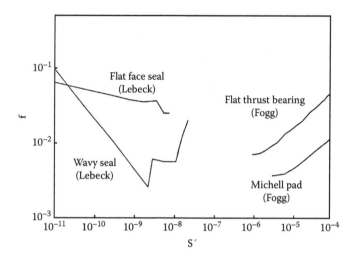

FIGURE 3.15 Comparison of parallel surface and inclined surface sliding $S' = \mu U/w'$.

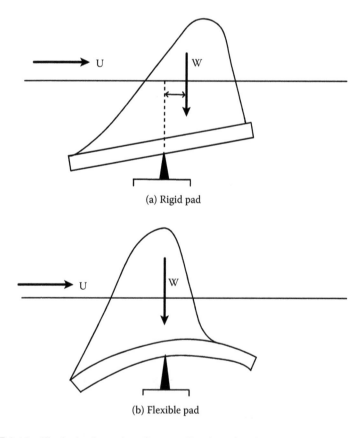

FIGURE 3.16 The hydrodynamics of a centrally pivoted pad.

where thermal effects are, likewise, excluded. Yet these bearing are superior to any other known bearing design at both low and high conditions of speed and load.

3.1.3 MIXED LUBRICATION

The Reynolds equation implicitly requires that a fluid film be present throughout the interface. This is seen most clearly from the presence of a viscosity term, which must be quantified in order to solve the differential equation; solids and asperity fields do not have this characteristic. Yet mixed regimes—in which a fluid film is present only over portions or only intermittently through the clearance space, with the rest subject to asperity contact—perform satisfactorily over a range of operating conditions. The most common analytical approach is to combine hydrodynamic action with elastic or plastic shear forces generated at the asperity contacts. It is, however, with great difficulty that such synthetic approaches can be reconciled with observed experimental facts (11), some of which are as follows:

- Even in heavily loaded or low-viscosity seals, the largest portion of the load is supported hydrodynamically with friction and wear much below what one would expect from direct asperity contact.
- No clear correlation can be found between load capacity and the geometrical characteristics—height, number, and shape—of surface asperities.
- Surface waviness was found to have little effect on the level of generated hydrodynamic pressures.

3.2 TRIBOPARTICULATES AS LUBRICANTS

Another family of phenomenological observations will later be shown as being related to those present in parallel surface operation. This refers to what is, in the literature, often called "third body" or wear debris lubrication. It refers to the observed ability of seemingly dry or semidry surfaces to operate in a manner reminiscent of fluid lubricants in that, in addition to being able to support a load, such seemingly dry interfaces exhibit many of the characteristics of hydrodynamic lubrication. Here, the substances that makes such operations possible will be termed *triboparticulates* because the particles that can form an interface layer must, in general, have certain characteristics not possessed by solid particles. The term triboparticulates thus delimits solid particles to those that can act as a nonfluid lubricant.

Godet and Berthier in France, followed by Heshmat, Pinkus, and Godet (Ref. 4, 5, & 7) in the United States, were the first to document the feasibility of the idea of using particulates as a lubricant. Figure 3.17 shows the change in the friction versus number of sliding cycles for three different metallic interfaces. These curves cover three stages of operation:

- Initial elimination of the natural screens that pollute the specimen surfaces
- Gradual passage from direct contact to the formation of a debris-covered interface
- Steady-state operation on a debris film and ejection of excess debris (equivalent to side leakage)

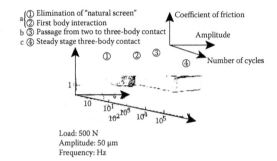

Load: 500 N
Amplitude: 50 μm
Frequency: Hz

Chemical composition (wt.%) of test materials

	C	Si	Mn	S	P	Ni	Cr	Mo	TT
I	0.378	0.33	0.38	$<10_3$	0.011	3.77	1.65	0.29	600°C
II	0.378	0.33	0.38	$<10_3$	0.011	3.77	1.65	0.29	200°C
III	<0.03					12.5	17	2.75	

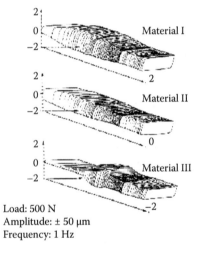

Load: 500 N
Amplitude: ± 50 μm
Frequency: 1 Hz

FIGURE 3.17 Frictional sliding data for three different metallic interfaces. (Godet, M. 1990.)

In all cases, steady-state operation on a debris interface layer was reached after roughly 1000 cycles.

Another series of tests using, initially, a chalk stick and, later on, various metals was conducted in which trace formation was recorded using a mirror placed at 45° against a semitransparent disk. The following behavior was noted:

- Different separation distances were obtained as a function of speed.
- Speed of particles varied across the film.
- Rate of wear decreased after the first few rotations.

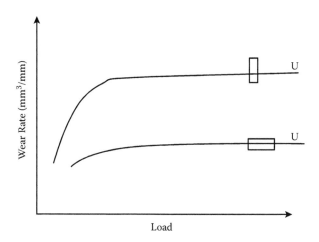

FIGURE 3.18   Wear rate for two different orientations of test specimen. (Godet, M. 1990.)

- Cutting a wedge at the entrance to the pin increased the load capacity.
- Wear rates varied with specimen orientation, as shown in Figure 3.18. The rate of wear dropped when the long side was parallel to the direction of motion. This corresponds to the behavior of a slider bearing, which would have a higher load capacity with a long rather than short dimension in the direction of motion.

Iordanoff and Berthier (8) conducted a study of triboparticulate lubrication by introducing a powder film into the interface. This was a rutile form of TiO_2 in the size range 0.1 μm to 0.2 μm. Tablets of this powder were compacted and placed, one in a hollow chamber and the other in a tight clearance, as shown in Figure 3.19. Figure 3.20 gives the results for five different shear rates. The curves reveal that a threshold yield stress exists before displacement sets in and that a maximum shear stress is, likewise, reached for the powder at high surface velocities. The values of yield and limiting stress, as well the overall stress levels, depend on the applied load, as shown in Figure 3.21.

A more direct demonstration of the semihydrodynamic behavior of powdered films was obtained by means of a specially designed rig (Figure 3.22). This apparatus consisted of one inclined and one horizontal plane with the powder placed between the two and subject to shear. The force reaction on the slider was measured with a dynamometer. To obtain a picture of the flow pattern inside the powder, ½ mm lead balls were dispersed throughout the powder and monitored with x-ray photography, as shown in Figure 3.23. Figure 3.24 shows the measured variation of load capacity as a function of traveled distance. The most striking results, most relevant to our present concerns, are the mapping of velocity profiles across the film as obtained from the photographic records. These profiles, given in Figure 3.25, show a velocity accommodation from the moving to the stationary surface and rupture at the large

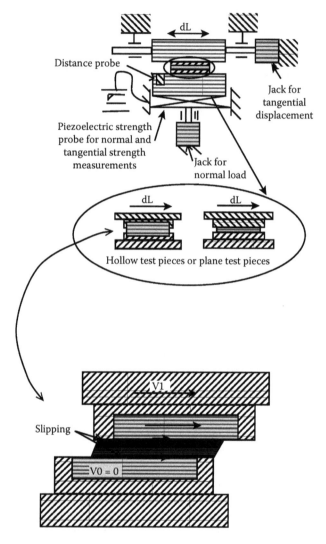

FIGURE 3.19 Testing arrangement used in Reference 8. (Iordanoff, I. and Berthier, Y.)

film height. These features are close to the behavior of a fluid film. The only new element here is the presence of slip at the moving boundary—though this, too, happens
with dilute hydrodynamic gas films. The only possible reservation one could have is
that the film height here was 1 cm, two or three orders of magnitude higher than in
hydrodynamic films.

Another series of tests on triboparticulates was conducted over a number of years
by Heshmat (7). The apparatus used is shown in Figure 3.26, the materials for both
the rotating and stationary parts being TiC cermet. A range of loads, speeds, and

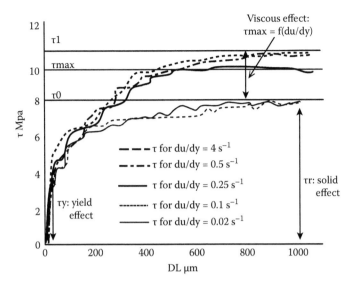

FIGURE 3.20 Stress curves for different rates of shear in tests of Reference 8. (Iordanoff, I. and Berthier, Y.)

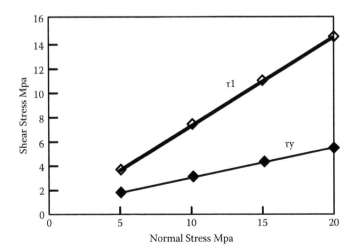

FIGURE 3.21 Values of overall, yield, and limiting stresses in powder films. (Iordanoff, I. and Berthier, Y.)

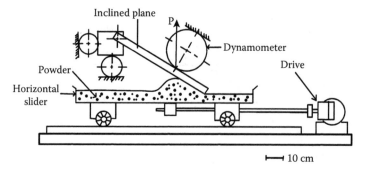

(a) Test apparatus

Powder	C1	F2	F1	G3	G2	G1
	Chalk $CaCO_3$	Magnetite Fe_3O_4	Hematite Fe_3O_3	Natural graphite	Synthetic graphite	Synthetic graphite
Size (μm)	1–4	20–30	5–10	15–30	50–60	0.5–2
Specific surface area $m^{2*}gr^{-1}$	2	--	12	1.1	2	25

(b) Powders used

FIGURE 3.22 Test apparatus and powders used in Reference 8. (Iordanoff, I. and Berthier, Y.)

ambient temperatures was employed, which changed the basic hydrodynamic parameter S'. The work concentrated on the behavior of the friction curve, the formation of
a wedge at the interface, and the behavior of the debris as a lubricant:

- Friction. Figure 3.27 shows a decrease in f with an increase in speed and
  temperature. If higher temperatures can be related to an increase in wear,
  then the figure shows a direct relationship between a drop in f and an
  increase in the amount of debris formation. As the rate of wear levels off,
  so, too, does friction.
- Wedge formation. After a brief running period, the mating surfaces, originally perfectly flat, were noted to have assumed a wavy pattern. Figure 3.28
  shows the measured waviness of the surfaces as a function of location and
  running time. The average wave height was about $8 \times 50 \times 10^3$ in. and the
  wave half-width 18×0.01 in. This yielded 2 mil of taper per inch in the circumferential direction, the very quantities used in designing tapered land
  bearings.
- Wear particles at the interspace. Figure 3.29 shows an ×1000 photograph of
  the particles found at the interface after disassembly. A three-dimensional
  photograph showed the particles to be of nearly spherical shape.

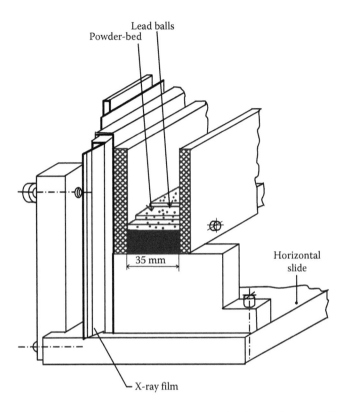

FIGURE 3.23 Method of measuring velocity distribution in powder films. (Iordanoff, I. and Berthier, Y.)

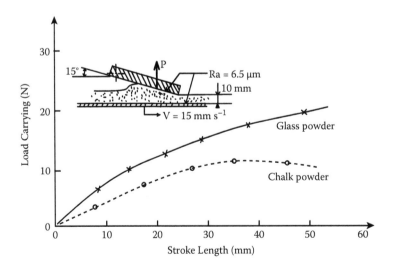

FIGURE 3.24 Load capacity of powder films versus stroke length. (Iordanoff, I. and Berthier, Y.)

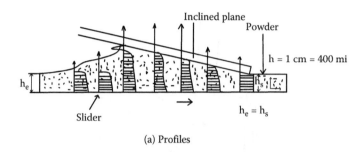

(a) Profiles

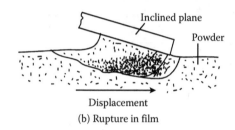

Displacement
(b) Rupture in film

FIGURE 3.25   Velocity profiles in powder films.

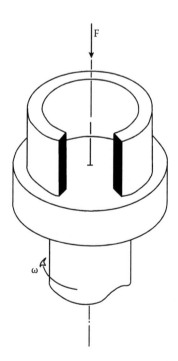

FIGURE 3.26   Test apparatus used in Reference 7. (Heshmat, H. 1991.)

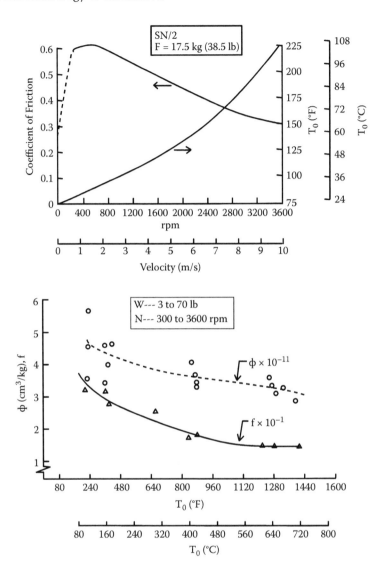

FIGURE 3.27 Friction as a function of speed and temperature. (Heshmat, H. 1991.)

3.3 GENERAL OBSERVATIONS

Based on the foregoing experimental evidence, operation of parallel plate devices is characterized by the following features:

1. A reduction of friction with the dimensionless grouping $S = (\frac{\mu U}{W'})$.
2. Levels of recorded friction are substantially below those expected from full or partial contact.

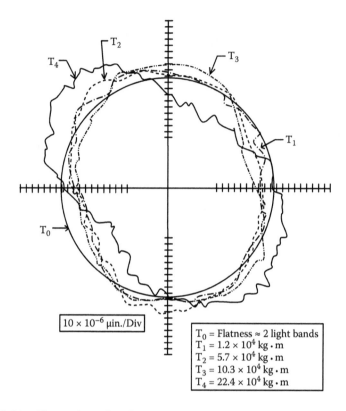

10×10^{-6} μin./Div

T_0 = Flatness ≈ 2 light bands
$T_1 = 1.2 \times 10^4$ kg · m
$T_2 = 5.7 \times 10^4$ kg · m
$T_3 = 10.3 \times 10^4$ kg · m
$T_4 = 22.4 \times 10^4$ kg · m

FIGURE 3.28   Changes in surface flatness due to wear. (Heshmat, H. 1991.)

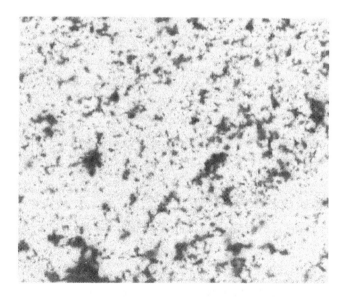

FIGURE 3.29   ×1000 Photograph of wear particles. (Heshmat, H. 1991.)

3. Satisfactory performance prevails with a wide variety of devices, such as plane sliders, thrust sectors of wide and narrow angular extent, flat face seals, centrally pivoted pads, and others.
4. In the mixed and boundary lubrication regimes, the levels of friction are below those based on interactions, whether elastic or plastic, of the interacting asperities.
5. The interface separation is smaller than in conventional hydrodynamic bearings or seals.
6. The imposed loads are carried only on a fraction of the bearing surface.
7. The level and shape of the generated pressures are close to conventional hydrodynamic films.

Experiments with triboparticulates revealed the following:

1. Films formed of triboparticulates, whether generated by wear or deliberately introduced, can support external loads.
2. Frictional levels in powder films are considerably below those expected from "dry" contacts.
3. The behavior of triboparticulate films—such as the shape of the velocity profile, presence of cavitation, and side leakage—bear a striking resemblance to the behavior of hydrodynamic films.

REFERENCES

1. Brix, V.H., Thin film lubrication, *Lubrication Engineering*, July 1962.
2. Fogg, A., Fluid film lubrication of parallel thrust surfaces, *Proc. Inst. Mechanical Engineers*, Vol. 155, 1946, pp. 49–53.
3. Ettles, C.M. and Cameron, A., The action of parallel surface thrust bearings, *Proc. Inst. Mechanical Engineers*, Vol. 180, Part 3K, 1966, pp. 177–91.
4. Godet, M., Third bodies in tribology, *Wear* 126, 1990, pp. 29–45.
5. Godet, M. and Berthier, Y., Continuity and dry friction; Osborne Reynolds approach, *Proc. 13, Leeds-Lyon Symposium, Tribology Series* 11, 1986, pp. 653–66.
6. Heshmat, H., Pinkus, O., and Godet, M., On a common tribological mechanism between interacting surfaces, *STLE Trans.*, Vol. 32, 1989, pp. 32–43.
7. Heshmat, H., The rheology and hydrodynamics of dry powder lubrication, *STLE Tribology Trans.*, Vol. 34, 1991, pp. 433–39.
8. Iordanoff, I. and Bertheir, Y., First Step for a Rheological Model for Solid Third Body, Lab. de Mec. de Contacts, Villeurbanne, Cedex.
9. Ishiwata, H. and Hirabayashi, H., Friction and sealing characteristics of mechanical seals, International Conference Fluid Sealing, BHRA, Paper D3, 1961.
10. Kanas, P.W., Microasperity lubrication in a boundary lubricated interface, MS at the University of New Mexico, May 1984.
11. Lebeck, A.O., Parallel sliding load support in the mixed friction regime, Part I—Experimental data and Part II—Evaluation of mechanisms, *ASME J. Tribology*, Vol. 109, January 1987.
12. Lebeck, A.O., Face seal waviness—prediction, causes and effects, *Proc. Int. Conf. on Fluid Sealing*, BHRA, April 1984.
13. Lenning, R.L., The transition from boundary to mixed friction, *Lubrication Engineering*, 1960, pp. 575–82.
14. Summers-Smith, D., Laboratory investigation of the performance of a radial face seal, *Int. Conf. on Fluid Sealing*, BHRA, April 1961.

4 Direct Contact– Hydrodynamic Continuum

For our present purposes, the hierarchy of interactions between two surfaces will start with dry contacts (such contacts always possess an unintended oxide or molecular layer) to deliberately applied coatings or solid lubricants, to those surfaces that operate on a combination of fluid film and asperity interaction (referred to either as boundary or mixed lubrication) and ending with the process of elastohydrodynamic lubrication (EHL). Proceeding along this spectrum, we can say that we are progressing from instances of high friction and high wear rates down to EHL of low friction and, ideally, no wear at all. Before proceeding to a discussion of the tribological aspects in each of these categories, we shall take a brief look at the overall phenomena of friction and wear as they are presently understood.

The most common approach to the phenomenon of friction is in terms of the adhesion theory (the Tabor school). Here, friction is caused primarily by the fusion of asperities of the opposing surfaces so as to bond or weld areas of each set of interacting asperities. When a metallic junction is formed, shearing may occur in four different ways. If the junction is weaker than the metals themselves, shearing will occur at the junction interface. The amount of metal removed will be small, even though the friction may be high. Such a junction is usually formed in the presence of oxide films. If the junction is stronger than one of the metals, shearing will occur in the weaker metal, with fragments of it adhering to the harder surface. This will gradually build up a film of softer metal on the opposing surface so that, ultimately, the sliding may simulate that of two similar metals. Friction, surface damage, and wear will be high. If the junction is stronger than both metals, shearing will occur inside the bulk of the weaker metal and, occasionally, also inside the stronger metals so that there will be considerable removal of the softer material and also some of the stronger material. In the case of two similar metals, the process of deformation and welding will usually work-harden the junction and increase its shear strength above that of the two parent materials. Shearing, therefore, will occur inside the bulk of the materials. Surface damage in both metals can thus be considerable. Figure 4.1 shows, schematically, the mechanics involved in these processes.

In terms of adhesion theory, the frictional force is, to a rough approximation, given by

$$F = A_a \tau$$

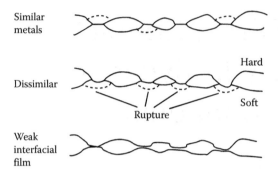

FIGURE 4.1 Loci of adhesive junction rupture.

where A_a is the contact area between asperities and τ the shear strength at the point of rupture.

The junction shear strength must be overcome before motion can commence. The tangential force must, therefore, be larger than the τ of the bulk material because work hardening has taken place in shearing the asperity junctions. Without such an assumption, the calculated values of F are considerably below experimental results. The wear volume, V_w, produced by the frictional force is proportional to the distance, S, traveled and the load, W, but inversely proportional to the hardness, H, of the material, or

$$V_w = \left(\frac{1}{3}\right) k_1 \left(\frac{WS}{H}\right) \tag{4.1}$$

where k_1 is a wear coefficient typical of a particular material. The factor 3, above, is a result of the particular formulation of the adhesion theory, which assumes the production of hemispherical particles when the weaker of the two materials ruptures.

Another reason for F being larger than that calculated from the adhesion theory is given by Tabor as follows: If W is increased, the total area of asperity contact will be proportional to W. Thus, in general, we have

$$A_a = \frac{W}{P_o} \tag{4.2}$$

where P_o is the stress at the contact area. When clean surfaces come into contact, the bond across the interface is the same as at a grain boundary in the metal. The force required to shear the junction will be a frictional force given by

$$F = A_o s = \left(\frac{W}{P_o}\right) s \tag{4.3}$$

where s is the interfacial shearing strength. The frictional force is, thus, proportional to load and is independent of the size of the bodies. The coefficient of friction then becomes $f = \frac{F}{W} = \frac{s}{P_o}$. For most materials, s is of the order of 0.2 P_o so that in the present model, $f = 0.2$. This, as pointed out before, is much below experimental values, which, for metals in air, is closer to unity.

The process, shown in Figure 4.2, is somewhat as follows: when bonding takes place at rest under an imposed external load, the junction is in a plastic condition. When a tangential force is applied to produce sliding, the plasticity condition at the junction is exceeded, the surfaces sink, and contact area is increased until a new state of plasticity over a larger area is created. If the original condition under a stationary load is an elastic junction, the process will be the same: contact area will increase until plastic conditions prevail. The only difference is that, in the elastic case, the tangential force required to produce motion will have to be larger, and so too the coefficient of friction.

A less common interpretation of friction and wear is that of abrasion (the Cavendish school), where plowing, or something similar to it, is seen as the mechanism leading to abrasive wear. This consists of removal of material by plastic deformation caused by hard protuberances or loose, hard particles

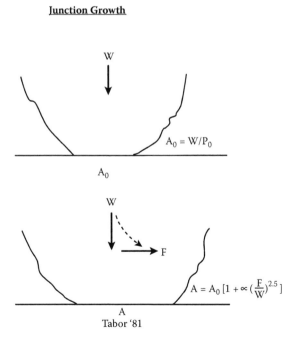

Junction Growth

$A_0 = W/P_0$

A_0

$$A = A_0 \left[1 + \propto \left(\frac{F}{W}\right)^{2.5}\right]$$

A

Tabor '81

FIGURE 4.2 Increase in adhesive contact area due to tangential force.

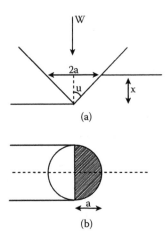

FIGURE 4.3   Model of abrasive wear protuberance.

forced against and sliding along a mating surface. In the first case, it is referred to as a two-body and in the second case as a three-body abrasive process. Figure 4.3 is a simple model of a protuberance, idealized as a cone, indented into and gouging a groove in an opposing surface. The normal load, W, is supported beneath the protuberance by plastic flow, producing a pressure, P, over the area of contact. With the cone moving, the active area is only in front of the cone, yielding the relationship

$$W = p\left[\frac{(\pi a^2)}{2}\right] = \frac{[px^2 \tan^2 \alpha]}{2}$$

As shown by Hutchings (11), the foregoing leads to an expression for the volume removed

$$V_w = k_2\left(\frac{WS}{H}\right)$$

which is the same as that obtained from adhesion theory, though it started from a completely different postulate. The value of the coefficient k_2 is given as between 5×10^{-3} and 50×10^{-3}.

In the abrasive model, the hardness parameter is of prime importance. The wear rate becomes very sensitive to the ratio of abrasive hardness H_a to abraded surface H_s when $\frac{H_a}{H_s}$ is less than one. According to this theory, significant plastic flow sets in when the mean contact pressure reaches about 3 times the uniaxial yield stress of the material, which is the indentation hardness of the surface. A sphere of hardness H_a will cause plastic indentation when $\frac{H_a}{H_s}$ is larger than 1.2. This is shown, schematically, in Figure 4.4.

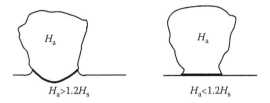

FIGURE 4.4   Effect of hardness ratio on asperity interaction.

The sizes of abrasive particles can cover a wide range: between 5 and 500 microns. Behavior such as that depicted in Figure 4.5 is typical of metals and other materials that exhibit plastic flow. Small particles cause proportionately less wear than large ones. Generally—whether by filtration or by centrifugal action—larger particles are more easily eliminated from the interface, so that most remaining particles are of the smaller sizes. This has some relevance to our subsequent treatment of debris or powder layers acting as lubricants.

Generally, friction is due to the combined effects of asperity deformation, f_d, plowing by the asperities and wear particles, f_p, as well as the adhesion between asperities, f_a. Their relative contribution to total friction depends on the history of sliding, the materials used, surface topography, and the environment. Asperity deformation determines the static friction coefficient and, to a minor degree, also contributes to the dynamic coefficient. On the other hand, plowing or abrasion is very important in sliding, for it not only raises the frictional force but also produces delamination wear, which in turn causes further wear at the interface. Plowing produces in the surface grooves lined on

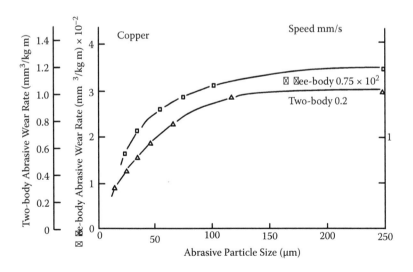

FIGURE 4.5   Relation between abrasive particle size and rate of wear.

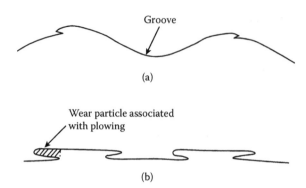

FIGURE 4.6 Mode of generation of loose wear debris due to plowing.

either side by ridges. When these ridges are subjected to repeated loading, their edges tend to break off to form loose wear particles, as shown in Figure 4.6.

According to Suh (20), the relative values of these several friction components are as follows:

- Asperity deformation: $f_d = 0.43$ to 0.75. This is applicable to asperities sloping at an angle of 4° to 20°, with the entire load carried by the asperities.
- Plowing: $f_p = 0$ to 1.0. The high values are associated with sliding of identical materials, while the low values apply when a soft surface slides against a polished harder surface. Normally, its value is on the order of 0.4.
- Adhesion: $f_a = 0$ to 0.4. The low values apply to mildly lubricated surfaces, while the high values pertain to identical metals bare of oxides or other contaminants.

We have seen from the foregoing that, whereas adhesion can only affect a transfer of material to one or both surfaces, the production of loose wear particles requires a plowing action. Two other mechanisms credited with the generation of loose wear debris are the formation and breakup of oxidized material and particle fatigue. In experiments (12) conducted with radioactive annealed steel pins running against inactive hardened steel rings, the transfer of the radioactive material produced curve A in Figure 4.7. Initially, transfer was rapid but then it leveled off. Wear of the ring, however, proceeded at the same high rate as the initial rate on the pin (curve B). The material initially transferred from the pin to the ring was a thin gray layer. This later became a brown powder, identified as hematite or hydrohematite. This was interpreted as resulting from the following sequence of events: (1) transfer of material from the pin to the ring, (2) oxidation of the transferred layer, and (3) breakup and removal of the oxide as loose wear.

In experiments conducted by Kerridge and Lancaster (13), it was found that adhesive transfer was the initiating process in the wear of 60/40 brass against tool steel or Stellite. Here, however, the loose wear debris was not oxidic but

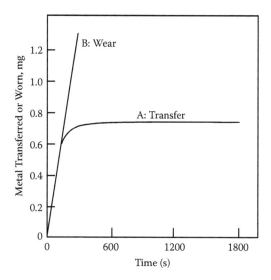

FIGURE 4.7   Formation of loose wear debris by tribo-oxidation.

metallic in character. The initial particle sizes transferred from the pin to the ring were only 5% of those produced at equilibrium; in addition, the loose wear particles at equilibrium were larger than the transferred particles. The mechanistic process consistent with these results is (1) initial adhesive transfer from pin to ring, (2) repetitive compression and agglomeration of the transfer particles by plastic deformation and adhesion and, finally, (3) detachment of loose particles due to fatigue as the pin repetitively loads the transferred material against the ring.

4.1 DRY FRICTION AND WEAR

A very comprehensive series of experiments on the process of friction and wear was conducted by Steijn in 1959. The parameters investigated included different materials, effects of time or distance traveled, load, and temperature. From our standpoint, his experiments carry the additional relevance of having two different test rigs; one was the familiar stationary pin against a ring, while the other used a rotating ring in place of the pin. While a pin is unlikely to retain loose debris at the interface, a ring would; the test results thus include the effect of a particle layer at the interface. In both tests series, the stationary ring was made of a 64 Rockwell B hardness steel polished to an rms of 10–12 μin.

4.1.1 Tests with a Spherical Pin

The pin was a soft annealed SAE 1018 sphere, 1/4 in. in diameter. Figure 4.8 gives the results in terms of sliding distance and load. The level of wear, though linear, does not stay constant; at some point there is a decrease in the rate of wear. Thus,

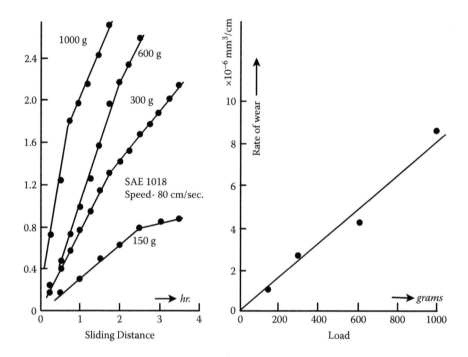

FIGURE 4.8 Adhesive wear rate as a function of sliding distance and load.

as postulated before, while friction is independent of the area of contact, its level has changed. Steijn attributed the reason for the change in the slope of the wear line after a passage of time to the onset of oxidation. It was noted that the wear products were not metallic but a reddish-brown powder identified by x-ray diffraction as $\alpha - Fe_2O_3$. From examining the pin surface, it was also concluded that oxidation occurred after the particles separated from the pin. It is thus postulated that, with time, an oxide film builds up between the two surfaces, reducing the rate of wear.

Wear tests were also conducted on two nonferrous alloys: a 60/40 free-cutting brass and a 24ST aluminum. These specimens were cone-shaped with a 45° semi-apex angle. The ring was 64 Rockwell C steel. The results of the tests are given in Figures 4.9 and 4.10. Here, the slope of the wear lines remained constant throughout the test period.

4.1.2 Tests with a Rotating Ring

The dimensions of both annular rings were 9/8 in. OD and 7/8 in. ID. The rotating ring was a 60/40 brass running against a 64 Rockwell C ring. The rings were manufactured with radial grooves, ranging from no slots to 1, 2, 3, and 4 slots, as shown in Figure 4.11. These escape ports facilitated the discharge of the accumulated wear debris at the interfaces. In all these tests, only the brass ring was slotted. Figure 4.12 gives the test results for various loads. While the wear remained linear

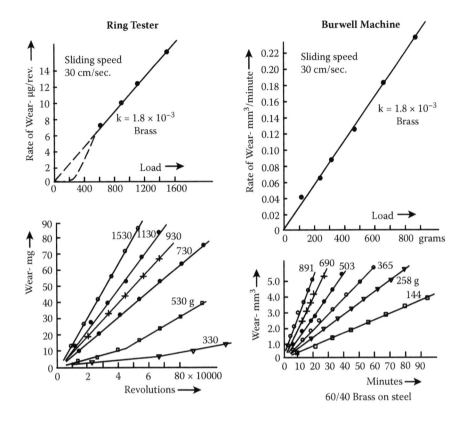

FIGURE 4.9 Comparative wear tests.

with sliding distance, when these results were plotted in terms of load, a very interesting plot was obtained. It can be seen that over a range of operation—up to loads of about 1 kg—there was a progressive reduction in rate of wear as the number of slots decreased, which is equivalent to progressively retaining more wear debris at the interface. This also shows up on the curves of Figures 4.9 and 4.10, where the two-ring tests are compared with those of the pin. As shown in Figure 4.13, no such advantage was obtained when similar metals were tested with slotted rings.

Using the data from both the brass and aluminum tests, the values of k contained in the wear formula $V_w = k(\frac{W}{p_o})$ were calculated. The obtained k values are given in Table 4.1. For brass they are of the order of 10^{-3} and for aluminum closer to 10^{-4}.

Of additional interest is the size of the produced wear particles because only those below a certain diameter can act as an interface layer. According to Sarkar

$$d \geq \left(\frac{W}{H}\right)6\times10^4 \tag{4.4}$$

where d is particle diameter. Experiments with steel on steel have produced a mean diameter of the order of 120 microns. The ratio $\frac{W}{H}$ in the foregoing equation can be

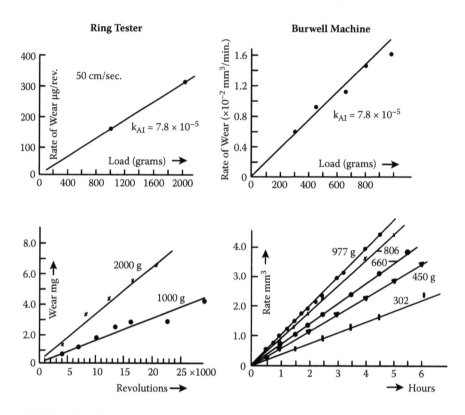

FIGURE 4.10 Wear tests for an aluminum coating on steel.

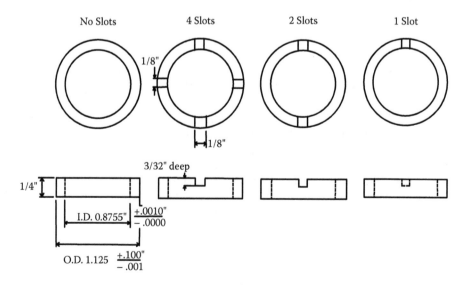

FIGURE 4.11 Configuration of slotted rotating rings.

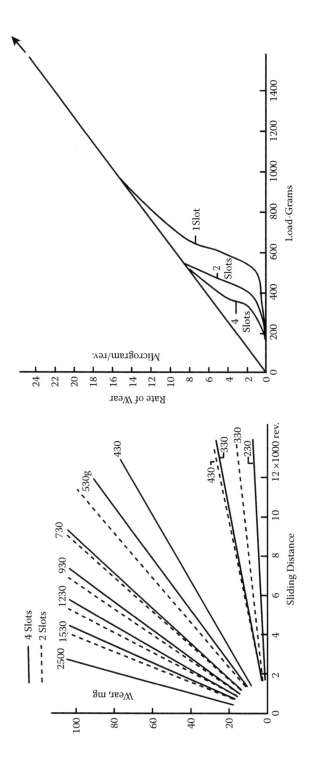

FIGURE 4.12   Wear tests with rotating brass ring.

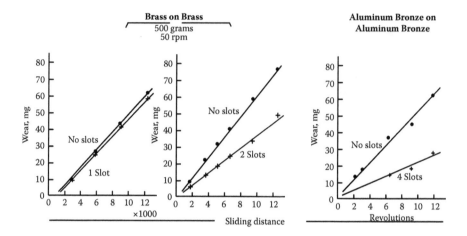

FIGURE 4.13   Wear tests with rotating ring for similar metals.

considered as the ratio of surface to volume energy of a wear particle. A high surface energy implies a large critical diameter before it can detach itself from the parent material; soft materials have a high surface energy, whereas hard materials have a high volume energy. Tests with an alloy containing 21% Si produced the size distribution of Figure 4.14, showing that small sizes are the most numerous. However, a more reliable load dependent distribution is given in Table 4.2.

The mechanisms of asperity deformation and plowing imposed on adhesion will also produce abrasive and cutting modes of wear. Czichos then separates wear into two categories: light and severe. A similar subdivision is proposed by Sasada (17). In Figure 4.15, the straight linear wear (vertical dimension) is plotted for a pin-ring test setup as a function of sliding distance and material hardness. As expected, wear decreases with hardness. Under an SEM examination, the surfaces showed that in the

TABLE 4.1

Values of Coefficient k in (V_m) Formula

		Ball-on-Plate Machine		Ring Machine		
Material	Hardness (10^6/g per cm^2)	Rate of Wear (cm^3 per cm per g)	k-Value	Rate of Wear (cm^3 per cm per g)	k-Value	Sliding Speed (cm per s)
60/40 brass (3.5% lead)	11.6	0.154×10^{-9}	1.8×10^{-3}	0.165×10^{-9}	1.9×10^{-3}	30
Aluminum alloy (24ST)	13.0	5.96×10^{-12}	7.8×10^{-5}	6.89×10^{-12}	8.9×10^{-5}	50

Source: Stejin, 1959.

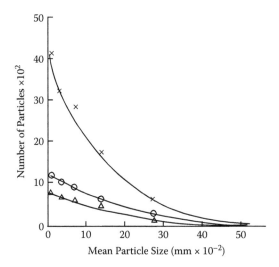

FIGURE 4.14   Debris size distribution due to wear for an Al–Si alloy surface.

high-wear region—a hardness range of 200 to 300 HV—adhesion is the dominating process, whereas in the low-wear mode—hardness of 450 to 600 HV—tribo-oxidation and abrasion dominate. Thus, an important aspect for the transition from severe to low wear is the formation of oxidative reaction layers. The great influence of the environmental atmosphere was further corroborated in tests performed in a vacuum. A comparison between the results of Figures 4.15 and 4.16 leads to the following differentiation in behavior in air and in a vacuum:

- Different ranking of wear as a function of hardness
- More loose wear particles for sliding in air
- Severe material transfer, particularly at the beginning, when running in a vacuum
- Different condition of the surfaces, with adhesion prevailing in a vacuum and abrasion in air

4.2   COATINGS AND DRY LUBRICANTS

In the previous section, it was pointed out that there are, in fact, no "dry" or clean surfaces but that, except in a near-perfect vacuum, they all carry some oxide or other compound film, or at the least, a monomolecular layer. This is one example of the overlapping nature of tribological operations—in the present case, that of dry surfaces with those that are deliberately coated with a tribological film. The inadvertent oxidation of most metals is a fast process, retarded only by the presence of the oxide layer itself. Freshly cleaned metals—copper, iron, aluminum, zinc, nickel, chromium, and others—will acquire 10 to 100 Å oxide layers in about 5 min. This will even take place at very low oxygen pressures. The process begins with physisorption and chemisorption, and proceeds all the way to the formation of a stable oxide on the surface.

TABLE 4.2

Wear Debris Sizes for Al–Si Alloys as a Function of Load (Pin-Bush Machine with Surface Speed 196 cm s^{-1})

Materials	Load (kg)	Percentage of Total Number of Particles of a Given Mean Particle Size (mm × 10^{-2}) for a Material at a Given Load								Total Sliding Distance (cm × 10^4)
		1.13	3.4	6.8	13.6	27.2	54.3	111.2	225.0	
Pure aluminum	0.5	31.0	22.4	19.8	16.7	8.2	1.65	0.12	0.00	96.191
	2.0	32.8	24.2	22.8	13.2	5.2	1.35	0.25	0.03	47.581
Al + 2% Si	0.5	33.3	24.5	20.5	15.1	5.6	0.80	0.13	0.00	75.471
	1.0	29.8	27.2	23.7	14.1	4.3	0.70	0.03	0.00	88.638
	1.5	31.0	27.5	21.9	13.1	4.8	1.20	0.10	0.02	90.074
	2.0	28.8	26.3	24.3	14.2	5.7	0.30	0.20	0.02	56.582
Al + 6.5% Si	0.5	32.8	25.1	20.9	14.3	5.4	1.20	0.20	0.00	89.811
	1.0	27.6	23.8	22.1	17.9	6.5	1.40	0.40	0.10	83.850
	1.5	31.8	27.3	23.5	12.2	3.8	1.00	0.25	0.03	103.564
	2.0	29.0	25.3	23.4	16.5	5.1	0.50	0.10	0.00	64.662
	2.5	30.3	24.8	21.7	13.6	6.3	2.30	0.13	0.00	7.876
Al + 12% Si	0.5	37.5	23.3	18.0	13.6	6.9	0.58	0.00	0.00	71.880
	1.0	36.2	28.2	20.6	10.6	3.6	0.62	0.00	0.00	87.381
	1.5	26.0	24.0	22.7	18.1	8.1	0.64	0.30	0.04	101.745
	2.0	26.7	24.2	23.2	18.2	6.5	1.00	0.06	0.00	71.856
	2.5	29.3	25.2	23.0	15.7	5.7	0.90	0.05	0.00	97.017
	3.0	27.7	24.5	21.8	15.9	7.6	2.30	0.13	0.00	83.826
Al + 21% Si	1.0	27.5	24.5	22.7	16.5	7.1	1.40	0.14	0.00	97.591
	2.0	28.7	25.1	23.3	15.5	6.0	1.10	0.14	0.02	101.769
	2.5	32.0	25.0	22.4	13.8	5.0	1.20	0.26	0.07	101.505

Source: Sarkar, Chapter 8. 16.

4.2.1 GENERAL NATURE OF COATINGS

From previous consideration, it is apparent that, in order to reduce friction between metals, it is necessary to have low values for either A, s, or both. This can be achieved only by depositing a thin film of soft metal on a hard substrate. In such an arrangement, the shear strengths will be of the soft layer while the contact surface, A, will change very little because the hard metal supporting the load will undergo little deformation. This, naturally, will hold only as long as there is no breakdown in the soft film. One must, thus, have a layer of appropriate thickness, neither too thick nor too thin. Indium film on steel is shown in Figure 4.17. The friction first decreases with the decrease of film thickness because this results in reduction in the track area. However, when the thickness reaches a level on the order of 10^{-6} cm (30 atomic layers), the film has been broken through and the friction rises again. This peculiar

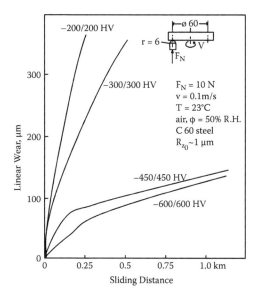

FIGURE 4.15 Effect of hardness on wear in air.

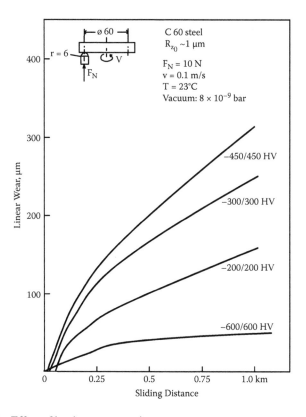

FIGURE 4.16 Effect of hardness on wear in a vacuum.

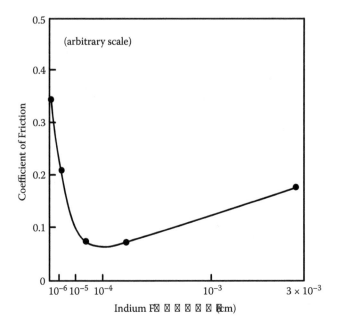

FIGURE 4.17 Effect of solid film thickness on friction coefficient : indium coating on steel.

knee-shaped frictional behavior occurs when solid film thickness decreases and has, indeed, its correspondence in liquid lubrication under conditions of starvation (4). An additional feature of metallic films is that, in cyclic operation, they are prone to be worn off due to the slider repeatedly traversing the same track.

The effect of a rise in temperature is shown in Figure 4.18. Friction decreases linearly with temperature, reaching a minimum at the melting point of the layer.

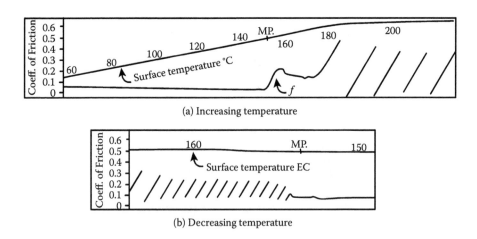

FIGURE 4.18 Effect of temperature rise on friction coefficient: indium coating on steel.

When melting is completed, friction rises again. With films of both indium and lead, the friction drops to nearly half its value in going from room to melting temperature. This decrease in friction when the solid liquefies has, of course, its correspondence in the use of lubricants under hydrodynamic conditions. Another continuum aspect between solid and liquid films is observed in the use of mercury films on silver. A minute quantity of mercury produces a marked reduction in friction because the mercury wets the silver and forms an amalgam, acting as a full-fledged lubricant. Similar phenomena are likely to occur in a number of other cases, including commonly used bearing materials, particularly under boundary lubrication conditions.

Whereas the coefficient of friction, f, for both unlubricated and thinly lubricated steel surfaces remains constant with load, when a coating is used, the value of f falls with an increase in load, as shown in Figure 4.19. In the first two cases (I and II), the area over which metallic contact occurs is proportional to load; hence, the constancy of f. With the use of a coating, an increase in load produces only a small increase in deformation of the underlying hard metal. The increase in the frictional force is likewise small, yielding a drop in f. In the case of indium on steel (III), the initial value was $f = 0.2$; at the heaviest load, it leveled off at $f = 0.04$. In this respect, too, solid lubricants simulate hydrodynamic operation, where an increase in load produces a drop in the coefficient of friction.

As was stated earlier, the thickness of the coating is an important factor in producing both low friction and low wear rates. The effect of changing the film thickness of a nickel film on an AISI steel is shown in Figure 4.20. As can be seen, a minimum occurs at about 1 micron film thickness, on either side of which the wear rates are higher. Table 4.3 gives wear rates and friction coefficients for a number of coatings on steel of various hardness. When a nickel film was used on very soft steel, there was immediate failure. On the other hand, when the coatings were much softer than the underlying steel, good results for both wear and coefficient of friction were

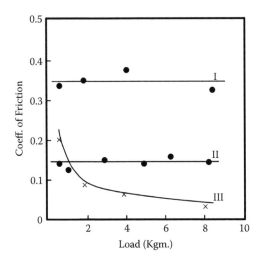

FIGURE 4.19 f As a function of W for coated and uncoated surfaces.

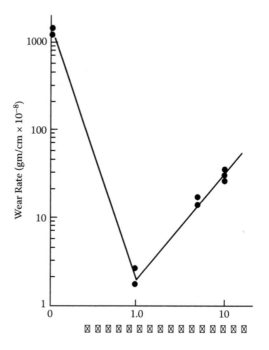

FIGURE 4.20   Optimum thickness of solid film: Ni on steel.

TABLE 4.3
Comparative Wear Resistance of 1 micron Coatings on Steel

Coating	Substrate and Hardness (kg/mm²)	Wear Rate ($\times 10^{-8}$ mg/cm)	Coefficient of Friction
Cd	AISI 1018 (84)	3.6	0.35
Cd	AISI 1095 (170)	1.8	0.25
Cd	AISI 4140 (270)	3.6	0.35
Cd	AISI 4140 (370)	3.6	0.25
Ag	AISI 1095 (170)	1.8	0.33
Au (diffused)	AISI 1095 (170)	1.8	0.85
Au (Ni underlayer)	AISI 1095 (170)	−1.8[a]	0.9
Ni	AISI 1018 (84)	Immediate failure	—
Ni	AISI 1095 (170)	Immediate failure	—
Ni	AISI 4140 (270)	Immediate failure	—
Ni	AISI 4140 (370)	Immediate failure	—
Ni	AISI 4140 (460)	1.8	0.45

Source: Cheng H. S. 1978, 4.

Note: Experimental conditions: wear-tested in argon, normal load 2.25 kg, sliding distance 108 m.

[a]   These specimens gained weight.

TABLE 4.4
Types of Solid Lubricant Coatings

Inorganic

Layer-lattice	MoS_2, graphite, graphite fluoride
Other inorganic solid compounds	CaF_2, PbO, TiO_2
Soft metals	Pb, Ag, Au, Pb-Su

Organic

Soaps, fats, waxes	Calcium stearate, dimethylstearamide
Polymeric plastics	Polytetrafluoroethylene (PTFE), polyimide
Layer-lattice solids	Phthalocyanines
Diamond-like carbon (DLC)	i-C a-CiH

obtained. Thus, the important factor, as postulated earlier, is the relative hardness of the two materials that represents the important parameter in the use of soft coatings.

4.2.2 KINDS AND CHARACTERISTICS OF COATINGS

Table 4.4 lists a number of commonly used coatings characterized, primarily, by their low shear strength, either because of their composition or due to the nature of the material. However, some films that do not possess low shear strength—diamond-like carbon (DLC) or TiN—are also capable of reducing friction due to their adhesion being of low strength. In addition to low shear strength, these coatings must also not be too thick and have a low wear rate. A wear rate of a monolayer (0.3 nm) per pass would remove 1 micron of material in 3000 passes (18).

The most commonly used solid lubricants are molybdenum disulfite and graphite. The lubricating properties of MoS_2 are ascribed to the ease of sliding of the lamellar layers of its structure. Each layer consists of two planes of sulfur atoms and an intermediate plane of molybdenum atoms. These layers are held together by weak van der Waals bonds, yielding the low shear strength desired. Its main disadvantage is its affinity to oxygen and water vapor, the latter tending to degrade its lubricating qualities. The lubricating effect of MoS_2 against AISI steel is shown in Table 4.5.

TABLE 4.5
Wear Rates for Dry and MoS_2-Covered Surfaces

	Wear Rate (cm^3/Nm)	
	Load: 58.8 N	**Load: 255 N**
Unlubricated	0.561×10^{-6}	0.561×10^{-6}
MoS_2 film	0.5405×10^{-10}	0.714×10^{-10}

Note: Speed: 0.60 m/s.
Source: Dorinson and Ludemea, 1985, Chapter 13.

Under conditions of the MoS_2 being constantly replenished, it reduced the wear rate by a factor of 4 or 5 orders of magnitude (22,23). In fact, graphites, on the other hand, need moisture to achieve their low friction and wear characteristics. Graphite is usually deposited by bonding it to a substrate with a resin binder or by rubbing layers of graphite powder onto the substrate. Graphite fluoride is a material with the generic formula of $(CFx), n$, resulting from the action of elemental fluorine on graphitic carbon. The value of x can range from 0.68 to 1.12, with n being a large number. Soft metal films such as Pb, Au, and Ag (and in some cases alloys such as Pb–Sn) act as lubricants by shearing at a lower stress level than either of the substrates to which they are deposited. However, to achieve long life they must meet other requirements: optimal thickness of 0.1 to 0.2 microns (18) and good adhesion.

When abrasive delamination or chemical instability presents a problem, it is often desirable, from the standpoint of wear, to have a hard and tough surface. However, with most metals, hardness can be obtained only at the expense of toughness. The solution, then, is to coat a tough material substrate, not with a soft but a hard film. The friction and wear will be reduced when this coating is sliding against a softer material. The role of the hard layer is to prevent plowing: the mechanism responsible for abrasion and delamination wear (6). Coatings for such applications are carbides, oxides, nitrides, borides, and amorphous glasses. Carbides have a higher hardness and higher melting point than the oxides and nitrides. In some cases, a combination of both soft and hard layers may be advantageous. Schematically, this is portrayed in Figure 4.21 (7,8). Figure 4.22 shows the wear resistance of surface coatings under both adhesive and abrasive wear. For an adhesive wear process, the nitrided coatings exhibit the best behavior; the poorest was vanadized steel. In contrast, for an abrasive process, the best results were obtained for the vanadized surface coating. Thus,

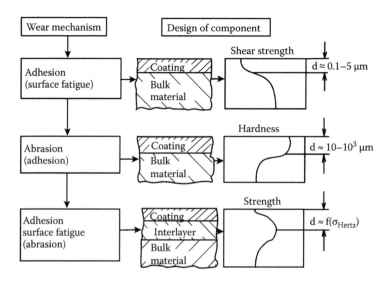

FIGURE 4.21   Topography of various coatings.

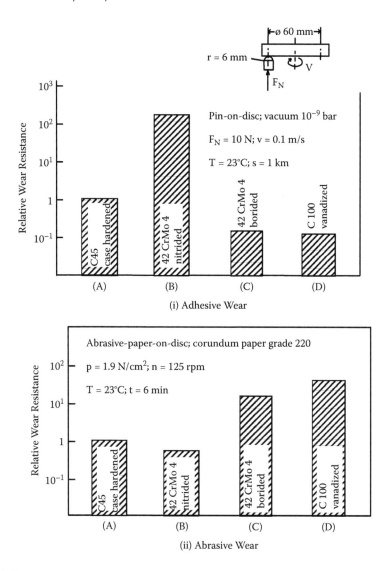

FIGURE 4.22 Response of various coatings in adhesive and abrasive wear modes.

the studies for adhesive and abrasive wear processes, using some diffusion surface coatings, yielded different rankings for the employed surface coatings, as follows:

For Adhesive Wear Resistance	For Abrasive Wear Resistance
1. Nitriding	1. Vanadizing
2. Case hardening	2. Boriding
3. Boriding	3. Case hardening
4. Vanadizing	4. Nitriding

4.3   BOUNDARY LUBRICATION

Quite often, a distinction is made between boundary and mixed lubrication. When this is done, there are no definitions attached to this distinction for good reason; if at all, they differ only in degree. Here, these two terms will be used interchangeably. Both refer to the tribological state; then, though a fluid lubricant is present, the operational conditions are such that extensive portions of the interface are subject to asperity contact. Thus, one may treat boundary lubrication as being closer to the regime of solid lubricants, whereas mixed lubrication would be contiguous with elastohydrodynamic operation.

4.3.1   THE NATURE OF BOUNDARY LUBRICATION

There is much that remains unknown about boundary lubrication because the subject involves fluid dynamics, plastic flow, thermal effects, chemical and molecular interaction, change of phase, and others. Even the elementary tasks of measuring film thickness or the pressure field are difficult. Perhaps the most vital piece of information—the relative areas either free of or under asperity contact—is still beyond experimental reach. Thus, unlike the previous sections where most of the information came from experiments, in the case of boundary lubrication, analytical treatments tend to dominate direct experimental approaches.

Early in the century, the main picture of boundary lubrication was that of Hardy. Treated mainly in terms of chemical and molecular interactions, it was postulated that when a lubricant is introduced, its molecules are physically adsorbed and oriented so as to form unimolecular layers on the mating surfaces. Slip over the asperities is, then, not between metals but between the molecular layers. In essence, the interacting asperities ride over each other like tiny lubricated sliders.

However, much of the available evidence makes this theory insufficient for providing a full explanation. Microscopic examination shows that the gouging resulting from wear is much deeper than the dimensions of a molecular layer. There is sufficient micrographic evidence attesting to the surfaces undergoing adhesion and abrasion. Thus, friction in boundary-lubricated metals cannot be due simply to the sliding of one monomolecular layer over another, as interpreted by Tabor and portrayed schematically in Figure 4.23. When a lubricant is present, it is trapped between the interacting surfaces and pressurized. At high pressures, the lubricant film is likely to break down, resulting in direct metal adhesion. Also, when sliding is appreciable, high temperature will lead to a breakdown of the trapped lubricant, causing additional metallic junctions. Resistance to motion, then, is offered by the metallic junctions as well as by the lubricant film, or

$$F = A[s_m + (1 - \alpha)s_L]$$

$$(4.7)$$

where
A = area supporting the load
α = fraction of A where lubricant film has broken down
s_m = shear strength of metallic junction
s_L = shear strength of lubricant film

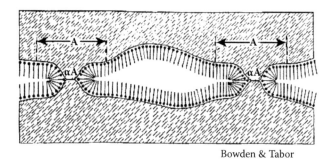

Bowden & Tabor

A - Area supporting the load
αA - Metallic junction

FIGURE 4.23   Tabor model for boundary lubrication.

Given that s_m is so much higher than s_L, even a small breakdown of the lubricant film will appreciably affect the friction between boundary-lubricated surfaces.

4.3.2   EXPERIMENTAL RESULTS

The first set of test results to be cited will be for near-zero velocities, thus eliciting the quasi-stationary character of boundary lubrication under load. The next two sets of experimental data will consider the use of different lubricants under sliding conditions.

The near-zero sliding tests were performed using stainless steel surfaces with two lubricants: stearic acid and cholesterol. These were deposited in either mono- or multimolecular layers. With the film placed on the upper surface and the lower moving at 1 cm/s, the resulting friction was as shown in Figure 4.24. While, initially, the friction for various molecular layers was the same, about $f = 0.1$, as sliding continued, friction proved to be less for the thicker layers. This was due to the gradual loss of

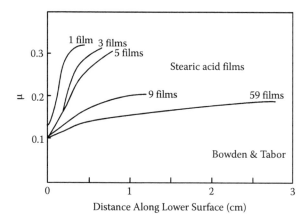

FIGURE 4.24   Coefficient of friction with upper surface lubricated and sliding.

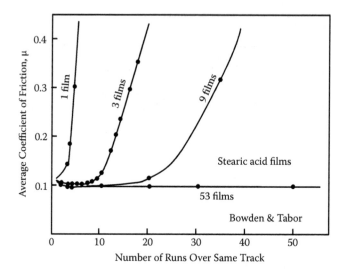

FIGURE 4.25 Coefficient of friction with lower surface lubricated and upper surface reciprocating.

the thin lubricant film on the upper moving surface. In the second set of experiments, the lower surface was lubricated while the upper surface remained unlubricated. Because, under these conditions, fresh layers of film came under sliding, there was no effect of the number of molecular layers used. However, when the upper member was cycled (that is, run repeatedly reciprocated over the lower surface), then the data of Figure 4.25 were obtained, essentially duplicating the previous results. The foregoing two figures are for the stearic acid; similar results were obtained for the cholesterol lubricant. The only difference with the latter was that, even on start-up, the thinner the film, the higher the initial friction coefficient. For both lubricants, the results in the speed range of 0.001 to 1 cm/s were independent of the velocity. Another general conclusion was that as the film thickness increased, there was a decrease in the rate of wear of the film (21).

The next series of experiments employed sliding velocities up to 100 cm/s and an alkyl-naphthalene oil as a lubricant (28–35 cSt at 38°C). The main measurement was wear rate, given in microns, of material worn away. The test rig was a pin-on-disk setup with an apparent contact area of 0.3 mm². The pin material was a steel of 64 HRC hardness and a surface roughness of 0.09 micron R_{max}. The disk material was steel with a 15 HRC hardness and a surface roughness similar to that of the pin. Figure 4.26 gives a representative plot of the rate of wear in terms of sliding distance. From this and similar plots, Akagaki et al. (1) concluded that the wear debris originated from the plastic flow at the asperity junctions, that amount of wear is a direct function of material hardness, and that the wear rate decreased with an increase in sliding velocity.

Another set of pin-on-disk experiments using an SAE oil was provided by Czichos (7). Material hardness and surface roughness of the steels used were the two basic variables: the first ranged from 300 to 600 HV and the second from 1 to 4 microns

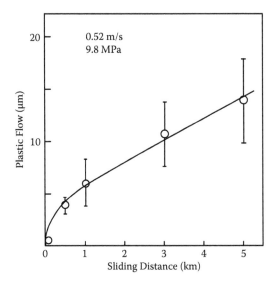

FIGURE 4.26   Level of wear versus sliding distance.

rms. Figure 4.27 provides profilographs obtained for the different hardness combinations employed. Essentially no wear in the lower surface occurred when its hardness was higher than that of the stationary pin. Table 4.6 gives the relative wear rates for the combinations of hardness and surface roughness used. One of the conclusions reached was that the lowest wear rate was obtained with a rough stationary pin sliding against a hard, smooth moving disk; reversing this condition resulted in a 25-fold increase in wear rate. The difference in wear rates between sliding and rolling motion is provided in the diagram of Figure 4.28.

4.3.3   THEORETICAL ANALYSES

As intimated previously, little experimental information is available on the inner workings of boundary-lubricated interfaces. This only tempts the analyst to formulate

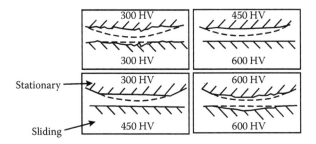

FIGURE 4.27   Modes of wear for different hardness ratios.

TABLE 4.6

Relative Wear Rates for Different Hardness and Surface Roughness Conditions (C60 Steel)

Pin (stat.)	Disk (mov.) R_{ms} (µm)	~300 HV		~600 HV	
		1	4	1	4
~300 HV	1	20	22	2	3
	4	25	22	1	3
~600 HV	1	19	19	14	14
	4	25	20	12	11

Source: Czichos et al., 1985.

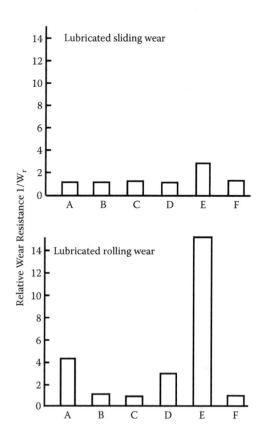

FIGURE 4.28 Wear resistance in rolling and sliding motion.

models and proceed toward a possible mathematical solution. Because these models are, for the most part, tentative, it would not be very fruitful to cover these analyses in detail. What will be done here is to present the approaches taken by several investigators and whenever an analysis is carried through to what may be considered a solution, this will be cited as well.

It is clear that a critical parameter in boundary lubrication is the ratio $\lambda = (\frac{h}{\beta})$, where h is the film height and β is the height of the asperities. Ideally, one would like to have a functional distribution of this ratio over the surfaces. But, given that the number of asperities is practically infinite, statistical or other methods have to be used to arrive at an estimate. This would still not provide the two domains of interest—one without and the other with asperity contact—nor, certainly, the degree of overlap or contact area of the individual protuberances. However, an attempt in this direction, using β as a basic parameter, is contained in a 1997 paper by Hua et al. (10). The problem is formulated as follows:

In the valleys, a sufficiently thick film exists to separate the two surfaces; otherwise metal-to-metal contact prevails. In addition, temperature effects are considered, including their impact on both lubricant and boundary film breakdowns.

A typical surface roughness of a gear tooth is shown in Figure 4.29a. By using a reference plane and a residual topography, the surface to be used in the

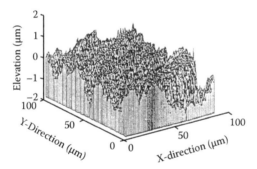

(a) Original surface (Gear tooth surface - Shaved)

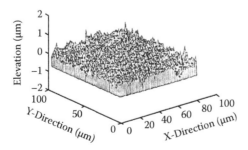

(b) Frequency-domain filter function (highpass Gaussian) <12_12>

FIGURE 4.29   Surface topography of gear tooth.

analysis is as shown in Figure 4.29b. Facing it is a smooth parallel plane. At the rim of asperity contact areas, the hydrodynamic pressures are made equal to the elastic pressures. Together, the Hertzian and viscous pressures then contribute to the developed total load capacity. An iteration procedure is, naturally, required to satisfy the foregoing boundary conditions. The flowchart for this procedure is given in Figure 4.30. Numerical solutions for contact pressures and for film thickness are given in Figure 4.31 for pure rolling and in Figure 4.32 for pure sliding. In the latter case, the asperities are stationary, and a pressure spike is generated at the trailing end due to the asperity blocking the outflow of lubricant. According to the foregoing analysis, the contact areas of asperities contribute about 20% of the total load. Finally, Figure 4.33 gives the temperature distribution in the interface.

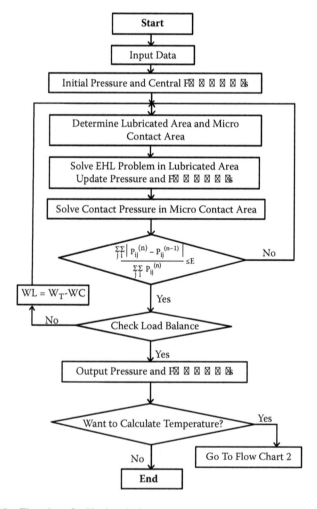

FIGURE 4.30 Flowchart for Hua's solution.

$U_r = 0.871$ (m/s) s = 0.0 ha = 0.193 (µm) W = 60 (N)

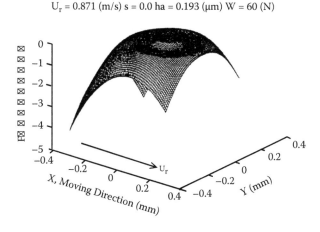

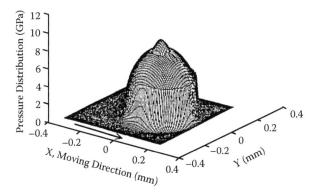

FIGURE 4.31   Film thickness and pressure distribution in rolling boundary lubrication.

A most comprehensive analysis is contained in a 1985 paper by Oh (14), in which the formulated equations are sufficiently general so that, by eliminating certain terms, they reduce themselves to either that of dry contact or hydrodynamic lubrication. The approach is of dividing the areas into three subregions: metal contact, hydrodynamic lubrication, and cavitation or zero pressure. The mathematical approach used is that of a generalized complementary set of equations. By using all or only some of the operators involved, separate lubrication regimes can be extracted from the general solution. Given the extensive mathematics involved, a specific case of a journal bearing on an elastic support, shown in Figure 4.34, is given as a sample solution. The data for the bearing are as follows:

$$D = 2.5\,\text{cm}; \quad \left(\frac{L}{D}\right) = 1; \quad C = 0.0125\,\text{mm}; \quad \mu = 0.0069\,\text{Pa} - s; \quad E = 19.2\ 10^{10}\,\frac{N}{m^2},$$

$$v = 0.3; \quad \text{OD of support} = 5\ \text{cm}; \quad N = 2,000\,\text{to}\,4,000\,\text{rpm}; \quad W = 1,200\ \text{to}\ 33,600\,\text{N}$$

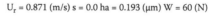

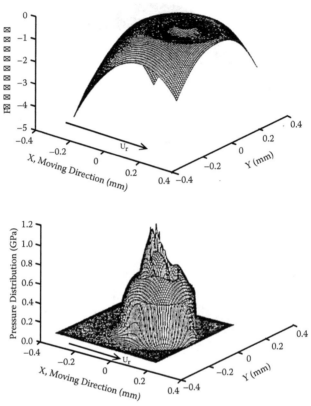

FIGURE 4.32   Film thickness and pressure distribution in sliding boundary lubrication.

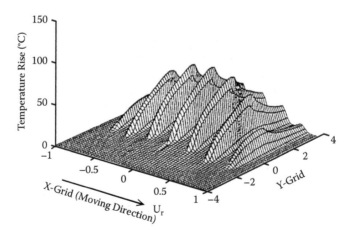

FIGURE 4.33   Temperature distribution in sliding for boundary lubrication.

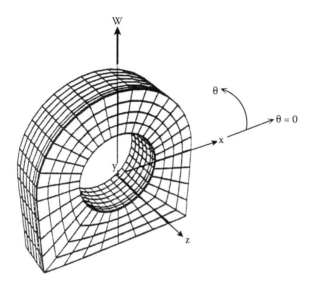

FIGURE 4.34 Journal bearing with elastic support.

Solutions for pressure distribution, frictional force, and eccentricity ratio as well as bearing regions under either asperity contact or under fluid film are given in Figures 4.35 to 4.40. The following comments can be made with regard to these figures:

1. Pressure profiles. Due to symmetry, Figures 4.35a to 4.35e show only half of the axial distribution. These are all for conditions of 2,000 rpm and loads ranging from 1,200 to 33,600 N.
2. Figure 4.36 shows the various portions of the surface under either asperity contact or hydrodynamic operation. Shading density and pattern reflect the contact conditions at the vertices of an element. A cross-hatched pattern denotes metal contact while a dot pattern relates to a fluid film. Cavitating regions show no shading. The shading density increases with the number of nodes subject to each of the three modes. Specifically, referring to Figures 4.35 and 4.36, we have:

   For the lightly loaded case, (a), elastic distortion is minimal, and the solution is typical of a conventional hydrodynamic case.

   In case (b), elastic distortion becomes noticeable and the bimodal mode is close to that of an EHL solution.

   For case (b), the deformation causes the bearing to wrap around the journal.

   Metal contact first occurs at a load of 15,600 N, primarily along the outer edge of the bearing. With increasing load, the contact area increases and moves inward, intruding upon the previously cavitated regions.

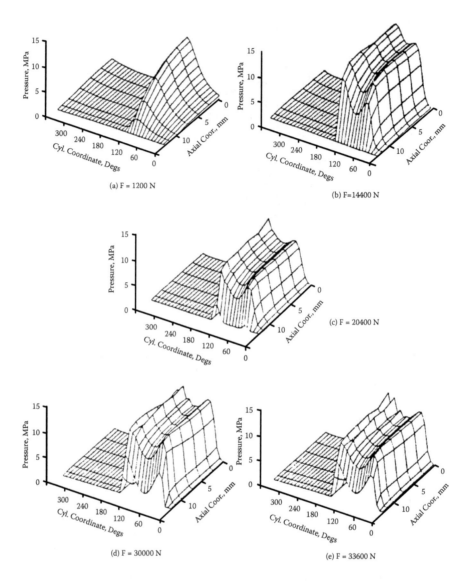

FIGURE 4.35 Pressure distribution in journal bearing operating in boundary lubrication regime.

3. Friction. In the hydrodynamic region, the friction was based on the viscous shear while in the contact areas on Coulomb friction with a friction coefficient of 0.08. The total coefficient of friction is plotted in Figure 4.37, which resembles the Stribeck curve. The knee in the curve is the locus of transition from hydrodynamic to mixed lubrication regime.

4. Journal eccentricity. The variation of ε and attitude angle with load is shown in Figures 4.38 and 4.39. Here, ε can be larger than unity due to

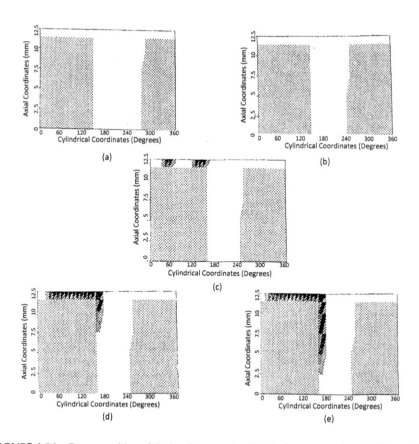

FIGURE 4.36 Decomposition of the bearing area, load levels the same as those in Figure 4.35 (Ref. 14).

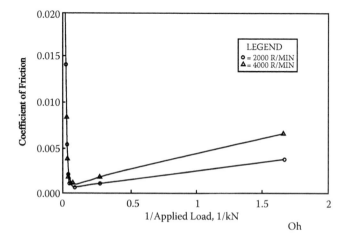

FIGURE 4.37 Coefficient of friction in journal bearing operating under boundary lubrication.

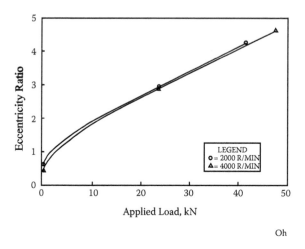

Oh

FIGURE 4.38   Eccentricity ratio in journal bearing operating under boundary lubrication.

the deflection of the bearing surface. As the load increases, the direction of journal displacement aligns itself more and more with the load vector—a consequence of the departure from purely hydrodynamic operation.

5. Proportion of metal contact area. These results are shown in Figures 4.40, 4.41, and 4.42. With an increase in load, both metal contact and the hydro-dynamic regions increase at the expense of the cavitated region. Of interest is the fact that the knee in the Stribeck curve—Figure 4.37—coincides with the points in Figure 4.42, where metal contact first appears. As noted earlier, as little as 6% of metal contact produces a significant rise, about 4 times, in the friction coefficient.

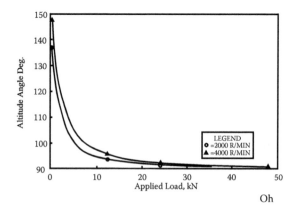

Oh

FIGURE 4.39   Altitude angle in journal bearing operating under boundary lubrication.

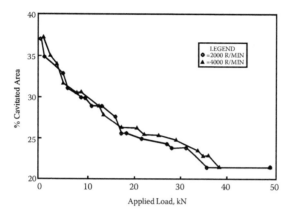

FIGURE 4.40 Extent of asperity contact area in journal bearing operating under boundary lubrication.

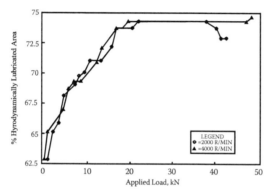

FIGURE 4.41 Region of hydrodynamic lubrication in journal bearing operating under boundary lubrication.

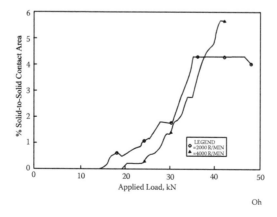

FIGURE 4.42 Region of asperity contact in journal bearing operating under boundary lubrication.

4.4 ELASTOHYDRODYNAMIC LUBRICATION (EHL)

Just as boundary and mixed lubrication constitute wings abutting the solid lubricants and the EHL regimes, so, too, we have partial as well as full EHL modes, which can be interposed between the mixed and hydrodynamic lubrication regimes. Furthermore, EHL can be extended to overlap the domain of solid lubricants. The pressures in EHL applications are often high enough to boost lubricant viscosity to extremes and convert the fluid lubricant into a solid; this reverses the process mentioned in Section 4.2 whereby, under high temperatures, solid lubricants are often liquefied into a fluid film.

As stated, the area of EHL can be subdivided into partial or full. The dividing line is whether the ratio of average film thickness to the rms of the combined surface asperities is larger or less than 3. In the partial EHL domain, it is visualized that, although a good fluid film exists, it is interrupted by small areas of contacting asperities. It is, in fact, believed that most EHL applications, such as rolling element bearings, seals, gears, and cams, operate in the partial EHL domain. Two features distinguish EHL from ordinary hydrodynamic lubrication: one is that the loads and geometry are such that elastic deflection occurs in one or both surfaces; and that the motion consists of both rolling and sliding, the latter due to the different angular velocities of the surfaces. As shown in Figure 4.43, the principal mating modes are line contact and point contact. A more general set of mating surfaces is that of Figure 4.43c, which, under load, yields an elliptical contact area.

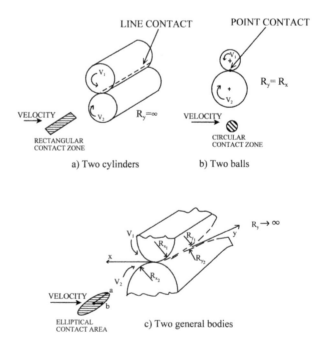

FIGURE 4.43 Principal mating modes in EHL.

A number of important parameters make their appearance in treating EHL. The major ones are as follows:

- Slide to roll ratio; $s = 2\frac{(U_1-U_2)}{(U_1+U_2)}$
- Viscosity as a function of pressure; $\mu = \mu_o e^{\alpha p}$
- Equivalent radius: $R = \frac{R_{x1}R_{x2}}{(R_{x1}+R_{x2})}$
- $\frac{1}{E_D} = \frac{1}{2}[\frac{\frac{(1-v_{12})}{E1+(1-v_{22})}}{E_2}]$
- Load: $W = \frac{W}{E_D R}$
- Velocity factor: $U = \frac{\mu_o U}{E_D R}$
- Velocity $U = \frac{1}{2}(U_1 + U_2)$
- Material factor: $G = \alpha E_D$

Above, the subscripts 1 and 2 denote the two surfaces.

In a most general way, one can say that the dynamics of an EHL film result from a superposition of hydrodynamic and Hertzian stresses, shown in Figure 4.44. Analyses of EHL involve a great deal of mathematics; even so, it has to be supplemented by empirical and experimental coefficients in order to obtain concrete solutions. The following material is taken from a review by Cheng (5), culled

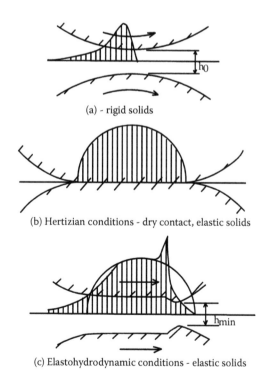

(a) - rigid solids

(b) Hertizian conditions - dry contact, elastic solids

(c) Elastohydrodynamic conditions - elastic solids

FIGURE 4.44 Hydrodynamic and Hertizian components in EHL.

from an extensive roster of analytical and experimental works including those of Cheng himself.

4.4.1   RESULTS FOR FULL EHL

In general, an EHL film has three subregions: a converging film at the inlet, a parallel section over most of the contact area, and a spike or protuberance near the outlet, as sketched in Figure 4.44c. Examples of film thickness in a line contact geometry are given in Figure 4.45. Pressure variations are shown in Figure 4.46 where, as noted, there are spikes at the outlet due to the blockage of flow near the exit. Contour plots for the film thickness in a circular contact are given in Figure 4.47. Finally, values of h_{min}, in terms of several relevant parameters, are plotted in Figure 4.48.

Of particular interest is friction in EHL contacts, often referred to as *traction*. Its ramifications are even more complex than the other sought-after quantities. One set of results is shown in Figure 4.49. In the low-slip region, the traction rises with slip as in a Newtonian fluid. As slip increases, traction levels off or even drops. As shown in Figure 4.50, elements of fluid deform due to the higher velocity at the bottom. Part of the strain is elastic and is recoverable; that due to viscous dissipation is not. If the ratio of the recoverable to the permanent strain is high, the friction is governed by elastic deformation; if low, friction will follow viscous behavior. At very high sliding speeds, the lubricant most likely shears like a plastic solid.

4.4.2   RESULTS FOR PARTIAL EHL

This, as defined previously, is the regime where the ratio of average film thickness is less than three times the composite surface roughness. Under these conditions, the film is interrupted by asperity interference. As would be expected, any presence of asperity contact introduces severe complications. Whereas, usually, surface

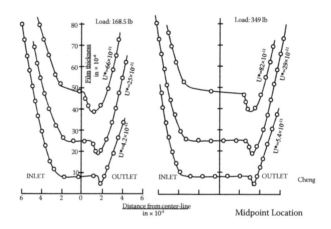

FIGURE 4.45   Film profiles in rolling EHL.

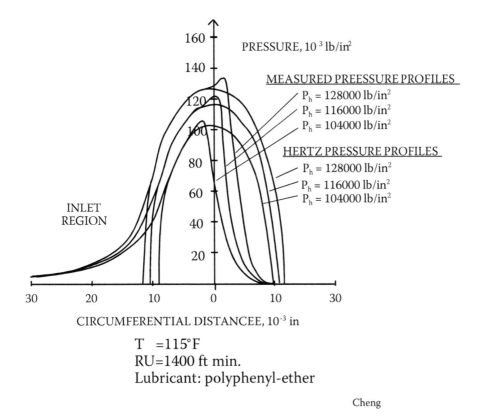

FIGURE 4.46 Pressure profiles in EHL.

roughness is determined by a single parameter—the standard deviation of the asperity heights from a mean plane—this is no longer sufficient for the present purposes. There are four notions involved here that need to be specified, namely:

1. The standard rms of surface roughness.
2. Height distribution function. This depends on whether the asperity height distribution is Gaussian or departs from it.
3. Autocorrelation function. This is a measure of the wave length structure of surface profile in a given direction.
4. x-y Correlation length given by a factor γ. This relates the ratio of wavelengths in the x and y directions. It is partly linked to the above autocorrelation.

Figure 4.51 depicts the flow pattern for longitudinally oriented ($\gamma > 1$) and isotropic ($\gamma = 1$) roughness. Results for the actual film thickness in partial EHL versus an assumed smooth film thickness are given in Figure 4.52. For an isotropic case, there is very little difference in the two film thicknesses. On the other hand, the effect for $\gamma = 6$ turns out to be less than for $\gamma \Rightarrow \infty$, which represents the case of longitudinal roughness only.

Dimensionless film thickness $H=h/R_x$		Dimensionless pressure $P=p/EN$	
A	4.3×10^{-6}	A	1.7×10^{-3}
B	4.6	B	1.6
C	5.0	C	1.5
D	5.5	D	1.4
E	6.0	E	1.2
F	6.6	F	1.0
G	7.4	G	0.7
H	8.2	H	0.3

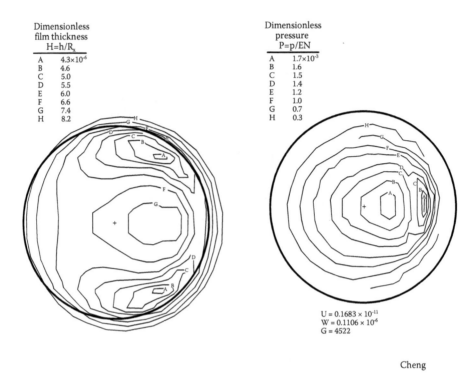

$U = 0.1683 \times 10^{-11}$
$W = 0.1106 \times 10^{-6}$
$G = 4522$

Cheng

FIGURE 4.47 Film thickness and pressure maps for circular contact area.

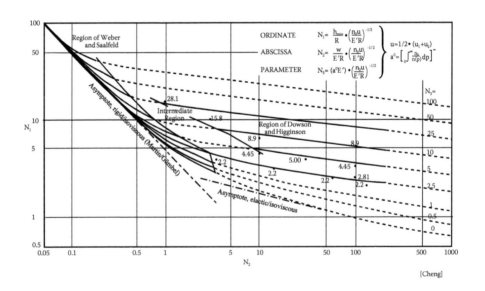

[Cheng]

JBW-081600

FIGURE 4.48 General performance of an EHL film.

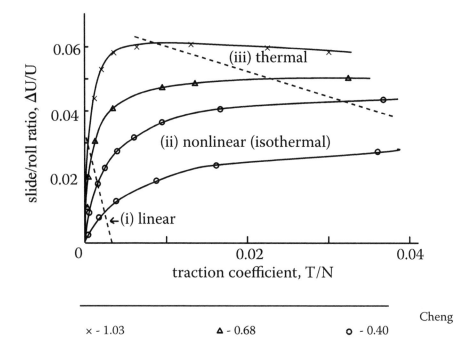

FIGURE 4.49   Traction coefficient in EHL for elliptical contact.

4.5 CONTINUUM ASPECTS

In this chapter, we have taken a look at four major modes of tribological operation, namely,

1. Dry surfaces, in practice, possible only under exceptional conditions such as a high vacuum
2. Surfaces with unintentional or intentionally introduced coatings and solid lubricants
3. Boundary and mixed lubrication
4. Partial and full elastohydrodynamic operation

It was seen that in almost any deliberate attempt to keep categories (1) and (2) apart, they merge into each other, the only difference being that in the "dry" condition the coating is most likely to be an oxide while in case (2) the films can cover a wide range of organic and inorganic materials. The boundary and mixed lubrication regimes operate under a wide range of relative proportions of asperity contact and fluid film operation. Spanning regimes (2) and (3) is the occurrence of high temperatures in solid lubricants, which may lead to the melting of the solid, duplicating conditions under boundary lubrication or even those of the hydrodynamic regime, known as melt lubrication, encountered, for example, in steel mill operations. The extreme edge of mixed

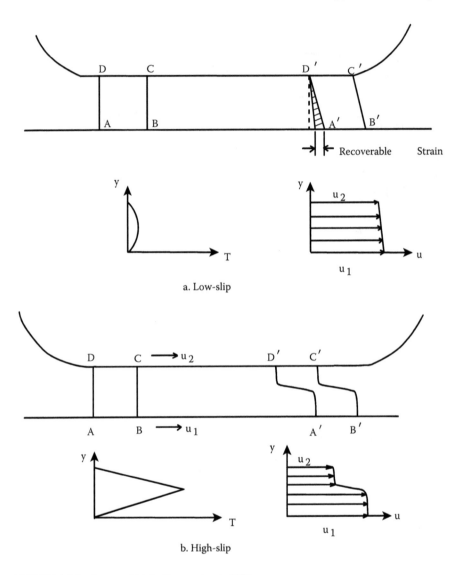

FIGURE 4.50 Low and high slip traction in EHL.

lubrication wherein fluid film covers only a small area can be said to coincide with the partial EHL mode, where there is only a small subregion of asperity interaction. On the other hand, the encountered very high stresses and pressures in elastohydrodynamic lubrication often convert the fluid into a solid, reaching all across the boundary and mixed lubrication regimes to link up with the region of solid lubricants. We thus have a wide spectrum of contiguous operations in which, at most, there is only a difference in degree as to which is the dominant mode prevailing at the interface.

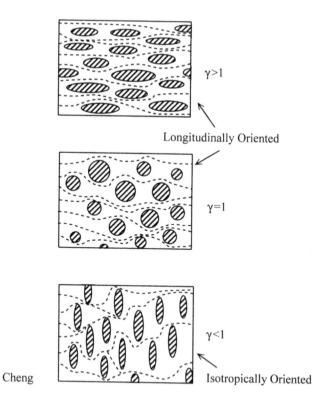

Cheng

FIGURE 4.51   Asperity contact areas in partial EHL.

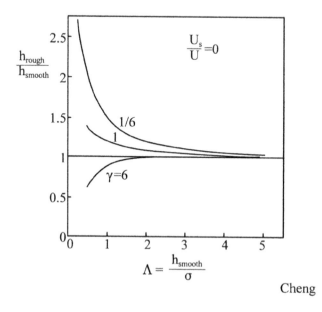

FIGURE 4.52   Film thickness in partial EHL for smooth.

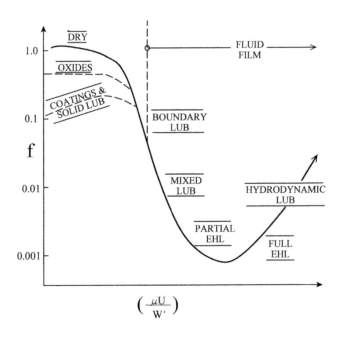

FIGURE 4.53 Coefficient of friction for several tribological regimes in terms of ($\mu U/W'$).

Not only is this overlap phenomenological and experimentally verified, but it can also be supported by some recent analytical treatments, where the classical Reynolds equation has been extended to cover simultaneously several regimes. As was shown in the Oh paper, one generalized differential equation can cover no fewer than three tribological regimes: dry lubrication, elastohydrodynamics, and pure hydrodynamic operation. By eliminating particular operators in this generalized equation, a particular mode can be obtained as a limiting solution. Otherwise, when solved in complete form, it provides a solution for the simultaneous presence of all three subregions.

When all six modes (including the subdivisions) are plotted against the familiar Stribeck curve, one obtains the plot of Figure 4.53, where the various regimes overlap one another along the span of the curve. In addition, there appears to be a relationship between the various modes of operation and film thickness. Figures 4.54 and 4.55 give values of some film thicknesses for tribological operations, ranging from monomolecular layers to full hydrodynamic conditions. Correlating these with the friction coefficient in Figure 4.56, one obtains another continuum, this time in terms of film thickness. Although the abscissas ($\frac{\mu U}{W'}$) and h in the two figures are in some way related, they are not quite identical. Figure 4.56 thus provides further support for the tribological continuity of the various lubrication regimes.

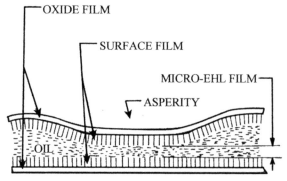

(a) Protective films for an asperity

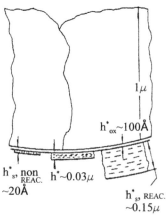

(b) Typical thickness of different surface films

Cheng

FIGURE 4.54   Film coatings on an asperity.

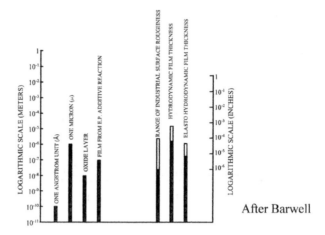

FIGURE 4.55   Thickness of film layers for various tribological regimes.

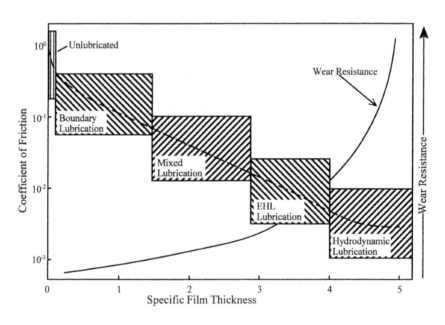

FIGURE 4.56   Coefficient of friction and wear rate for several tribological regimes in terms of film thickness.

REFERENCES

1. Akagaki, T., Kato, K., and Kayaba, T., Flow wear at sliding contact in boundary lubrication, *Proc. JSLE Int. Tribology Conf.*, May 1985 (Tokyo).
2. Barwell, F.T., Tribological application of surfaces, *Proc. Inst. Mechanical Engineers*, Vol. 182, Part K, 1967/8.
3. Bowden, F.P. and Tabor, D., *The Friction and Lubrication of Solids*, Clarendon Press, Oxford, 1986.
4. Heshmat, H., Starved bearing technology: theory and experiment, Ph.D. thesis, Mechanical Engineering department Rensselaer Polytechnic Institute, Troy, New York, December 1988.
5. Cheng, H.S., Fundamentals of elastohydrodynamic contact phenomena, *Fundamentals of Tribology*, Chapter 9, Suh, N.P. and Saka, N., eds., MIT Press, Cambridge, MA, 1978.
6. Jahanmir, S., A fundamental study on the delamination theory of wear, Massachusetts Institute of Technology, Department of Mechanical Engineering, Ph.D. thesis, 1977.
7. Czichos, H. and Habig, H.K., Wear of medium carbon steel: a systematic study of the influences of materials and operating parameters, *Proc. JSLE International Tribology Conf.*, May 1985 (Tokyo).
8. Czichos, H., *Tribology*, Elsevier Scientific Publishing Company, New York, 1978.
9. Dorinson, A. and Ludema, K.C., *Mechanics and Chemistry in Lubrication*, Chapter 13, Elsevier, Amsterdam, 1985.
10. Hua, D.Y., Qiu, L. and Cheng, H.S., Modeling of lubrication in micro-contact, *Tribology Lett*, 3 (1997), pp. 81–86.
11. Hutchings, I.M., *Tribology: Friction and Wear of Engineering Materials*, CRC Press, Boca Raton, FL, December 1992.
12. Kerridge, M., Metal transfer and the wear process, *Proc. Physical Society*, London, *Series B*, Vol. 68, pp. 400–407, 1955.
13. Kerridge, M. and Lancaster, J.K., The stages in a process of severe metallic wear, *Proc. Royal Society of London, Series A*, Vol. 236, 1956.
14. Oh, Kong Ping, The formulation of the mixed lubrication problem etc. as a generalized nonlinear complementarily problem, *Trans. ASME J. Tribology*, 108 (1986), pp. 598–604.
15. Rabinowicz, E., *Friction and Wear of Materials*, 2nd ed., John Wiley & Sons, New York, 1995.
16. Sarkar, A.D., *Friction and Wear*, Chapter 8, Academic Press, New York, 1985.
17. Sasada, T., Fundamental analysis of the adhesion wear of metals; severe and mild wear, *Proc. JSLE Int. Tribology Conference*, May 1985 (Tokyo).
18. Singer, I.L., Solid lubricating films for extreme environments, Material Research Society Symposium, Vol. 140, 1989.
19. Steijn, R.P., An investigation of dry adhesive wear, *Trans. ASME*, March 1959.
20. Suh, N.P., *Tribophysics*, Chapter 9, MIT Press, Cambridge, MA.
21. Tabor, D., Friction—the present state of our understanding, *ASME J. Tribology*, Vol. 103, April 1981.
22. Higgs, C.F., III, Heshmat, C.A., and Heshmat, H., Comparative evaluation of MoS_2 and WS_2 as powder lubricants in high speed, multi-pad journal bearings, presented at the ASME/STLE Tribology Conference, October 1998 (No. 98–TRIB–41), *ASME Transactions, J. Tribology*, July 1999, Vol. 121, pp. 625–630.
23. Higgs, C.F. III and Heshmat, H., Characterization of pelletized MoS_2 powder particle detachment process, 2000–TRIB–46, presented at the 1999 ASME/STLE Joint Tribology Conference and published in the *ASME J. Tribology*, July 2001, Vol. 123, No. 3, pp. 455–461.

5 Hydrodynamic–Solid Films Continuum

In dealing with the subject of tribological continuum, Chapter 4 had concluded with EHL operation: the first mode of widely recognized hydrodynamic operation. The present chapter will start with what may be considered the core of tribology—the traditional hydrodynamic lubrication of a wide spectrum of mechanical devices—and then go on to extend the continuum beyond it to such areas as grease, slurry lubrication, and powder films, just encroaching on the rarest of operations: kinetic flow. It will be seen that, even what was termed here the hydrodynamic mode, abuts at one end (namely, that of compliant surface bearings the discipline of EHL), while its other end leads us into powder film lubrication.

5.1 HYDRODYNAMIC FILMS

In dealing with the wide field of hydrodynamic lubrication, we intend here to concentrate on the essentials so that the other modes—from dry contacts to kinetic flow—can be contrasted against this traditionally central field of tribology. In particular, we would like to expand on the variety of boundary conditions possible for two reasons: because they are sparsely treated in the literature; and also because these nonconventional boundary conditions are essential features in grease and powder lubrication, forming an additional link in the body of tribological processes.

It is instructive, first of all, to position the subject of hydrodynamic lubrication within the overall framework of fluid dynamics, thereby highlighting the basic proposition of hydrodynamic theory. If we look at the three common fields of fluid dynamics shown in Table 5.1, we see that what distinguishes hydrodynamic lubrication from the other two disciplines is the neglect of inertial effects in the fluid film. The validity of this approach emerges from even a qualitative analysis (1), where the inertia terms turn out to be two orders of magnitude smaller than the viscous and pressure terms. Another general feature in hydrodynamic lubrication is that the clearance space between the interacting surfaces is some three orders of magnitude smaller than the surface dimensions; even when the film is only two orders of magnitude smaller, the hydrodynamic forces become insignificant.

The basic element of a tribological device is shown in Figure 5.1. The thing to note is that the fluid film has, generally, both converging and diverging regions. To produce hydrodynamic forces, there must be relative motion between the surfaces, which can be either in the longitudinal direction, normal direction, or both. From a fluid dynamics standpoint, the flow in a hydrodynamic film consists essentially of a combination of Couette flow, due to the velocity component U, and of a Poiseuille

TABLE 5.1
Common Branches of Fluid Dynamics

Area	Inertia Forces	Viscous Forces	Pressure Forces
Potential flow	☒	○	☒
Boundary layers	☒	☒	○
Fluid film tribology	○	☒	☒

flow, due to a pressure gradient. These component flows vary along the film in both the longitudinal (x) and axial (z) directions.

5.1.1 INCOMPRESSIBLE LUBRICATION

5.1.1.1 Basic Equations

Most conventional bearings use an incompressible fluid as a lubricant. The most common is petroleum oil, occasionally a synthetic liquid and, still more rarely, water. The two basic bearing designs are journal bearings, which support radial loads, and thrust bearings, which accommodate any axial loads exerted by the rotor.

The basic differential equation for incompressible lubricants is the familiar Reynolds equation, which in its simplest form is as follows:

$$\frac{\partial}{\partial x}\left[\frac{h^3}{\mu}\left(\frac{\partial p}{\partial x}\right)\right] + \frac{\partial}{\partial z}\left[\frac{h^3}{\mu}\left(\frac{\partial p}{\partial z}\right)\right] = 6U\left(\frac{dh}{dx}\right) \tag{5.1}$$

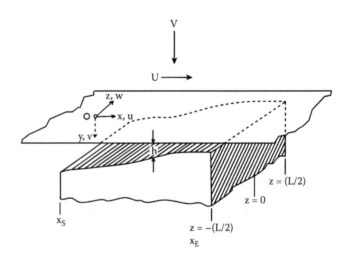

FIGURE 5.1   Basic tribological element.

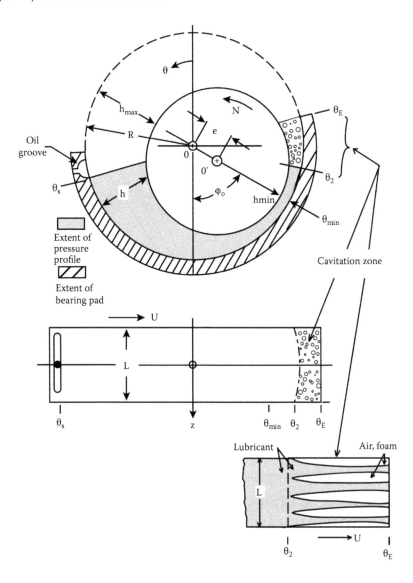

FIGURE 5.2 Extent of fluid film in a journal bearing pad.

For a journal bearing, shown in Figure 5.2, we use the normalizations

$$\bar{x} = R\theta; \quad \bar{z} = \frac{z}{(L/2)}; \quad \bar{h} = \left(\frac{h}{C}\right) = 1 + \varepsilon\cos\theta; \quad \bar{p} = \frac{p(C/R)^2}{\mu\omega}; \quad \bar{\mu} = \frac{\mu}{\mu_o}$$

The Reynolds equation then becomes

$$\frac{\partial}{\partial\theta}\left[\frac{\bar{h}^3}{\bar{\mu}}\left(\frac{\partial\bar{p}}{\partial\theta}\right)\right] + \frac{\partial}{\partial z}\left[\frac{\bar{h}^3}{\bar{\mu}}\left(\frac{\partial\bar{p}}{\partial z}\right)\right] = 6\left(\frac{d\bar{h}}{d\theta}\right) \tag{5.2}$$

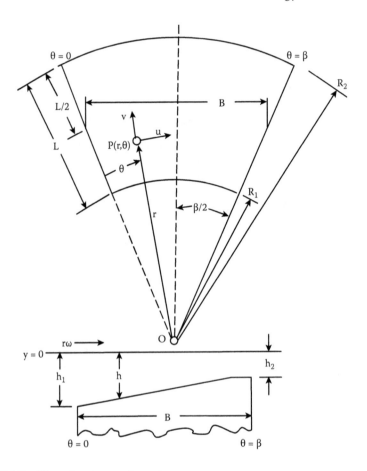

FIGURE 5.3   Thrust bearing configuration.

For a thrust bearing, shown in Figure 5.3, consisting of sectorial pads with coordinates (r, θ), the normalizations used are

$$\bar{r} = \left(\frac{r}{R_2}\right); \quad U = 2\pi r N; \quad \bar{h} = \left(\frac{h}{\delta}\right); \quad \bar{p} = \left(\frac{p}{\mu N}\right)\left(\frac{\delta}{L}\right)^2; \quad \delta = (h_1 - h_2)$$

and the Reynolds equation takes the form:

$$\frac{\partial}{\partial \bar{z}}\left[\frac{\pi \bar{h}^3}{\bar{\mu}}\left(\frac{\partial \bar{p}}{\partial \bar{z}}\right)\right] + \frac{1}{\bar{r}}\frac{\partial}{\partial \theta}\left[\frac{\bar{h}^3}{\bar{\mu}}\left(\frac{\partial \bar{p}}{\partial \theta}\right)\right] = 12\pi\bar{r}\left(\frac{R_2}{L}\right)^2\left(\frac{d\bar{h}}{d\theta}\right) \tag{5.3}$$

To solve journal bearing problems, we need to know two parameters: (L/D) and ε; similarly, for thrust bearings: (L/R_2) and $\delta = (h_1 - h_2)$ or the taper of the sector. When solved, the Reynolds equation provides the velocity and pressure fields, which by proper integration enables one to obtain the basic performance parameters of load, dissipation, flow, and other relevant quantities. These are all obtained in normalized form. For example, the total hydrodynamic force for a journal bearing is a quantity called the Sommerfeld number, given by

$$S = \left(\frac{\mu N}{P} \right)\left(\frac{R}{C} \right)^2$$

and for thrust bearings its equivalent is

$$T = \left(\frac{\mu N}{P} \right)\left(\frac{L}{\delta} \right)^2$$

In practice, both journal and thrust bearings consist of two or more pads, together completing an arc of 360°. The essential features of each pad (28) can be summarized as follows:

- In a journal bearing pad of sufficient angular extent, the pressure profile starts at the beginning of the pad, θ_s, and ends at some θ_2 ahead of the physical end of the pad. The region $(\theta_E - \theta_2)$ is a region of cavitation, where the lubricant breaks up into streamlets, as shown in Figure 5.2. The part where cavitation occurs does not contribute to the load capacity, the pressures over that region being close to or slightly below ambient. This follows simply from the continuity requirements of the incompressible lubricant.
- With a variation of operating conditions, the journal center moves along a locus, which in generalized form looks like that of Figure 5.4. At very light loads, when the eccentricity is close to the geometric center of the bearing, the attitude angle ϕ is close to 90°; this angle keeps decreasing with an increase in eccentricity until, at $\varepsilon = 1$, the attitude angle equals zero. These angular quantities are in reference to the direction of the load vector, which can be oriented at any arbitrary angle with respect to the boundaries of the pad.
- In thrust bearings, where the pads usually have a converging film throughout the sector, the pressure profile extends from the start to the end of each pad and there is no cavitation. Exceptions occur when a thrust bearing pad has a crowned geometry or when a tilting pad bearing is sufficiently flexible to undergo bending under the load. The films over such sectorial pads are then similar to journal bearing pads with diverging films.

The Reynolds equations cited earlier contain the term $\bar{\mu} = (\mu/\mu_o)$. Most analyses simply assume that $\mu = \mu_o$, whether this μ_o represents the inlet or some average viscosity, so that $\mu = 1$ and the viscosity term is eliminated from the differential equation. However, in nearly all cases, particularly when liquids are used, the viscosity is a strong function of the temperature, and the latter varies considerably throughout the fluid film. In particular, there is a strong temperature variation across the height of the film (y direction), as shown schematically in Figure 5.5. Thus, to complete the formulation of hydrodynamic lubrication equations, one would have to provide at least two additional expressions: the energy equation for calculating temperature distribution in the circumferential, axial and normal directions; and an expression for the variation of viscosity with temperature. Should the situation be nonadiabatic, which again is the most prevalent case, conductive

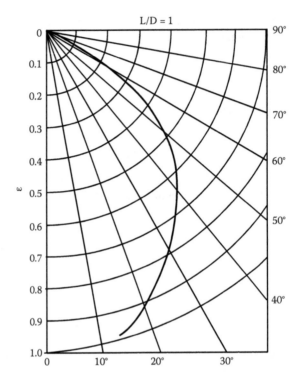

FIGURE 5.4   Locus of shaft center for circular bearing.

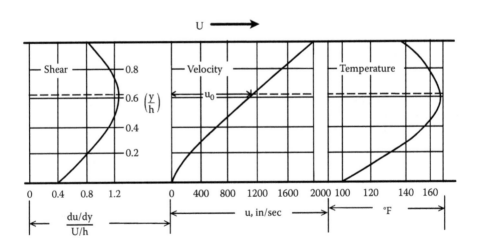

FIGURE 5.5   Shear strain, velocity, and temperature profiles across the fluid film.

and convective heat transfer expressions would have to accompany the energy equation. These important aspects of hydrodynamic lubrication are somewhat beyond the present scope and are treated in great detail in Reference 28.

5.1.1.2 Boundary Conditions

The variety of boundary conditions arising in hydrodynamic lubrication, particularly in journal bearing configurations, result from several possible physical arrangements in the practical usage of bearings, dampers, piston rings and similar devices. A journal bearing, regardless of how many individual pads it has, forms a closed 360° shell. Lubrication of this axially finite shell involves two important variables: the angular location or locations where the lubricant is admitted; and its pressure level—in practice, ranging from half to several atmospheres (5 to 50 psig). The geometry where the lubricant is introduced is either in the form of an axial groove of arbitrary axial extent—L_G, often only a circular hole which is merely a special case of an axial groove; or along what is called a circumferential groove, which runs either along the sides or in the bearing center. Equivalent to such a circumferential mode of lubrication is the "flooded" condition where a bearing is merely submerged in a pool of lubricant. Under all the latter conditions, the fluid is expected to enter the bearing clearance axially from the sides ($z = 0, L$). These several structural arrangements are shown in Figure 5.6. The implications of such different physical arrangements result

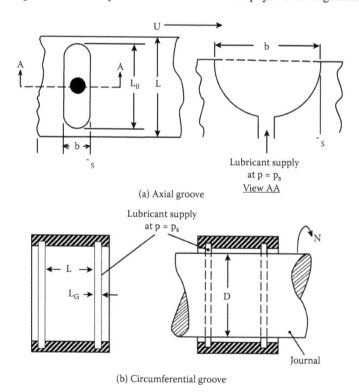

(a) Axial groove

(b) Circumferential groove

FIGURE 5.6   Basic forms of lubricant admission.

in different boundary conditions attending the solution of the hydrodynamic equations, which are described in the following text:

1. Standard conditions. The simplest boundary conditions apply to cases where the lubricant is admitted via an ample groove at high inlet pressure in the converging section of the clearance (see Figure 5.2) so that the film is full at both the start and end of the bearing pad. In that case, the simple requirement is

$$p_1 = p_2 = 0 \tag{5.4}$$

the zero condition denoting ambient pressure.

2. Cavitating conditions. When the pad is extensive enough that $\theta_E\theta_2$ produces a region of cavitation over the trailing portion of the pad, the downstream boundary condition, as derived in Reference 28, is given by

$$p_2 = \left(\frac{dp}{d\theta}\right)_2 = 0 \tag{5.5}$$

The cavitation region is sketched in Figure 5.2, and a photograph of the cavitating streamlets is shown in Figure 5.7 (30). Equation 5.5 is adequate for most practical problems. When the exact geometric behavior of these streamlets in the region $\theta_E\theta_2$ is required, the downstream boundary

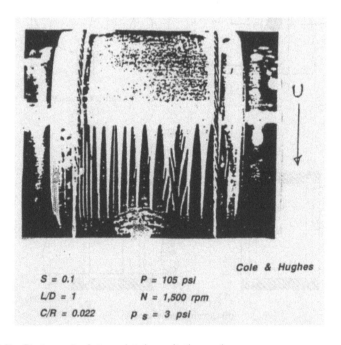

Cole & Hughes

S = 0.1	P = 105 psi
L/D = 1	N = 1,500 rpm
C/R = 0.022	p_s = 3 psi

FIGURE 5.7 Photograph of streamlets in cavitating region.

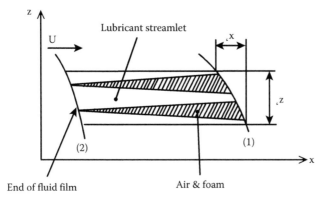

FIGURE 5.8 Streamlet geometry in a downstream cavitating region.

condition, as derived in Reference 28, is given by

$$p = \left(\frac{\partial p}{\partial x}\right)\Big|_{\theta_2} = O \tag{5.6a}$$

$$\frac{U h_2}{2} = \frac{U h_1}{2} - \frac{h_1}{12\mu}\left[\frac{\partial p}{\partial x} - \frac{\partial p}{\partial z}\left(\frac{d x}{d z}\right)\right]_1 \tag{5.6b}$$

and portrayed in Figure 5.8. When these equations are solved for the envelope $g\,(\theta, z)$, one obtains the width of the cavitating streamlets until the point where the fluid film is again complete.

Boundary conditions (Equation 5.6) apply equally to cases where the lubricant is admitted to a diverging section of the bearing. Here, we will have an upstream region of cavitation until convergence of the film makes the streamlets merge into a continuous film, as shown in Figure 5.9. In the general case, there could be cavitating regions in both the upstream and downstream portions of the pad. The commencement and termination of the full film, as a function of the (L/D), eccentricity ratio, and point of lubricant admission, are shown in Figures 5.10 and 5.11.

3. Restricted lubricant supply. Several geometric features may lead to having an incomplete starting film, even when admission is at a converging section of the film. This can be caused by an insufficient axial extent of the groove, too small an inlet orifice, or an insufficient inlet pressure p_s of the lubricant. A photograph of such incomplete axial extent of the inlet films due to too low a value of p_s is shown in Figure 5.12. A solution for such a partial film extent is given by the relationship (5)

$$U\lambda\left(\frac{h - h_2}{2}\right) = \frac{U L_o h_o}{2} - \frac{h_o^3}{6\mu}\int_0^{(L_o/2)}\left(\frac{\partial p}{\partial x}\right)_o d z + \frac{h^3}{6\mu}\int_0^{(\ell/2)}\left(\frac{\partial p}{\partial x}\right)_o d z \tag{5.7}$$

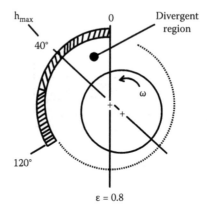

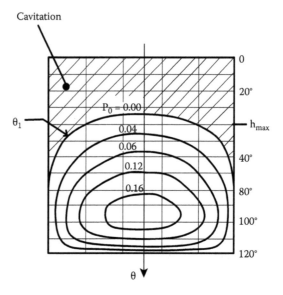

FIGURE 5.9 Cavitation in upstream portion of journal bearing.

where λ_s is the local width of the film, a function of the angular coordinate. Some solutions for this relationship are shown in the plots of Figure 5.13.

4. Circumferential groove or flooded condition. It should be intuitively apparent that lubrication via a circumferential groove is a poor alternative to an axial groove. While, in the latter case, the rotating journal sweeps continuously into the clearance space, a sheet of lubricant, the fluid filling the circumferential groove, has great difficulty entering the tight clearance space, which is only mils in height. Unless the pressure is such that a groove is extremely high, the fluid will penetrate only a small axial distance into the clearance space, leaving large portions of the bearing surface dry. Following the analysis of Reference 29, the boundary conditions for this

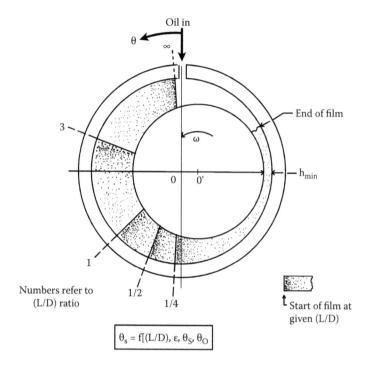

$$\theta_s = f[(L/D), \varepsilon, \theta_S, \theta_O]$$

FIGURE 5.10 Start of full film for upstream cavitation.

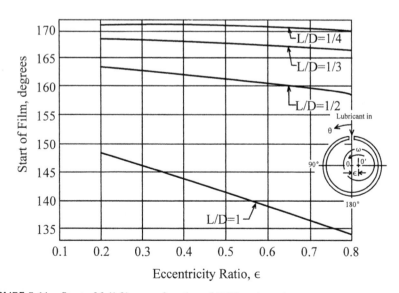

FIGURE 5.11 Start of full film as a function of (*L/D*) ratio and ∈.

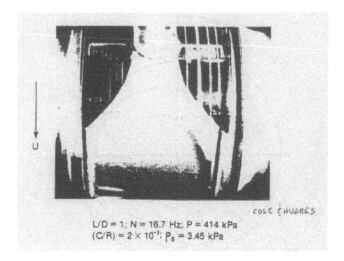

COLE & HUGHES

L/D = 1; N = 16.7 Hz; P = 414 kPa
(C/R) = 2 × 10^{-3}; P$_s$ = 3.45 kPa

FIGURE 5.12   Photograph of incomplete inlet film.

form of lubrication are given by

$$\frac{\partial p}{\partial x} = 0; \quad \frac{\partial p}{\partial z} = 0 \qquad \qquad \text{Region 1} \quad (5.8a)$$

$$\frac{U h_2}{2} dz = \left[\frac{U h_1}{2} dz - \frac{h_1^3}{12\mu} \frac{1}{R} \left(\frac{\partial p}{\partial \theta} \right) dz + \frac{h_1^3}{12\mu} \left(\frac{\partial p}{\partial \theta} \right) R d\theta \right] \qquad \text{Region 2} \quad (5.8b)$$

where the nomenclature is given in Figure 5.14 and the dimensionless inlet pressure is

$$\bar{p}_s = \left(\frac{2 p_s}{3 \mu \omega \varepsilon} \right) \left(\frac{C}{L} \right)^2 \qquad \qquad (5.9)$$

Figure 5.15 gives a sample solution for a full (360°) journal bearing at an (L/D) = 1 and $\varepsilon = 0.5$, lubricated via circumferential grooves. As can be seen, the higher the value of p_s, the fuller the film; however, even at its highest value there is still a cavitating bubble at the center of the bearing. At low values of inlet pressure, the cavity extends almost along the entire circumferential centerline of the bearing.

5.1.2 COMPRESSIBLE LUBRICATION

It would seem that, in discussing overlapping regimes in tribology, liquid lubrication would be more akin to EHL on the one hand and powder lubrication on the other, than would compressible fluid films, but such is the interwoven nature of tribological phenomena that, in many ways, vapors have more in common with these two disciplines than do incompressible films. It is compliant surface air bearings that provide

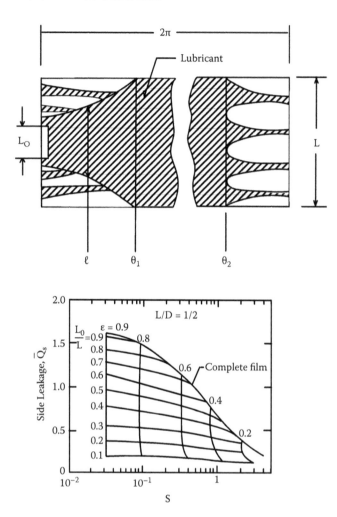

FIGURE 5.13 Reduced side leakage due to incomplete inlet film.

a backward link to EHL, with which we concluded Chapter 4, and several properties of gas lubrication lead us on to powder lubrication, dealt with later in the chapter. First, there is compressibility that, in the form of compaction, is also a feature of tri-boparticulate films. Then, under some conditions, gas lubrication induces slip at the boundaries—also a basic characteristic of powder flow in bearings.

5.1.2.1 Rigid Surfaces

In rectangular coordinates, the Reynolds equation that includes a variable density assumes the following form:

$$\frac{\partial}{\partial x}\left[\frac{h^3}{\mu}\rho\left(\frac{\partial \rho}{\partial x}\right)\right]+\frac{\partial}{\partial z}\left[\frac{h^3}{\mu}\rho\left(\frac{\partial \rho}{\partial x}\right)\right]=6U\left[\frac{d(\rho h)}{dx}\right] \tag{5.9a}$$

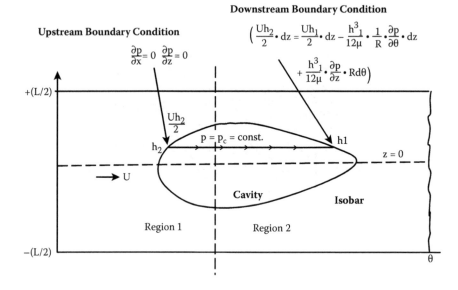

Downstream Boundary Condition

$$\left(\frac{Uh_2}{2} \cdot dz = \frac{Uh_1}{2} \cdot dz - \frac{h^3_1}{12\mu} \cdot \frac{1}{R} \cdot \frac{\partial p}{\partial \theta} \cdot dz \right.$$

$$\left. + \frac{h^3_1}{12\mu} \cdot \frac{\partial p}{\partial z} \cdot Rd\theta \right)$$

Upstream Boundary Condition

$$\frac{\partial p}{\partial x} = 0 \quad \frac{\partial p}{\partial z} = 0$$

FIGURE 5.14 Cavity formed due to circumferential groove or flooded lubrication.

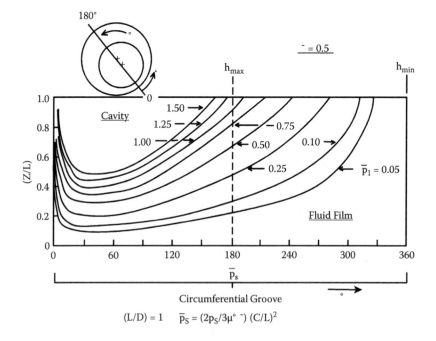

Circumferential Groove

$(L/D) = 1 \qquad \bar{p}_S = (2p_S/3\mu^\circ)(C/L)^2$

FIGURE 5.15 Cavity boundaries as a function of circumferential groove pressure.

Applying the perfect gas equation $p = \rho RT$, the density term for isothermal conditions becomes proportional to pressure, and we have

$$\frac{\partial}{\partial x}\left[\frac{h^3}{\mu}P\left(\frac{\partial p}{\partial x}\right)\right] + \frac{\partial}{\partial z}\left[\frac{h^3}{\mu}P\left(\frac{\partial p}{\partial x}\right)\right] = 6U\left[\frac{\partial(ph)}{\partial x}\right] \qquad (5.9b)$$

Normalizing the terms for a plane slider of extent B, we can write

$$\bar{p} = \left(\frac{p}{p_a}\right); \quad \bar{h} = \left(\frac{h}{h_2}\right); \quad \bar{x} = \left(\frac{x}{B}\right); \quad \bar{z} = \left(\frac{z}{B}\right)$$

yielding

$$\frac{\partial}{\partial \bar{x}}\left[\bar{h}^3\frac{\partial \bar{p}^2}{\partial \bar{z}}\right] + \frac{\partial}{\partial \bar{z}}\left[\bar{h}^3\frac{\partial \bar{p}^2}{\partial \bar{z}}\right] = 2\Lambda\frac{\partial(\bar{p}\bar{h})}{\partial \bar{x}} \qquad (5.9c)$$

where $\Lambda = [\frac{6\mu UB}{p_a h_2^2}]$ is the basic parameter for gas lubrication, the way the Sommerfeld number is the basic parameter for incompressible cases.

Some of the characteristics that differentiate gas from liquid-lubricated bearings are the following:

- Gas bearing performance is a function of the ambient pressure p_a. The higher the ambient pressure, the higher the load capacity.
- There is no cavitation. Even when a bearing has a divergent film, the pressure in that region will be subambient but can never fall below absolute zero pressure. For this reason, the boundary conditions in gas bearings are simply

$$p_o = p_B = p_a \qquad (5.10)$$

- The general shape of the pressure profile in gas bearings is given in Figure 5.16a. Compared to the incompressible case, the peak pressure is displaced closer to the trailing end of the pad ($x = B$).
- The load capacity of gas bearings is considerably lower than in liquid-lubricated bearings, mainly because of the lower viscosity of gases and vapors. As shown in Figure 5.17, they are two orders of magnitude below water and three orders of magnitude below petroleum oils (9). Their power requirements, consequently, are reduced by similar levels.

The standard requirement in hydrodynamic bearings is for the fluid film to have at its boundaries the same velocity as the adjacent surfaces. This, however, is true for flows where the molecular mean free path of the gas, λ, is much smaller than the film thickness. Gas bearings generally have thin films, and the foregoing does not always hold, as shown in Table 5.2. If the mean molecular path becomes comparable to h, there will be slip at the boundaries, and a new parameter enters the equations, namely,

$$m = \left(\frac{\lambda}{h}\right) = \left(\frac{\text{Mean Free Path}}{\text{Film Thickness}}\right) \qquad (5.11)$$

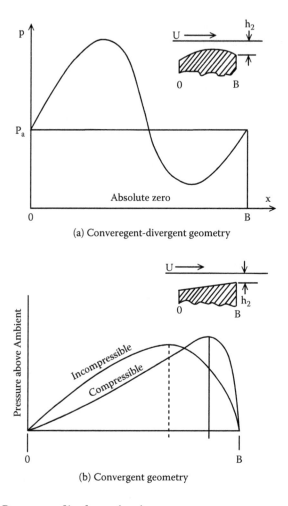

FIGURE 5.16 Pressure profiles for gas bearings.

The analysis for this case, as developed in Reference 4, requires that the continuity equation be solved with the following boundary conditions:

$$u_o = U + \lambda \left. \frac{\partial u}{\partial y} \right|_o ; \quad u_h = -\lambda \left. \frac{\partial u}{\partial y} \right|_h ; \quad w_o = \lambda \left. \frac{\partial w}{\partial y} \right|_o ; \quad w_h = -\lambda \left. \frac{\partial w}{\partial y} \right|_h \qquad (5.12)$$

With these boundary conditions on the fluid, the resulting Reynolds equation becomes

$$\frac{\partial}{\partial x}\left[ph^3 \frac{\partial p}{\partial x}\left(1 + 6\frac{\lambda}{h}\right)\right] + \frac{\partial}{\partial z}\left[ph^3 \frac{\partial p}{\partial z}\left(1 + 6\frac{\lambda}{h}\right)\right] = 6\mu U \frac{\partial}{\partial x}(ph) \qquad (5.13)$$

which, due to the assumptions in the analysis, is valid for the range of $0 \le m1$.

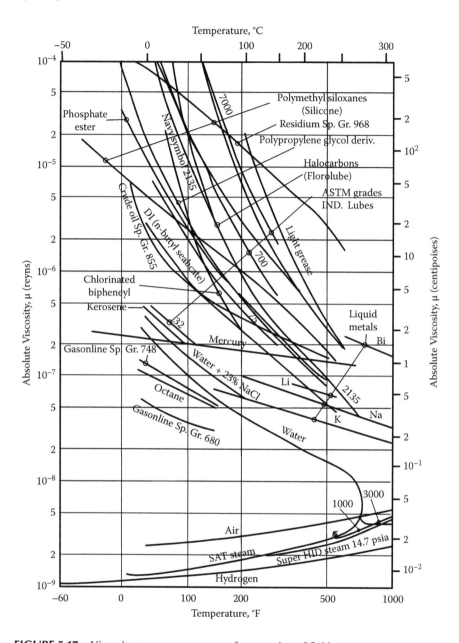

FIGURE 5.17 Viscosity–temperature curves for a number of fluids.

TABLE 5.2
Values of *m* for a Selection of Gas Bearings

Radial Clearance (Micro-Inch)	Ambient Pressure (Psi)	Ambient Gas	Molecular Mean Free Path, λ (Micro-Inch)	Mol. Mean Free Path Radial Clearance
130	14.7	Air	2.52	0.019
210	14.7	Air	2.52	0.012
72	14.7	Air	2.52	0.035
135	14.7	Air	2.52	0.019
135	4.	Air	9.26	0.069
135	14.7	Helium	7.32	0.054
135	3.	Helium	35.90	0.266
135	14.7	Neon	5.20	0.039

A sample solution for a plane slider with the characteristics

$$\left(\frac{h_1}{h_2}\right) = 2 \quad \text{and} \quad \Lambda = 7.5$$

is given in Figure 5.18. As can be seen, the effect of the ratio *m* on the performance of gas-lubricated bearings is as follows:

- Values of *m* greater than 0.01 have a noticeable effect on load capacity and, thus, on the friction coefficient.
- Load-carrying capacity decreases with a rise in *m*. This is most pronounced at low speeds, whereas at very high speeds the effect of *m* tends to vanish.
- Friction coefficient decreases with a rise in *m*.

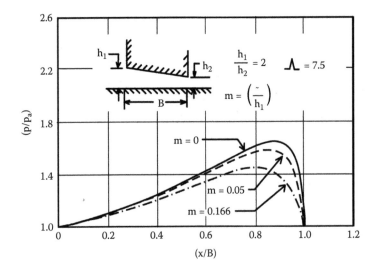

FIGURE 5.18 Pressure profiles for plane slider with slip at the boundaries.

5.1.3 COMPLIANT SURFACE FOIL BEARINGS (CSFB)

While the elastohydrodynamic lubrication (EHL) operation discussed in Chapter 4 dealt with unintended surface deflections due to extremely high loadings, compliant surface foil bearings are designed with the deliberate intent of producing structural deflections. This is achieved by placing a matrix of flexible supports under the load-carrying foil. The basic structure of a CSFB is shown in Figure 5.19. The top foil is usually anchored at the trailing end with the upstream end free and the bumps under the foil acting as springs. As compared with a conventional gas bearing, the CSFB has the following advantages:

- Higher load capacity. In terms of the customary W-h relationship, the CSFB will carry a higher load for a given film thickness.
- Lower power loss. This is due to the larger average film thicknesses in a CSFB.
- Ability to absorb shocks. Due to the ability of the foils to deflect radially, the bearing can tolerate runner excursions, misalignment, or external shocks.
- Stability. As shown in Figure 5.20, by using multiple top foils and a variable matrix of bump foils, the spring and damping characteristics can be varied in the circumferential, axial, or both directions, in accordance with a preferred design scheme.

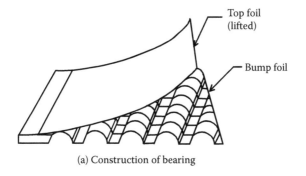

(a) Construction of bearing

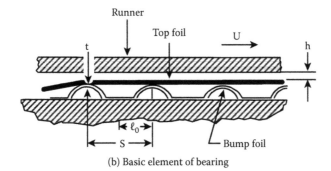

(b) Basic element of bearing

FIGURE 5.19 Basic construction of a compliant surface (foil) bearing.

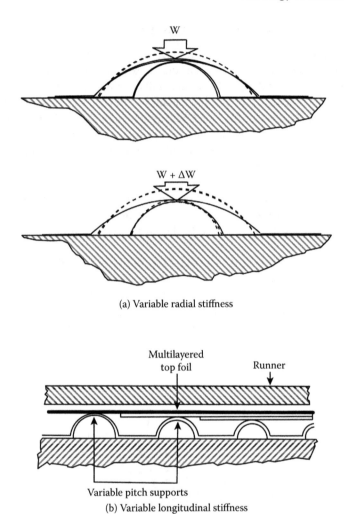

(a) Variable radial stiffness

(b) Variable longitudinal stiffness

FIGURE 5.20 Arrangements for a variable stiffness matrix.

- Endurance of high temperatures. The bearing exhibits superior qualities in operating at high temperatures, primarily in that it is not subject to thermal distortions.
- Tolerance of contamination. Due to its higher film thickness and ability to conform, such a bearing can tolerate the incursion of foreign matter without undue harm to the mating surfaces (13–23).

A CSFB operates with a dual film: one hydrodynamic between runner and top foil and an elastic one underneath the top foil. A journal bearing constructed on these principles is shown in Figure 5.21. To arrive at a mathematical solution, one must simultaneously solve a set of hydrodynamic and elastic equations in order to obtain

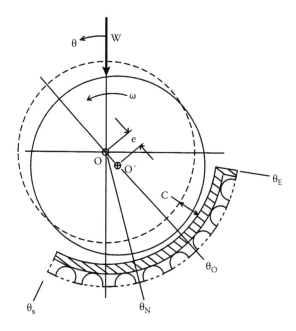

FIGURE 5.21   Compliant surface journal bearing.

the total deflections, pressures, and stability coefficients. The compressible fluid equation is the same as that given by Equation 5.9. Following the analysis given in Reference 9, the film thickness variation $h\,(\theta)$ is written as due to both hydrodynamic eccentricity e and deflection of the foil, the latter assumed to be proportional to local pressure, yielding

$$h(\theta) = C + e\cos(\theta - \Phi_o) + K_1(p - p_a) \tag{5.14}$$

where K_1 is a constant reflecting the structural rigidity of the bumps. K_1 is given by

$$K_1 = (\alpha C/p_a) \quad \text{where} \quad \alpha = (2p_a s/CE)(l_o/t)(1 - v^2) \tag{5.15}$$

In the preceding equation, α is the compliance of the bearing with the structural quantities s, l_o, and t defined in Figure 5.19. Consequently, the normalized film thickness is given by

$$\bar{h} = \left(\frac{h}{C}\right) = 1 + \varepsilon\cos(\theta - \Phi_o) + a(p - p_a) \tag{5.16}$$

One of the other advantages of CSFBs is that they do not allow for the generation of subambient pressures. Whenever diverging portions of the film tend to produce subambient pressures, the external pressure p_a underneath the top foil lifts it up until the pressures on both sides of the foil are equalized. This fact has different implications for diverging films at the start or at the end of the pad. For an upstream divergence, hydrodynamic pressures will start forming again at a point θ_1 where the film thickness h_1 equals the film thickness h_s at the start of the pad. Thus, the entire pad portion $\theta_s\theta_1$ is ineffective as a bearing surface. For divergence at the end of the pad, the foil will lift off to align itself parallel to the runner. This is similar to the trailing end of a cavitating liquid-lubricated bearing. The boundary conditions in such a CSFB are then

$$at\,\theta_s\;\;\bar{p}=\left(\frac{p}{p_a}\right)=1 \tag{5.17a}$$

$$at\,\theta=\theta_2\;\;\;p=\left(\frac{p}{p_a}\right)=1;\;\;\;\frac{\partial\bar{p}}{\partial\theta}=0 \tag{5.17b}$$

$$at\,z=\left(\frac{L}{2}\right)\;\;\;\bar{p}=\left(\frac{p}{p_a}\right)=1 \tag{5.17c}$$

Some comment is required about the notion of minimum film thickness in CSFBs. Unlike in rigid surface bearings, the absolute minimum film height is not a good criterion for load capacity. Because, here, the hydrodynamic pressures cause a proportional structural deflection, the film thicknesses in the interior of the bearing—where pressures are high—will be larger than at the edges, where the pressure is ambient. Pictorially, the deflected bearing surface under load assumes the shape of a "basket" (Figure 5.22a). The axial film thickness at $\theta=\theta_o$ has a value of h_{min} at the edges, but only there; elsewhere along the line $\theta_o=$ constant, the film thicknesses are larger. A "nominal film thickness," h_N, is therefore defined as the criterion for load capacity. This h_N is taken as the minimum film thickness along the bearing centerline, $z=0$ (Figure 5.22b). In Figure 5.22c, the variation of h along this centerline is plotted for various values of a. While for rigid surfaces $(a=0)/h_{min})$ occurs at 180° with increasing values of a this h_{min} or our h_N shifts downstream and increases in magnitude. Thus, it is quite possible that the value of h_N may increase with an increase in load—a reflection of the advantages previously attributed to CSFBs.

5.2 RHEODYNAMIC LUBRICANTS

5.2.1 GREASES

The behavior of a grease as a lubricant holds a particular interest because much, if not most, of its behavior is also common to powder lubrication, a central interest in the present text. The structure of the substance, its rheological properties, response to shear, slip at the walls, formation of cores, and proneness to fracture all have their correspondence in a lubricant film consisting of compacted micron-sized solid particles.

5.2.1.1 Rheology of Greases

Greases consist of mineral oils thickened with metal soaps. These are produced by reaction between metal oxides or hydroxides and various fatty substances such as fatty acids. The most commonly used metal soaps are 10%–15% concentrations of

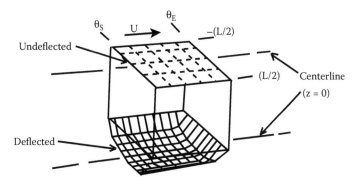

(a) Qualitative shape of surface

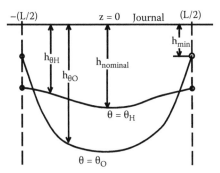

(b) Minimum and nominal film thickness

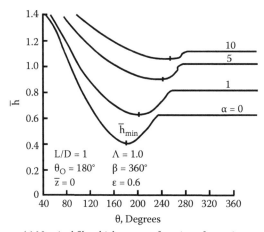

(c) Nominal film thickness as a function of compiance

FIGURE 5.22   Film thickness in a deflected compliant bearing surface.

lithium, calcium, sodium, or aluminum. Greases can hold their shape under their own weight but are soft enough to flow under shear, which is the basis of their utility as a lubricant under conditions that cannot accommodate a liquid. The solid phase of a grease may consist of elongated, spherical, or intermediate shaped particles. In some greases, the bonds between the particles are weak and the structure becomes layered under shear; in others, the bonds are strong. Some bonds are reversible in that after a time they reform, a property called thixotropy. Under severe shearing, not only the structure but also the particles themselves may be ruptured or degraded to yield a permanently softened substance.

Like many plastic substances, greases exhibit edge effects. In experiments conducted with two sets of disks (9)—one with equal size disks and clean edges and another in which one disk was twice the size of the other with uncleaned edges—the results were as shown in Figure 5.23. As can be seen, the torque due to the exposed edge was as large as that due to the gap. With concentric cylinders, there is also a centrifugal effect in that it causes an outward migration of particles. More serious is the fact that boundary layers are formed adjacent to the walls so that, at the rotating surface, the boundary layer will be low in particle concentration; at the extremes, it may consist of the liquid phase alone. This effect would be most pronounced in small gaps when the boundary layers constitute a large fraction of the film thickness.

From a lubrication standpoint, greases exhibit two important features: the presence of a yield stress, below which there is no flow and, as shown in Figure 5.24, a rapid drop in viscosity as the rate of shear rises. Greases are most often characterized as a Bingham plastic. Both Newtonian fluids and Bingham plastics exhibit a linear relationship between stress and shear rate except that the latter have a stress yield point that must be exceeded before flow takes place. Also, during a shear cycle, greases exhibit a hysteresis effect. Hence, in those parts of a grease layer where the yield point is not exceeded, a stagnant portion of the lubricant, called cores, will be

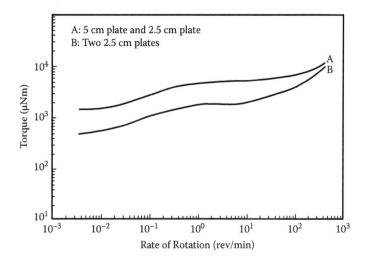

FIGURE 5.23   The effect of an exposed film edge in grease lubricated disks.

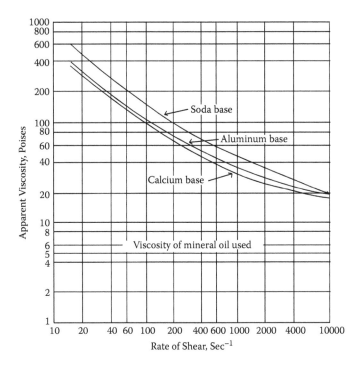

FIGURE 5.24   Variation of grease viscosity with shear.

formed. This feature of a grease is mathematically expressed by the relationship

$$\tau = \tau_o + \mu\left(\frac{\partial u}{\partial y}\right) \qquad (5.18)$$

where τ_o is the yield point, equal to zero in Newtonian fluids. These effects are sketched in Figure 5.25. Equation 5.18 yields a relationship $(\frac{dp}{dx}) = \mu$, which is similar to that of a Newtonian fluid except that it does not hold throughout the entire grease layer but only in those regions where the shear exceeds the yield point.

5.2.1.2   Grease-Lubricated (Rheodynamic) Bearings

Here we shall consider the performance of a grease-lubricated infinitely wide slider (25). The assumption of $(\frac{L}{D}) = \infty$ is a much more reasonable approach here than it is with liquid lubrication because side leakage—the main reason for resorting to finite bearing analysis—is radically reduced with grease or powder lubricants. Denoting by h_a and h_b the lower and upper limits of the stagnation core (where τ_o is not exceeded), we have, after considering the equilibrium of a one-dimensional element of the core, the relationship

$$\left(\frac{dp}{dx}\right)(h_a - h_b) = 2\tau_o \qquad (5.19)$$

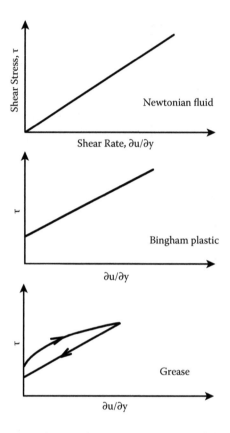

FIGURE 5.25   The stress-strain curve in a grease vs. a newton fluid and a Bingham plastic.

Applying this equation to a slider in which the core moves with some velocity u_c, and integrating for the three different regions of the film, we have

$$u = -\frac{1}{u}\frac{\partial p}{\partial x}(h_a - y)\frac{y}{2} + \left(\frac{U - u_c}{h_a}\right)y - U \qquad O \le y \le h_a \qquad (5.20a)$$

$$u = -u_c \qquad h_a \le y \le h_b \qquad (5.20b)$$

$$u = -\frac{1}{u}\frac{\partial p}{\partial x}\left(\frac{h + h_b}{2}y - \frac{y^2}{2} - \frac{h h_b}{2}\right) - u_c\frac{h - y}{h - h_b}h_b \le y \le h \qquad (5.20c)$$

The volume flow across any section of the one-dimensional slider must be constant or

$$\int_0^h u\,dy = q = -\frac{1}{12\mu}\frac{\partial p}{\partial x}\left[h_a^3 + (h - h_b)^3\right] - \frac{Uh_a}{2} - \frac{u_c}{2}(h + h_b - h_a) \qquad (5.21)$$

At the edges of the core, the velocity gradient must be zero; hence, by differentiating Equation 5.20 and equating the derivatives to zero, we obtain

$$\frac{1}{2\mu}\frac{\partial p}{\partial x} = \frac{u_c - U}{h_a^2} \tag{5.22a}$$

$$\frac{1}{2\mu}\frac{\partial p}{\partial x} = \frac{u_c}{(h - h_b)^2} \tag{5.22b}$$

By eliminating u_c we obtain from the foregoing and from Equation 5.19

$$\frac{1}{\mu}\frac{\partial p}{\partial x} = \frac{2U}{(h - h_b)^2 - h_a^2} \tag{5.23}$$

$$\pm \tau_o = \frac{\mu U (h_b - h_a)}{(h - h_b)^2 - h_a^2} \tag{5.24}$$

Equations 5.22 and 5.24 are sufficient to determine h_a, h_b, and u_c.

It follows from Equation 5.23 that if $(\frac{dp}{dx})\frac{2\mu U}{h^2}$, both h_a and h_b are unreal, and no cores will be formed. With the velocity gradient different from zero everywhere in the film, Equation 5.21 reduces to

$$q = -\left(\frac{h^3}{12\mu}\right)\left(\frac{dp}{dx}\right) - \frac{Uh}{2},$$

the same flow as in Newtonian films.

In those cases where h_a and h_b are real, cores will be formed in the fluid film. In the inlet region, $(\frac{dp}{dx})$ is negative, and if $(\frac{\tau_o}{\mu})$ is large enough, then it is possible to make the upper boundary of the core adhere to the stationary bearing surface; that is, $h_b = h$. At the outlet, $(\frac{dp}{dx})$ is positive and, with a sufficiently low value of $(\frac{\tau_o}{\mu})$, it is possible to obtain $h_a = 0$, and the core will adhere to the runner. If $h_a = 0$, then $u_c = U$ and the flow becomes, from Equations 5.21 and 5.22,

$$q = \frac{-U(2h + h_b)}{3} \tag{5.25a}$$

If $h_b = h$, then $u_c = 0$ and we have

$$q = \frac{-Uha}{3} \tag{5.25b}$$

There will, thus, be three regions in the bearing: (1) an inlet region with $h_b = h$ and $h_a = const.$, (2) an intermediate region with no core, and (3) an outlet region with $h_a = 0$ and $h_b = (const. - 2h)$. These regions are shown in Figure 5.26a.

It is possible for the values of $(\frac{\tau_o}{\mu})$ to be such that the core adheres to neither the bearing nor the runner. In that case, the cores will be detached and moving within the film with some velocity u_b, as shown in Figure 5.26b. A treatment of this case would be more complicated than that of the attached cores.

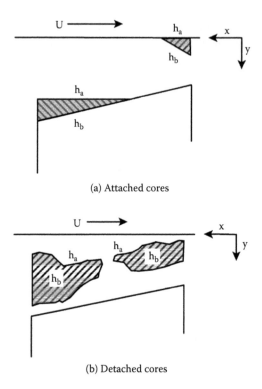

(a) Attached cores

(b) Detached cores

FIGURE 5.26   Grease lubricated slider.

For the case of attached cores, we can now calculate the performance of a plane slider with a film geometry given by

$$h = h_2 \left[1 + (a-1)\left(\frac{x}{B}\right) \right]$$

By integrating Equation 5.22, we have, for the inlet core,

$$h_a = h_2 \left[\frac{3a}{(1+a)} \right] \tag{5.26}$$

which shows that the inlet core height is proportional to the minimum or exit film thickness. Because h_a must always be less than h_1 (or ah_2), a real value for h_a is obtainable only if $a \geq 2$. Thus, a bearing with an inlet-outlet film thickness of less than 2 will operate without cores. Complete core formation can be shown to be given by the condition

$$\frac{\tau_o h_2}{\mu U} \geq \frac{(a-2)(a+1)}{a} \tag{5.27}$$

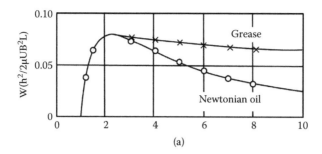

FIGURE 5.27 Load capacity of a rheodynamic vs. a hydrodynamics slider.

The integration of Equations 5.21 and 5.22 for pressure distribution and then for the total load yields

$$W = \frac{2\mu U B^2 L}{h_2^2} \frac{1}{(a-1)^2} \left(3 \ln 2 - \frac{1}{9} \ln \frac{a}{2} - \frac{20}{9} + \frac{a^2}{18} \right) \qquad (5.28)$$

The frictional force calculated from the tangential stresses at the moving surface yields

$$F = L \left[\tau_o B + \frac{2\mu B}{h_2} \frac{1}{a-1} \left(2 \ln 2 + \frac{2}{9} \ln \frac{a}{2} - \frac{13}{9} + \frac{2a}{9} \right) \right] \qquad (5.29)$$

The results of Figures 5.26 to 5.28 show that there is an increase in both load capacity and friction in a rheodynamic bearing. Detached cores would tend to somewhat reduce the load capacity and appreciably reduce friction, although in all cases they would still be higher than with a Newtonian fluid. The increase in load capacity is due to the throttling action of the cores, which essentially produces smaller clearances.

In journal bearings, as in the case of sliders, cores would tend to form near the maximum and minimum film thickness with the former attached to the bearing and the latter to the runner, as sketched in Figure 5.29.

5.2.2 SLURRIES

Slurries can exist in two basic forms, as shown in Figure 5.30. In part (a), solid particles in a liquid are attracted to and contact each other to flocculate into a gel structure. Shearing a slurry of this sort requires a stress to overcome the interparticle attraction plus a component to shear the viscous fluid. The critical shear flow required to bring about motion is spent in breaking the particle bond, and it is a major part of the yield stress τ_o. Dispersed slurries in which the particles remain separated either due to Brownian motion or by the application of a dispersant coating on the particles are shown in part (b) of Figure 5.30.

5.2.2.1 Viscosity of Slurries

In slurries in which there is no flocculation, the main parameter determining viscosity is the volumetric (solid/fluid) ratio, called either solid fraction or concentration.

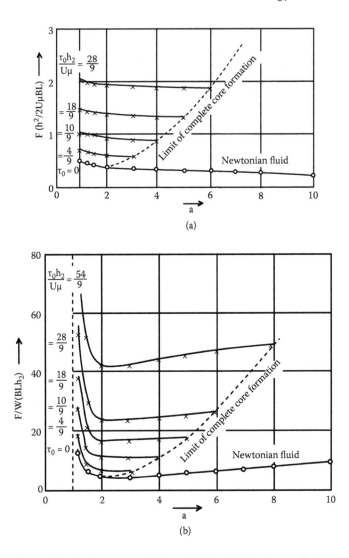

FIGURE 5.28   (a) Friction in a grease lubricated slider. (b) Friction coefficient.

Albert Einstein (10), in his study of Brownian motion in 1906, was the first to offer a formula for the viscosity of solutions, namely in a grease lubricated slider the simple expression

$$\mu_R = \left(\frac{\mu}{\mu_L}\right) = 1 + k_1 \Phi \tag{5.30}$$

where μ_L is the viscosity of the liquid or solvent, Φ the concentration, and k_1 a constant equal to 2.5. The main restriction on this equation is that the solution has to be dilute, that is, $\Phi = 1$. This was followed by a number of efforts to extend the applicability of this expression, of which the work of Vand in 1948 (27) is perhaps the

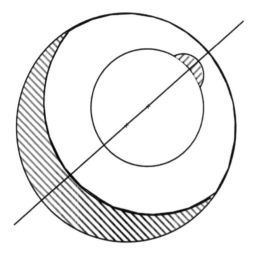

FIGURE 5.29   Location of attached cores in a grease-lubricated journal bearing.

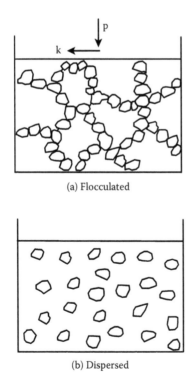

FIGURE 5.30   Two kinds of slurries.

most comprehensive. He expanded the foregoing equation to account for a number of physical conditions, as follows:

1. For concentrations up to 8%:

$$\mu_R = \frac{(1 + 0.5\Phi - 0.5\Phi^2)}{(1 - 2\Phi - 9.6\Phi^2)} \tag{5.31}$$

From Einstein's formula, the relative viscosity at 8% is 1.2; from Equation 5.31 it is 1.32.

2. No limitation on concentration (a formula proposed by S. A. Arrhenius, a Swedish physicist and chemist)

$$\mu_R = e^{(2.5\Phi)} \tag{5.32}$$

3. Effect of proximity of walls. This represents a case where a slurry flows in a very narrow channel when the nearby walls affect the flow. This effect can be accounted for if the spacing between the walls in a Couette type of flow is reduced by a factor of $K_w = (2.6a)$, where a is the radius of the particles.

4. Effect of particle interactions. Due to the mutual interaction between the particles, such as when collisions are taken into account, Equation 5.32 is modified to

$$\mu_R = \frac{e^{(2.5\Phi)}}{(1 - k_2\Phi)} \tag{5.33}$$

with $k_2 = 0.61$. For this case, an 8% solution would have a relative viscosity of 1.23.

When all of the foregoing effects are combined, the resulting equation for the relative viscosity of slurries becomes

$$ln(\mu_R) = \frac{(2.5\Phi + 2.7\Phi^2)}{(1 - 0.61\Phi)} \tag{5.34}$$

The variation of a slurry viscosity as a function of concentration based on Equation 5.34 is given in Figure 5.31. If the carrier fluid is water, whose viscosity at room temperature is approximately 1 cp, this figure provides the actual viscosity of the slurry in centipoises.

Vand (27) conducted viscosity measurements on a saturated solution of zinc iodine in a carrier of water and glycerol in which the fluid viscosity was 60 cps. Figure 5.32 gives the relative viscosities obtained from Equation 5.34 and from the experiments. As can be seen, the theoretical values hold up well up to concentrations of 30%–40%. Another set of data conducted on a slurry made up of magnetites suspended in three different carriers produced test data identical with the theoretical curve up to concentrations of 40%–50% (27). The variation of slurry viscosity with temperature is reproduced in Figure 5.33. The temperatures in these plots ranged

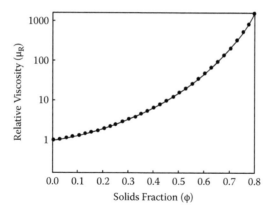

FIGURE 5.31 Variation of slurry viscosity with concentration (Equation 5.34).

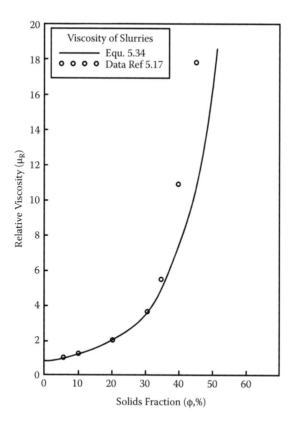

FIGURE 5.32 Theoretical and experimental values of slurry viscosity.

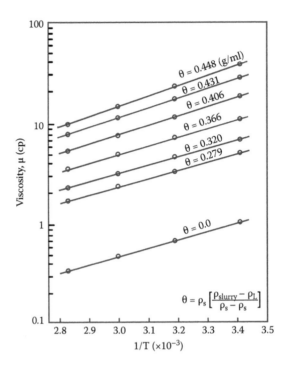

FIGURE 5.33 Experimental variation of slurry viscosity as a function of temperature and particle density.

over 70°C with the viscosity varying by factors of 3 to 5, depending on the relative density of the fluid and particles.

Clearly, slurry viscosity could not be a function solely of concentration but also of rate of shear and at least some particle properties such as size and shape. The authors of References 2 and 3 investigated the effect of some of these parameters by use of a method called Dissipative Particle Dynamics (DPD), consisting essentially of boxed-in models of slurries. They specified the viscosity to be dependent on the Peclet number, which contains a number of particle properties, as well as on the rate of shear

$$P_e = \gamma \frac{a^2}{D_o} = \frac{6\pi\mu_L a^3 \gamma}{k_B T} \tag{5.35}$$

where D_o is Einstein's diffusion coefficient, γ the shear rate, and $(k_B T)$ the thermal energy. The Peclet number expresses the ratio between the hydrodynamic forces due to shear and the forces due to Brownian motion, tending to restore the fluid to equilibrium; or the ratio between the time needed to deform the slurry by shear $(\frac{1}{\gamma})$ and the time for Brownian motion to restore equilibrium $(\frac{a^2}{D_o})$. The level of the Peclet number reflects the degree of departure from equilibrium.

With this approach, the relative viscosity is now a function of the rate of shear or

$$\mu_R(\gamma) = 1 + 2.5\Phi + k_2(\gamma)\Phi^2 \tag{5.36}$$

The dependence of the relative viscosity on the shear rate will now become more pronounced with an increase in concentration and will be contained in the factor k_2 (γ). This function has two limiting values: (1) for $\gamma = 0$, where Brownian motion dominates, $k_2 = 6.2$ and (2) for the limit of γ, where hydrodynamic forces dominate, $k_2 = 5.2$.

A transition was observed from the low shear viscosity $\mu(O)$ to the high shear viscosity $\mu(\infty)$. At this point, the Peclet number is given by

$$P_e = \frac{1}{2}[\mu(O) + \mu(\infty)], \tag{5.37}$$

its value being dependent on the concentration.

In order to set up a DSD model for, say, a 30% concentration, 45 spheres with a radius of 3.5 were positioned in a cube $30 \times 30 \times 30$—dimensions are all relative and, thus, dimensionless. For other concentrations, corresponding changes in the box size and sphere radii can be made. Only Brownian and hydrodynamic forces were made to act on the spheres, the shears varying from 3×10^{-6} to 0.05 with a ($k_B T$) value of 0.0033.

Figure 5.34 shows the obtained viscosity for spherical particles as a function of Peclet number (P_e). It follows a typical shear-thinning curve. Low Peclet numbers are fairly constant at a level of 4.5, which corresponds with a first Newtonian plateau. At higher shear rates, 1, the Pe of 10, at viscosity-thinning regime occurs until a second Newtonian plateau is reached at a value of 3.0. The characteristic viscosity, as defined in Equation 5.37, was found to be equal to 3.7 (for 3.77 and 4.5 limit values). Using a smaller box of $20 \times 20 \times 20$, the same values were obtained for the high-shear range

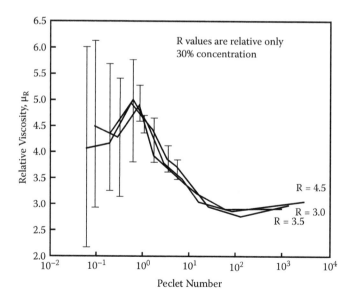

FIGURE 5.34 Variation of slurry viscosity as a function of Peclet number for spherical particles.

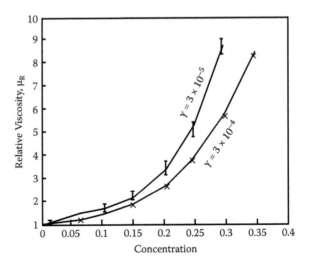

FIGURE 5.35 Variation of slurry viscosity as function of concentration for slender rod particles.

but not for the low end when the value turned out to be significantly larger, namely, 5.5. The impact of a smaller envelope was attributed to the effects of wall proximity.

Similar modeling was done for slender rods and disk-shaped particles. The dependence of the slurry viscosity with these shaped particles is represented in Figures 5.35 and 5.36.

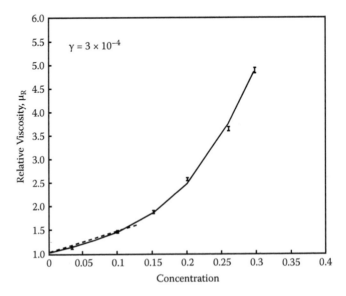

FIGURE 5.36 Variation of slurry viscosity as a function of concentration for disk-shaped particles.

The results produced by these modeling experiments can be summarized as follows:

- For 30% concentration of spherical particles, the values of both high and low shear thinning behavior agrees with existing experimental data.
- Denser suspensions at 40% give good results for the high shear viscosity but for the low shear rates, the size of the envelope makes itself felt and produces different results.
- For a constant shear rate, dilute suspensions of rods and disks show a linear increase in viscosity with concentration; at higher concentrations, the viscosity variation is proportional to the third power of the concentration.

5.3 TRIBOPARTICULATE LAYERS

Powders—that is, bodies of compacted solid particles separated by air voids—have many applications in science and industry. Our particular interest is in their possible use in tribological processes. As such, we need to know something about their physical and rheological properties, as not all powders (and not in all their variegated forms) are suitable as lubricants. While it is clear that a powder layer can be deliberately fed into the clearance between mating surfaces, our interest extends to the process of wear, which, although unintended, may provide the surfaces with a powder-like lubricant. Next, we want to know something about their rheology, to form a bridge or extension to the viscosity, compressibility, flow, etc., associated with liquid and gaseous lubricants. Knowing their physical and rheological properties, one can attempt to obtain analytical solutions for a number of tribological devices—a subject treated in Chapter 7.

The behavior of a powder is controlled by two mechanisms. One is the interparticle friction and attraction; the other is the solid–gas interaction, which involves viscous forces and physicochemical processes affecting the cohesion between particles. Thus, to characterize a powder, one must specify the properties of the particles, the entire assembly, and, also, often, the medium in which they are suspended. Table 5.3 gives a listing of some of the variables involved (HHP16 and HHP26). Only small portions of these are of interest here, but that still leaves enough of them to be considered.

5.3.1 POWDER MORPHOLOGY

- Size and shape. Of the numerous features that characterize a compacted powder, size and shape of the solid particles are the most elementary. It is, perhaps, arguable whether size can at all be specified by a single number; not only conceptually, but experiments, too, yield different values depending on the measurement technique used. Most commonly, the notion of an equivalent spherical diameter is used. At least half a dozen of these can be formulated, namely,
  - Equivalent volume diameter: $d_v = [6\,\pi/V]^{1/3}$
  - Equivalent surface diameter: $d_s = [\frac{S}{\pi}]^{1/2}$

TABLE 5.3
List of Powder Variables

a.   **Characteristics of a Powder Particle**
Material Characteristics

1.	Structure
2.	Theoretical density
3.	Melting point
4.	Plasticity
5.	Elasticity
6.	Purity (impurities)

Characteristics Due to the Process of Fabrication

1.	Density (porosity)
2.	Particle size (particle diameter)
3.	Particle shape
4.	Particle surface area
5.	Surface conditions
6.	Microstructure (crystal grain structure)
7.	Type and amount of lattice defects
8.	Gas content within a particle
9.	Absorbed gas layer
10.	Amount of surface oxide
11.	Reactivity

b. **Characteristics of a Mass of Powder**

1.	Particle characteristics
2.	Average particle size
3.	Particle size distribution
4.	Average particle shape
5.	Particle shape distribution
6.	Specific surface (surface area per 1 g)
7.	Apparent density
8.	Tap density
9.	Flow of the powder
10.	Friction conditions between the particles
11.	Compressibility (compactibility)

c. **Characteristics of Porosity in a Mass of Loose Particles**

1.	Total pore volume
2.	Pore volume between powder particles
3.	Pore volume within powder particles
4.	Number of pores between powder particles
5.	Average pore size
6.	Pore size distribution
7.	Pore shape

- Equivalent surface-volume diameter: $d_{sv} = [\frac{6V}{S}]$
- Equivalent projected area diameter: $d_a = [\frac{4A}{\pi}]^{1/2}$
- Equivalent perimeter diameter: $d_p = [\frac{P}{\pi}]$
- Martin diameter: $d_M = \frac{[d_{max}+d_{min}]}{2}$
- Sieve diameter: d_{si}

In the foregoing, V = particle volume, S = surface area, A = projected area, P = particle perimeter, and d_{max}/d_{min} = the maximum and minimum particle diameters.

Particle shapes are usually irregular, appearing as fibrils, rods, platelets, disks, or spheres; some of these are shown in Figure 5.37. Shape is often specified by a quantity called sphericity Ψ or the Wadell index, which is the ratio of surface area of a spherical

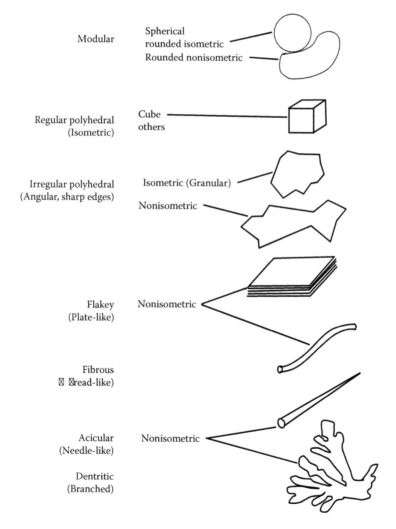

Modular — Spherical
rounded isometric
Rounded nonisometric

Regular polyhedral
(Isometric) — Cube
others

Irregular polyhedral
(Angular, sharp edges) — Isometric (Granular)
Nonisometric

Flakey
(Plate-like) — Nonisometric

Fibrous
(Thread-like)

Acicular
(Needle-like) — Nonisometric

Dentritic
(Branched)

FIGURE 5.37   Some typical powder particle shapes.

particle of the same volume to the surface area of the nonspherical particle or

$$\text{Wadell index} = \left[\frac{d_v}{d_s}\right]^2$$

where d_v is the diameter of a sphere having the same volume and d_s is the diameter of a sphere having the same surface area as the irregular particle being measured. Because a sphere has the smallest surface area for a given volume, the maximum value of sphericity is 1. For a cube, it is equal to 0.806; for round sand particles, it is in the range of 0.8–0.9; for an elliptical shape, this ratio is given by the cubic root of the ratio of the minor to major axes. One of the shortcomings of the sphericity approach is that it is not unique; a number of different shapes can yield the same sphericity number.

The shapes of some of powders are shown photographically in Figures 5.38 to 5.41. Figures 5.38a and 5.38b show optical micrographs of a spheroidal nickel powder, while Figure 5.39 shows a light micrograph of a packed nickel oxide powder with spherical and irregular particles. A rutile form of titanium dioxide powder that has been isostatically compressed at 10,000 psi is given in Figure 5.40. The boron nitride powder shown in Figure 5.41 had particles of irregular polyhedral, isomeric, and nonisomeric shapes; the flakes are 1 to 2 microns thick and 10 to 20 microns in diameter. In looking at these photographs, one must keep in mind that a high-resolution micrograph provides only a two-dimensional image, which does not reveal the spatial dimensions of the particles.

- Compaction and density. Under freely assembled conditions, the density of a powder assembly is the total mass of the particles divided by the bulk volume. The volume of the pores is usually measured by a technique called mercury displacement. When poured into a powder assembly, mercury will, under atmospheric conditions, penetrate the voids down to 15 microns. Under packed conditions, however, the problem is more complex. This is due partly

(a) Prior to exposure at 1500°F   (b) After exposure at 1500°F

FIGURE 5.38 Pure nickel particles at room and at 1500°F temperatures.

FIGURE 5.39   NiO powder compacted under 300 psi.

to the fact that, in addition to the voids between the particles, there is also particle porosity, which under pressure is reduced along with the interparticle spaces. Powder compaction involves several processes, that is:

1. Particle rearrangement leading to closer packing
2. Formation of arches and vaults enabling the voids to withstand the imposed pressure

FIGURE 5.40   Rutile form of TiO_2 particles compressed at 10,000 psi.

FIGURE 5.41   Boron nitride powder particles.

3. A complex process of deformation, cold welding, fragmentation, and percolation resulting in elastic and plastic deformation of the particles
4. Bulk compression of the entire assembly

Some characteristics of packed powders are given in Table 5.4 (34).

• Rheology

Powder compaction, as achieved first by compaction and then by dilatation during motion, and the stress–strain relationship exhibited during fluidlike shearing are two of the prominent rheological features of powder films. Others, too, make their appearance, such as an equivalent viscosity and a form of cavitation due to fracture of the powder substance. Quantitative expressions for these characteristics will be offered in the subsequent chapters when formulating equations for the solution of powder-lubricated bearings. Here, only a qualitative picture of powders as a quasi-fluid will be offered to link powder layers with the fluid films discussed previously.

A powder can withstand certain stresses without permanent distortion so that it returns to its original shape when the stress is released; this gives it certain elastic properties. Upon reaching a sufficiently high level of stress, the powder will yield and begin to flow. This critical stress level depends on the normal stress exerted on the substance. Thus, in general, the stress–strain relationship will look something similar to Figure 5.42. Upon start of shearing, the porosity of the powder—that is, its bulk density—will increase. The yield locus, thus, depends also on the initial porosity of the powder. Its level rises with a reduction in powder porosity, as shown in Figure 5.43. As intimated earlier, under some conditions a powder layer is likely to fracture. When a blade advances upon a layer of compacted powder, as shown in Figure 5.44, there is an

TABLE 5.4

Characteristics of Some Compacted Powders

Type of Packing	Points of Contact or Coordination Number	Volume of Unit Cell	Porosity	Pore Volume Per cm³ of Spheres cm³/cm³	Pore Volume Per Unit Cell[a]	Radius of Sphere Inscribed in the Cavities[b]	Radius of Circle Inscribed in the Throats Connecting Cavities[b]
	n		e				
Hexagonal close packed (rhombohedral)	12	$0.71d^3$	0.260	0.350	$0.18d^3$	$0.225R$—octahedral $0.414R$—tetrahedral	$0.155R$
Body-centered tetragonal	10	$0.75d^3$	0.302	0.430	$0.23d^3$	$0.291R$	$0.265R$ $0.155R$
Primitive hexagonal (orthorhombic)	8	$0.87d^3$	0.395	0.654	$0.34d^3$	$0.527R$	$0.414R$ $0.155R$
Primitive cubic	6	d^3	0.476	0.910	$0.48d^3$	$0.732R$	$0.414R$
Tetrahedral	4	—	0.660	1.94	—	$1.00R$	$0.732R$

[a] Diameter of particle.
[b] Radius of particle.

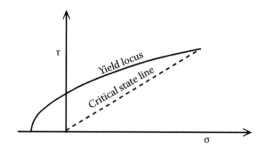

FIGURE 5.42   Yield locus for powder layer.

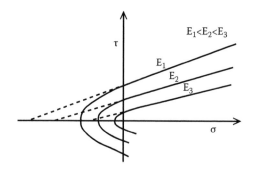

FIGURE 5.43   Yield locus for powder layer as function of porosity.

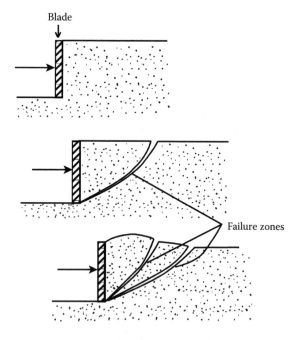

FIGURE 5.44   Formation of fractures in powder layer under pressure.

initial compression of the substance with only a small amount of displacement. With the blade pressure continuing, failure zones form in the bed and the longer the pressure is maintained, the larger and more numerous these rift zones will be (31,33).

5.3.2 Tribological Wear

It has been apparent from a number of investigations that the heretofore used notions of friction and wear in the "dry contact" of materials are both qualitatively and quantitatively at odds with experimental evidence. These important phenomena have always been considered intrinsic to the materials employed—that is, a function of the nature of the interacting surfaces only. Thus, much effort was spent on investigating the effects of surface roughness, environmental conditions, the formation of chemical layers, and specimen orientation relative to motion and other surface-related factors. However, in a fairly large number of works, a fresh approach has been taken to the subject in that the interacting surfaces are, in addition, subject to a process akin to a lubrication mechanism in which the lubricant is some sort of particle stratum. While it is clear that a powder layer can be fed externally to a set of mating surfaces and can be selected to meet the tribological requirement of a satisfactory interface layer, our interest here extends to the natural formation of such powder beds due to the ongoing wear of the interacting surfaces. This should explain a number of physical phenomena, which are a result of such unintended lubrication of the two surfaces.

Many tribological devices are covered with solid coatings to facilitate machine operation, of which one of the most common is a polytetrafluoroethylene (PTFE) layer. In experiments with this layer (24), it is shown that the friction and the rate of wear itself are dependent on the very process of wear and that the tribological effect depends on both the duration of the sliding and external load. In the tests, the PTFE counterface was an AISI stainless steel surface, subject to unidirectional sliding at a speed of 0.1 m/s. Three series of tests were run as follows:

1. Effect of time of sliding at a constant load and given surface roughness.
2. Effect of different counterface roughness within a range of 0.05–0.22 micron R_a at constant load and sliding time.
3. Effect of load at a counterface R_a roughness of 0.07 microns, the load ranging from 1.61 to 9.64 kN.

The results of the tests are given in Table 5.5, from which the following observations were deduced:

- The PTFE fibers were transferred to the steel counterface and then began to aggregate and flow in the interface.
- The rate of wear is initially rapid but stabilizes earlier than the coefficient of friction. While the rate of wear reaches a steady limit in 4–6 h, it takes friction 20–30 h to reach a constant value.

TABLE 5.5
Results of Wear Tests

| | | Conditions | | | | | Results | | | | | | | |
| | | | | | | | Temperature, θ (°C) | | | Wear depth μm | Weight Change (mg) | | Steady-State Wear Rate | |
	Test No	Time (h)	Load (kN)	Speed (m/s)	Axial Roughness (μm R_a)	Friction μ (Final)	Measured (2 mm Deep)	Calculated 2 mm Deep	Calculated Surface	Bearing	Counterface	Bearing	10^{-3} μm/m	10^{-7} mm³/Nm
Time	1	1	4.82	0.1	0.06	0.16	45	55.6	60.2	4	+0.3	−0.2	—	—
	2	2			0.047	0.23	59	67.2	73.5	7	−0.2	−1.2	—	—
	3	5			0.066	0.22	62	65.6	71.6	19	+1.8	−2	—	—
	4	50			0.025	0.25	71	70.5	77.3	40	+2.7	−12.1	1.36	1.06
Roughness	5	50	4.82	0.1	0.053	0.195	64	61.4	66.9	21.7	—	−9.5	1.28	1.0
	6				0.056	0.180	65	58.9	64	39	—	−14.7	1.67	1.3
	7				0.134	0.162	60	56	60.6	29	—	−12.2	1.99	1.5
	8				0.22	0.185	62	59.8	65	18.7	—	−11	1.81	1.4
Load	9	50	1.61	0.1	0.07	0.215	40	41	43.4	7	—	−3.8	0.33	0.75
	10		4.82		0.069	0.20	65	62.2	67.8	21	—	−13.8	1.23	0.94
	11		9.64		0.067	0.110	76	75.5	81.6	55.7	—	—	3.99	1.5

Source: Lancaster 1980, Reference 24.

- The value of the limiting rate of wear increases more rapidly than the increase in load; a 6-fold increase in load increases the wear rate by a factor of 12.
- When there is a reduction in the rate of wear with time, it is associated with the observed accumulation of the PTFE debris at the interface; the friction coefficient, on the other hand, is not reduced by the presence of the PTFE debris.
- The structure and composition of the PTFE debris in the film depends on the duration of sliding and the magnitude of the applied load. It is, however, independent of the counterface roughness, as shown in Figure 5.45, which is explained by having its roughness filled in by the PTFE fibers.

Overall, the tests seemed to verify the process of self-generation of a powder film at the interface. Moreover, the process seems to be self-regulating in that it leads to equilibrium conditions as far as the ongoing rate of wear and level of friction are concerned.

Not all powders can act as lubricants and one of the requirements is size of the solid particles, which in gross terms should be of the order of 1 micron. The following examination, using a range of sophisticated scanning methods (Fraunhofer diffraction, scanning electron microscope, ICP emission spectroscopy, and neutron

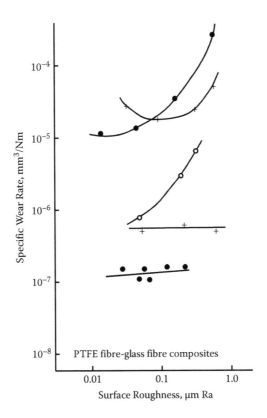

FIGURE 5.45 Variation of wear rate of PTFE layer as a function of surface roughness.

TABLE 5.6

Particle Size Distribution in Used Oil after Centrifuging

Cast Iron

Samples (N)	Compressor (N)	Iron Concentrations (ppm)	Size Distribution (%)		
			$d < 0.1$ (µm)	$0.1 \leq d \leq 0.5$ (µm)	$d > 0.5$ (µm)
1	5	43	55	22	23
2	18	37	57	15	28
3	18	110	35	19	46
4	21	60	40	25	35

Bronze

Samples (N)	Compressor (N)	Copper Concentrations (ppm)	Size Distribution (%)		
			$d < 0.1$ (µm)	$0.1 \leq d \leq 0.5$ (µm)	$d > 0.5$ (µm)
1	5	7.1	80	9	11
2	18	5	83	4	13
3	18	8.9	71	14	15
4	21	11	65	15	20

Babbitt

Samples (N)	Compressor (N)	Tin Concentrations (ppm)	Size distribution (%)		
			$d < 0.1$ (µm)	$0.1 \leq d \leq 0.5$ (µm)	$d > 0.5$ (µm)
1	5	6.7	70	4	26
2	18	4.3	80	5	15
3	18	6.5	50	10	40
4	21	7.8	50	13	37

activation analysis), provides the range of particles obtained from an engine oil covering cast iron, bronze, and babbitt materials. Table 5.6 gives the range of particle sizes obtained for all three materials, while Figure 5.46 plots a particular size distribution obtained from a reciprocating gas compressor. Most particles seemed to lie in the range of 0.1 to 0.3 micron; when the larger particles are included, the mean diameter was about 1 micron. It was also concluded that the size of the wear debris was of the same magnitude as surface roughness after the running in phase.

M. Godet, assisted by his French colleagues, was the first to introduce the notion of wear acting as a tribological layer—which he called third bodies—and he put the foregoing process in a more systematic form (31). A three-dimensional relationship, portraying friction as a function of the duration of sliding and amplitude of motion, is shown in Figure 5.47. The plot represents the behavior of glass sliding against steel, which, at the start, produces three regions:

- Elimination of the natural screens, which cover the two surfaces
- Debris accumulation and formation of an intermediate layer
- Steady-state condition involving a balance between formation and ejection of the generated debris

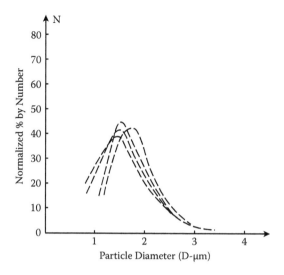

FIGURE 5.46 Wear particles size distribution in used oil from gas compressor.

Different materials, frequencies, and heat treatment will produce different shapes for the three regions and some such differences are shown in Figures 5.48 and 5.49. However, in all cases, a steady-state condition is reached after about 1000 cycles.

5.3.3 POWDER AS A LUBRICANT

While there remains little doubt about the presence of a debris film between interacting surfaces, two of the cardinal issues to be demonstrated are: that these films are capable of supporting an external load and, because the two interacting surfaces move at a different velocity—in most cases, one is stationary—how this difference

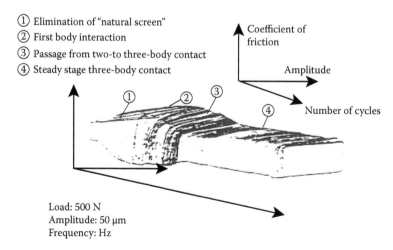

FIGURE 5.47 Frictional behavior of steel on steel.

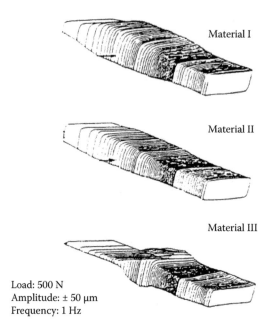

FIGURE 5.48 Frictional behavior of three different metallic materials.

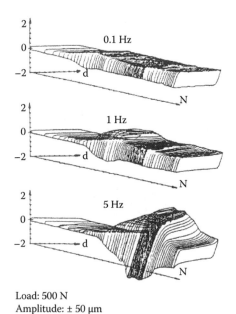

FIGURE 5.49 Effect of frequency on frictional behavior of stainless steel.

in velocity is accommodated across the solid film. The first task was undertaken by employing steel, glass, and chalk surfaces. When steel on steel was used, the load was 500 N; with glass and chalk, the load was 160 N. The progression of testing went through the following steps:

A. Standard fretting tests were run, and the formation of wear debris was monitored by periodically examining the surfaces during which the accumulation of oxide beds, accompanied by a decrease of wear rate, was verified.
B. Oxide beds were fed externally, and the behavior of the surfaces was monitored.
C. The third step consisted of establishing whether the formation of a load-carrying particle bed is specific to the oxides used in steps A and B or that all granular beds—regardless of the particle composition—are capable of carrying a load.

5.3.3.1 Tests A

- Depending on the initial specimen geometry, after about 5,000 to 50,000 cycles, the specimens were uniformly covered with red particles (oxides). The mechanism underlying this final state went through several steps, which are schematically portrayed in Table 5.7.
- At the start, adhesion occurs, and the material is work-hardened, leading to the formation of a new material phase.
- The work-hardened zones are brittle, leading to fragmentation and formation of debris of about 1 micron in size.
- Crushing of particles proceeds, further reducing their size. They then begin migrating across the surface, undergoing further plastic deformation. They are then oxidized mostly into red oxides, Fe_2O_3.

Table 5.8 shows the tests under conditions where the particles are completely contained in the interface or are free to leave at the sides of the specimen; the wayside leakage occurs in conventional bearings. The amount of wear was independent of the foregoing two operating conditions.

5.3.3.2 Tests B

In the external feeding, both gray Fe_3O_4 and red oxides were used. There was little difference from the results obtained from the in situ generated debris (A tests).

5.3.3.3 Tests C

One of the great advantages in running glass-on-glass and glass-on-chalk specimens was the use of visualization techniques to observe the ongoing process. Particle ejection was noted in these surfaces within 20 passes. They appeared as platelets and rolls, as well as in the form of granules and were usually three times the size of initial beads, forming almost immediately after motion starts.

While in fluid films, the velocity distribution across the interface is the familiar combination of Couette and Poiseuille flows (with no slip at the boundaries), the

TABLE 5.7
Metal-Metal Wear Process

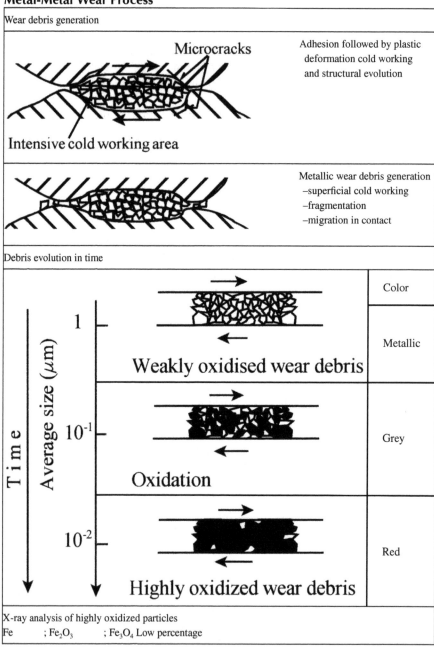

Wear debris generation	
Microcracks / Intensive cold working area	Adhesion followed by plastic deformation cold working and structural evolution
	Metallic wear debris generation −superficial cold working −fragmentation −migration in contact
Debris evolution in time	
Average size (μm) / Time	Color
1 — Weakly oxidised wear debris	Metallic
10⁻¹ — Oxidation	Grey
10⁻² — Highly oxidized wear debris	Red

X-ray analysis of highly oxidized particles

Fe   ; Fe_2O_3 ; Fe_3O_4 Low percentage

TABLE 5.8
Confined and "Side Leakage" Tests with Powder Films

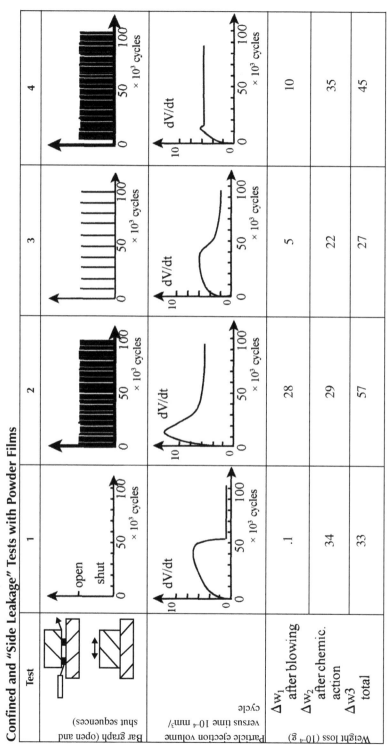

Test		1	2	3	4
Bar graph (open and shut sequences)		open / shut			
Particle ejection volume versus time 10^{-4} mm³/cycle	dV/dt				
Weight loss (10^{-4} g)	Δw_1 after blowing	.1	28	5	10
	Δw_2 after chemic. action	34	29	22	35
	$\Delta w3$ total	33	57	27	45

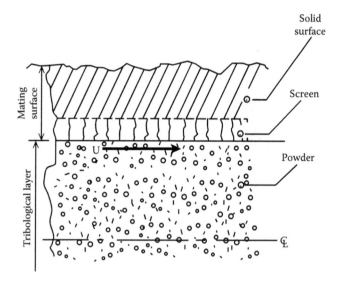

FIGURE 5.50   Velocity accommodation across powder layer.

velocity distribution across a powder film is much more problematic. The group of earlier researchers, as mentioned earlier, proposed the following picture. Referring to Figure 5.50, we have, for the case of symmetry, two primary zones. One is the screen, which consists of the thin top layer of the surface. The other is the bulk of the powder. Under the impact of high tangential stresses induced by the relative velocity, the tensile stresses generate cracks normal to the surface. The result is, as shown in Figure 5.51, a comblike set of short teeth capable of flexing in either direction. This is one zone where the surface velocity is first modified. The process, thus, involves first the fracture of the top layer of the surface and then the elastic flexing of the formed toothlike protrusions.

With regard to the velocity accommodation across the bulk of the powder film, the following test was conducted. A transparent glass specimen was run against steel. It

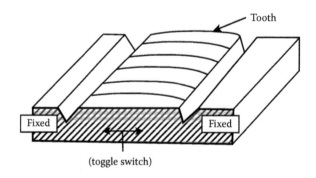

FIGURE 5.51   Surface cracks in polycarbonate specimen.

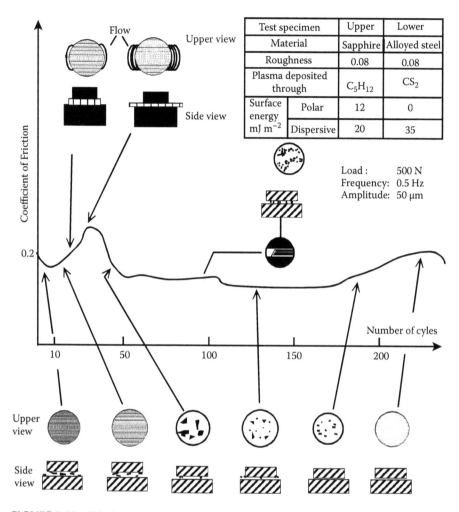

FIGURE 5.52 Friction–time relationship during initial velocity accommodation.

produced a white glass powder, which adhered to the glass, and a reddish powder, which adhered to the steel. In between was the bulk of the intermediate layer, consisting of a mixture of both powders. This bulk was sheared continuously, as well as being locally compacted. The primary importance of the natural film formation of the two interacting surfaces was observed in the following experiment. The effect of time on the velocity accommodation was obtained by running sapphire against steel, separated by a thick graphite film (35). The contact was observed through the sapphire in both top and side views. The process at the interface from the beginning onwards is portrayed in Figure 5.52, divided into phases:

- Phase I: 0 to 8 cycles. Film adhesion to sapphire is low, but high to steel. Accommodation occurs mostly at the screen.

- Phase II: 8 to 12 cycles. Accommodation sets in both at the screen and in the bulk of the graphite.
- Phase III: 12 to 25 cycles. The graphite film is fractured, both in the front and back of the contact—the tears being normal to the displacement. Accommodation has occurred through fracture of the graphite film. First sign of "side leakage."
- Phase IV: 25 to 50 cycles. The graphite film continues to fracture, forming parallel strips. Much of the graphite is ejected on the forward stroke but does not return to the interface on the return stroke.
- Phase V: Up to 175 cycles, the process of the previous phase persists until the graphite is completely eliminated. Beyond 175 cycles, contact occurs between the sapphire and the steel. White powder is produced by the sapphire and metallic debris by the steel. This is the onset of forming a natural debris film between the two surfaces.

5.4   GRANULAR FLOW LAYERS

While powder flow consists of a quasi-continuous substance made up of micron-sized solid particles, in kinetic flow we are dealing with a layer of discrete granules of roughly spherical shape, which may span the range of 1 to 100 mil (0.025 to 2.5 mm). We are interested in kinetic flow layers that may function as interface layers simulating—as shown in Figure 5.53—familiar hydrodynamic patterns of quasi-Couette and Poiseuille flows and combinations of the two.

When in a shear-dominated granular flow, gravity can be ignored, Bagnold (1954) (2) delineated three regimes tied to the rheological properties of the medium. Bagnold used a rotating cylinder Couette-flow-type apparatus in which he tested spherical droplets made of wax and lead stearate having a diameter of 1.32 mm (53 mil) suspended in water. The solid concentrations varied from 13% to 62%, and the value of the fluid viscosity was altered by adding a mixture of glycerin and methylated spirit; at a concentration of 55.5%, a buoyant state was achieved for the solid particles. The three regimes Bagnold defined were the macroviscous, Ba 40, where the fluid viscosity predominates and the shear is linearly related to the shear rate; a transitional regime 40 Ba 450; and the inertial regime Ba 450, where the interparticle collisions dictate the flow pattern and shear stress is proportional to the shear rate squared. This Bagnold number is defined as

$$Ba = \lambda^{1/2} \left[\frac{a_p^2 \rho_p}{\mu} \right] \left(\frac{du}{dy} \right)^2 = \frac{Inertial\ Force}{Viscous\ Force} \qquad (5.38)$$

In the foregoing, λ is related to the solids fraction and subscripts p relate to particle diameter (a) and density (ρ). As can be seen, there is, in the formulation of Equation 5.38, a striking analogy to Reynolds number, delineating the regimes of laminar, intermediate, and turbulent fluid flow.

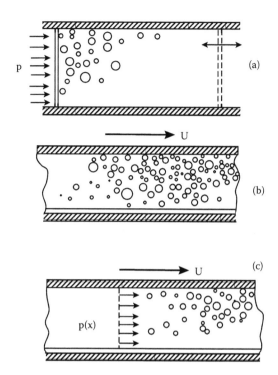

FIGURE 5.53 Simulation of Couette and Poiseuille flows in a granular layer.

The validity of the foregoing subdivision was evaluated by a number of researchers, much of which is summarized in a paper by Chen et al. (6). In place of γ, a viscosity parameter

$$\mu^* = (1 - KC)^{-(B/K)} \tag{5.39}$$

was defined in which $K = (\frac{1}{C_m})$, with C_m being the maximum concentration of the solid particles in the suspension and B the intrinsic viscosity. The dimensionless expression for correlating the rheological behavior of granular flows was given the form of

$$\frac{\tau}{\left[\rho_s a^2 \left(\frac{du}{dy}\right)^2\right]} = a_1 \mu^* \tag{5.40}$$

where a_1 is a coefficient to be determined experimentally.

Figure 5.54 shows a plot of the experimental results correlated via Equation 5.40. As shown in the figure, there was some dependence on the shear rate $(\frac{du}{dy})$ but, in general, the correlation is quite adequate.

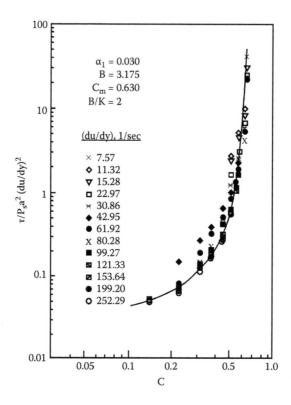

FIGURE 5.54 Fitted theoretical curve for granular flow versus Bagnold's experimental data (Chen).[6]

The major parameters affecting granular flow in the kinetic regime are

- Shear modulus
- Poisson ratio
- Interparticle friction coefficient
- Porosity or void ratio
- Solid fraction
- Particle diameter and density
- Viscosity of suspending fluid
- Effect of walls (boundary conditions)
- Initial and subsequent particle arrangement

Of these parameters, the stress–strain relationship in the flow of a granular layer is probably the most important. This involves a quantity called dilatation, which is a function of not only the elastic properties of the granules but also the geometric freedom the particles have in responding to an applied stress. Much of the available experimental work is, thus, concerned with this granular layer characteristic.

The relationship between an applied force F and the compression Δh of a granular assembly can be expressed by a power law

$$F = k(\Delta h)^m \qquad (5.41)$$

where k is some proportionality constant and m an exponent different from what it is when the foregoing expression is applied to two individual particles. This is due to a particle's multiple interactions with its neighbors. In obtaining a stress–strain relationship, Ammi et al. (1) used a box of Plexiglas cylinders, 4 mm (160 mil) in diameter, with a movable upper plate subject to a force F. The results conformed with Equation 5.41, modified slightly as follows:

$$F = F_o \left(\frac{\Delta h}{h_o} \right)^m$$

In the foregoing, the reference quantities F_o and h_o were the first indication of force required to compress the particles. The measured value of the exponent m was 3.9 ± 0.3, which held up well to the upper test limit of $F = 10,000$ N. A plot of the relationship between F and m is given in Figure 5.55. The surface roughness of the granules—100 microns—was higher than the local deformations, which at 2000 N were only 40 microns. When Plexiglas cylinders (with a better surface finish of 40 microns) were employed, the result was an $m = 3.4 \pm 0.4$. For rubber cylinders (for which the deformation is much higher than any surface nonhomogeneities), the measured value of m was 1.4 ± 0.1. The latter was very close to the value of individual granule interaction.

As said earlier, one reason for the departure of the exponent m from individual particle behavior was the multiplicity of interactions that particles undergo in an

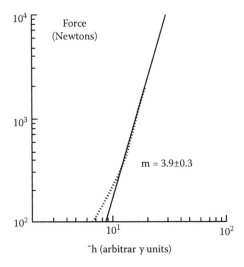

FIGURE 5.55 Stress–strain curve in a granular layer in a packing of 48×44 cylinders.

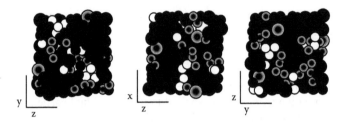

FIGURE 5.56 Three orthogonal views of stress intensity in an assembly of compressed balls.

assembly. This is much compounded by the particles' possible spatial distribution. Its extent can be seen from the experiments conducted by Cundall et al. (8), who used an assembly of 432 glass spheres with the following characteristics:

- Shear modulus of glass—2.9×10^{10} Pa
- Poisson ratio of glass—0.2
- Friction coefficient—0.3
- Radii of particles—40 of 0.1825 mm (7.3 mil) and 392 of 0.1075 mm (4.3 mil)
- Isotropic stress— 1.38×10^5 Pa
- Initial porosity—0.368

Compression/extension shear tests were performed, and stress levels on all three sides of the assembly were recorded. Figure 5.56 shows the results in which surface shading represents the mean stress carried by a particle, white areas indicating completely unloaded particles. The volume change of the assembly under both compression and extension is given in Figure 5.57, showing the presence of a hysteresis or damping effect, much as fluids and powders do.

The complexity of particle migration under loading was also exhibited in computer simulations. Solutions for a given granular assembly were obtained for different interparticle coefficients of friction. The effect of unidirectional dilatation

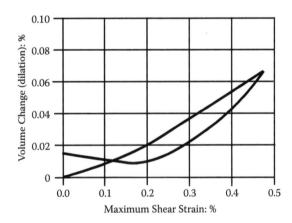

FIGURE 5.57 Volumetric changes of a granular assembly during compression and expansion.

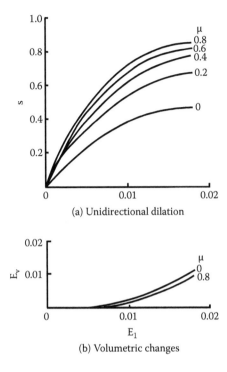

FIGURE 5.58   Stress–strain relationship in a granular assembly as a function of friction coefficient.

is shown in Figure 5.58a; the higher the friction coefficient, the higher the stress required to produce a given amount of shear. In going from zero friction to $f = 0.8$, twice the stress was required to produce a given displacement. But in Figure 5.58b we see that, while the effect was quite prominent in a linear direction, when volume change was considered, the effect of a high coefficient of friction was negligible, dilation in one direction being compensated by a contraction normal to it.

5.5   CONTINUUM ASPECTS

In examining the various tribological regimes, behavioral patterns common to seemingly disparate modes of operation became apparent. These can be listed as follows:

- Presence of incomplete films. In both hydrodynamic liquid-lubricated bearings and in powder and granular flows, the interface layer is frequently incomplete either in the axial, longitudinal, or both directions. This affects the performance of these devices, as well as the mathematics involved, particularly in formulating the correct boundary conditions.
- Variable viscosity. In nearly all the tribological regimes, there is a variation of viscosity in both the axial and longitudinal directions. In powder and granular films, this arises either because of variation of carrier fluid

viscosity or effective viscosity, due to variation in spatial density of powders and granular layers.

- Slip at the surfaces. Frequently, in hydrodynamic gas-lubricated bearings (and in most cases with grease, powder, and granular films), there is slip at the boundary. For the latter, there is also the phenomenon of core formation at either the surfaces or within the interface layer itself.
- Compressibility effects. These prevail in all gas- and vapor-lubricated bearings and to various degrees in slurry, powder, and granular films.
- Elastic effects. Deflection of the bearing surfaces occurs in compliant foil bearings, in elastohydrodynamic lubrication, and in rolling element bearings, gears, and cams.
- Inertia effects. Fluid inertia makes itself felt in determining the laminar and turbulent regimes in hydrodynamic bearings; it is always present in hydrostatic bearings and dampers and determines the level of the Bagnold number, thus delineating whether the regime is macroviscous or purely kinetic.

The concept of tribological surfaces as being separated by some sort of intermediate layer—be it molecular, oxides coatings, wear particles, or deliberately introduced lubricants—has serious implications on material testing and the attempts to generalize results obtained from a specific set of experiments. Looked at from a tribological perspective, there is no such thing as an intrinsic coefficient of friction or rate of wear for a pair of materials. One could almost claim that there would be as many frictional values and wear rates as sets of experiments. Speed, load, temperature, specimen shape, and environment will produce different interactions between the interface layers with significant effects on the resulting behavior of the materials tested.

The foregoing observations lead to the basic proposition that the various tribological regimes—whether they employ gases, liquids, powders, or even aggregates of granules—overlap in their ability to separate the mating surfaces and support an external load. A qualitative representation of this contiguity and overlap is portrayed in Figures 5.59 and 5.60. The dimensionless parameters that govern the

Incompressible		Compressible			Incompressible
Mixed	Liquids	Gases	Powders	Granular	Solids
$\dfrac{\sigma W'}{\mu U}$	$\dfrac{\mu U}{W'}\left(\dfrac{R}{C}\right)^2$	$\dfrac{\mu \omega}{Pa}\left(\dfrac{R}{C}\right)^2$	$\dfrac{U}{\tau}\left(\dfrac{L}{h}\right)^2$	$\dfrac{Pa^2}{\mu}\left(\dfrac{U}{B}\right)^2$	

FIGURE 5.59 Contiguity of various tribological regimes.

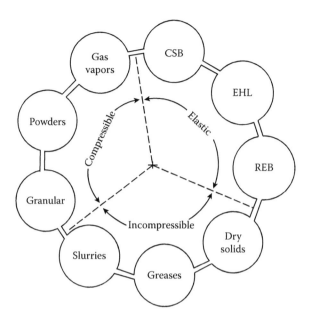

FIGURE 5.60   The closed circle of tribological regimes.

performance of the various segments do vary, but this is true only of the central core of a particular regime. At the extremities, these regimes blend into modes of operation governed by different sets of variables. Thus, in the hydrodynamic domain when the Sommerfeld number becomes sufficiently small, the mixed regime takes over. When the fluid becomes extremely viscous, it would shift close to a Bingham plastic and powder. An attempt to incorporate this tribological continuity into the familiar Stribeck curve would tentatively look like Figure 5.61. Although the various regimes combine differently, only a small number of parameters make their appearance, namely:

- Viscosity
- Density
- Surface roughness
- Loading
- Film thickness
- Surface dimensions

All dimensionless quantities are merely combinations of the foregoing six parameters. The ultimate aim of constructing such a tribological continuum would require the formulation of a "universal," dimensionless grouping, wherein with an asymptotic elimination of one or another of its terms, this overall construct would yield a particular tribological regime. This subject is taken up in Chapter 10.

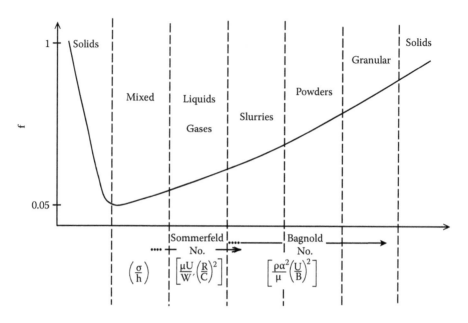

FIGURE 5.61 Tribological continuum on the extended Stribeck curve.

REFERENCES

1. Amni et al., Powders and grains, *Proc. Int. Conf. on Micromechanics of Granular Media*, Clermont-Ferrand, September 1989.
2. Bagnold, R.A., Experiments on a gravity free dispersion of large solid spheres in a Newtonian fluid under shear, *Proc. Royal Soc. London*, Serial A, 225, 49–69, 1954.
3. Boek, E.S. et al., Simulating the rheology of dense colloidal suspensions using dissipative particle dynamics, *Proc. Royal Soc. London*, Serial A, 225, 87–93, 1954.
4. Burgdorfer, A., The influence of molecular mean free path on the performance of hydrodynamic gas lubricated bearings, *Trans. ASME, J. Basic Eng.*, March 1959.
5. Cameron, A., *The Principle of Lubrication*, Longmans, London, 1966.
6. Chen, Cheng-lu and Chu, Y., Granular flow rheology: Role of shear rate number in transition regime, *J. Eng. Mechanics*, Vol. 122, pp. 192–198, May 1996.
7. Cole, J.A. and Hughes, C.J., Oil flow and film extent in complete journal bearings, *Proc. Inst. Mechanical Eng. (London)*, Vol. 170, No. 17, 1956.
8. Cundall et al., Evolution of elastic moduli in a deforming granular assembly, *Powders and Grains*, Balkema, Rotterdam, 1989.
9. Dobson, G.R. and Tompsett, A.C., The rheometry of lubricating greases, *The Rheology of Lubricants*, ed. Davenport, T.C., Applied Science Publications, U.K., 1973.
10. Einstein, A., On a new determination of molecular dimensions, *Annalen der Physik*, (4), 19, pp. 289–306, 1906.
11. Etsion, I. and Pinkus, O., Analysis of finite journal bearings with incomplete films, *ASME J. Tribology*, January 1975.
12. Floberg, L., On hydrodynamic lubrication with special reference to sub-cavity pressures and number of streamlets in cavitating regions, *Acta Polytech. Scand.*, Mechanical Engineering Series 16, pp. 3–35, 1965.

13. Heshmat, H., Walowit, J., and Pinkus, O., Analysis of gas-lubricated compliant thrust bearings, *J. Lubrication Technology, Trans. ASME* 105, No. 4, (1983), pp. 638, 646–655.

14. Heshmat, C.A. and Heshmat, H., An analysis of gas lubricated, multi-leaf foil journal bearings with backing springs, *Trans. ASME, J. Tribology*, Vol. 117, No. 3, 1995, pp. 437–443.

15. Ku, C.-P.R. and Heshmat, H., Compliant foil bearing structural stiffness analysis: Part I—Theoretical model including strip and variable bump foil geometry, *J. Tribology, Trans. ASME*, Vol. 114, No. 2 (1992), pp. 394–400.

16. Ku, C.-P.R. and Heshmat, H., Compliant foil bearing structural stiffness analysis: Part I—Experimental investigation, *J. Tribology, Trans. ASME*, Vol. 115, No. 3 (1993), pp. 364–369.

17. Heshmat, H., Role of compliant foil bearings in advancement and development of high-speed turbomachinery, Publication of ASME, Pumping Machinery, ed. by P. Cooper, FED—Vol. 154 (1993).

18. Heshmat, H., Advancements in the performance of aerodynamic foil journal bearings: High speed and load capability, *Trans. ASME, J. Tribology*, Vol. 116, No. 2, April 1994, pp. 287–295.

19. Walton, J.F. II and Heshmat, H., Compliant foil bearings for use in cryogenic turbopumps, *Proc. Advanced Earth-to-Orbit Propulsion Technology Conference held at NASA/MSFC*, May 17–19, 1994, NASA CP3282, Vol. 1, September 1994, pp. 372–381.

20. Heshmat, H., Chen, H.M., and Walton, J.F., II, On the performance of hybrid foil-magnetic bearings, The 43rd ASME Gas Turbine and Aeroengine Congress, Stockholm, Sweden, ASME Paper No. 98-GT-376, June 2–5, 1998, *J. Eng. Gas Turbines Power*, Vol. 122, No. 1, pp. 73–81, January 2000.

21. Heshmat, C.A., Xu, D., and Heshmat, H., Analysis of gas lubricated foil thrust bearings using coupled finite element and finite difference methods, *J. Tribology, Trans. ASME*, 122 (1), pp. 199–204 (2000).

22. Heshmat, H., Operation of foil bearings beyond the ending critical mode, *Trans. ASME, J. Tribology*, Vol. 122, pp. 192–198, January 2000.

23. Salehi, M., Heshmat, H., and Walton, J.F., II, On the frictional damping characterization of compliant bump foils, *ASME J. Tribology*, October 2003, Vol. 125, No. 4, pp. 804–813.

24. Lancaster, J.K et al., Third body formation and wear of PTFE fibre-based dry bearing, *ASME J. Lubrication Technology*, Vol. 102, April 1980.

25. Pinkus, O. and Sternlicht, B., *Theory of Hydrodynamic Lubrication*, Chapter 3, McGraw-Hill, New York, 1961.

26. Pinkus, O., *Thermal Aspects of Fluid Film Tribology*, ASME, New York, 1990.

27. Vand, V., Viscosity of solutions and suspensions, *J. Physical Chem.*, Vol. 52, pp. 227–314, 1948.

28. Heshmat, H., Starved bearing technology: theory and experiment, Ph.D. thesis, Mechanical Engineering Department, Rensselaer Polytechnic Institute, Troy, New York, December 1988.

29. Heshmat, H. and Artiles, A., Analysis of starved journal bearings including temperature and cavitation effects, *J. Tribology, Trans. ASME* 107, No. 1 (1985), pp. 11–13.

30. Heshmat, H., The mechanism of cavitation in hydrodynamic lubrication, *STLE Trans.*, Vol. 34, No. 2 (1991), pp. 177–186.

31. Heshmat, H., Pinkus, O., and Godet, M., On a common tribological mechanism between interacting surfaces, *STLE Trans.* 32, No. 1 (1989), pp. 32–41.

32. Heshmat, H. and Drewe, D.E., On some experimental rheological aspects of triboparticulates, *Proc. 18th Leeds-Lyon Symposium on Wear Particles: From the Cradle to the Grave*, Lyon (September 3–6, 1991), Elsevier Science Publishers, *Tribology Series* 18, 1992, NASA Contractor Report CR-189043, Volume I & II, October 1991.

33. Heshmat, H., Godet, M., and Berthier, Y., On the role and mechanism of dry triboparticulate lubrication, presented at the *49th STLE Annual Meeting*, Pittsburgh, Pennsylvania, May 1–5, 1994, *STLE Lubrication Eng.*, Vol. 51, No. 7 (1995), pp. 557–564.
34. Heshmat, H. and Brewe, D.E., On the cognitive approach toward classification of dry triboparticulates, *Proc. 20th Leeds-Lyon Symposium on Dissipative Processes in Tribology*, Lyon, France, September 7–10, 1993, Dowson et al., eds., published by Elsevier Science B.V., Tribology Series 27 (1994), pp. 303–328.
35. Iordanoff, I., Berthier, Y., Descartes, S., and Heshmat, H., A review of recent approaches for modeling solid third bodies, presented at the *ASME/STLE Joint Tribology Conference*, October 2001, San Francisco, CA and published in the *ASME J. Tribology*, October 2002, Vol. 24, No. 4, pp. 725–735.

6 Experimental Performance of Powder Layers

Chapters 6 and 7 deal with powder layers, my central concern in the present work. It is the latest technology making an appearance in the tribological spectrum and, in one form or another, is part of all tribological interactions, regardless of whether its presence is intended or not. It modifies—and hopefully clarifies—many of the tribological processes in which, if one ignores the wear process difficulties in interpreting much of the accumulated tribological experience result. The present chapter is confined to experimental work but forms the basis for the next chapter, which attempts to formulate an analytical approach to the field of powder lubrication. Some of the seminal testing on the effects of wear debris between interacting surfaces has been briefly discussed in Chapter 3. In tests with chalk layers observed through one transparent surface, it was seen that, indeed, a particle layer separates the surfaces and that, not unlike fluids, it has a variable velocity across the film. Parallel to the effect of the length-to-width ratio in hydrodynamic bearings, the orientation of a rectangular slider made a difference in powder layer performance. Moreover, forming a geometrical wedge at the slider entrance improved load capacity considerably. Much of the qualitative behavior of chalk was duplicated with other materials, such as metals and plastics. Evidence of the nature of powder flow was obtained by dispersing small lead balls in the film and observing their migration from inlet to outlet. A Couette-like velocity distribution across the film was observed, as well as film rupture reminiscent of fluid cavitation.

Based on these qualitative results, it was concluded that a close kinship exists between powder and fluid films. While a particulate film has no inherent viscosity, it exhibited a stress–strain relationship with a constant proportionality factor, which can be considered an "effective" or "equivalent" viscosity. In essence, given the appropriate conditions, a powder film can be looked at from a continuum standpoint and its behavior in a tribological device termed quasi-hydrodynamic. Powder lubrication, thus, forms an inter-regime mode of lubrication joining the "dry" and fluid films lubrication zones. Moreover, friction is deemed to be more than a simple function of the interacting materials but also a function of the accumulating wear debris.

While on one extreme—such as when the particles are of micron size—the particulate regime is contiguous with hydrodynamic lubrication, on the other end—should the particles be discrete and of relatively large size—powder lubrication abuts kinetic flow. The dynamic properties of layers of discrete particles have had their origin, interestingly enough, with Osborne Reynolds as early as 1885 (7,23). Such kinetic models involve spatial dilatation, which

239

can be looked at as a form of density variation and, as will be seen subsequently (in addition to compressibility), kinetic flows also have much in common with both the hydrodynamic and quasi-hydrodynamic regimes of lubrication.

While the foregoing generic tests and the conceptualization of powder lubrication as a quasi-hydrodynamic process have been more or less established, the present chapter goes beyond these qualitative efforts to generate a quantitative base for the application of this new technology. The experimental results described here offer data on the morphological, thermodynamic, and rheological characteristics of powder layers, as well as their special tribological features. The powders tested range from the familiar molybdenum disulfide to graphites, ceramics, and others. The comprehensive tests include the actual performance of devices operating with powder lubricants such as sliders, journal bearings, dampers, and others.

6.1 THE MORPHOLOGY OF POWDERS

A brief morphological description of powders was given in the Chapter 5. The intent there was to fit them into the tribological continuum from "dry" contacts to granular flows. Here, a more comprehensive and systematic characterization of powders will be given as it emerges from a series of experiments aimed specifically at evaluating them as viable lubricants in the operation of powder-lubricated devices. Powders are an assembly of particles that vary in size, shape, and material properties (17). We refer to powders as triboparticulates when they have suitable tribological characteristics and can act as interface layers. Thus, their size, for example, would have to be such as to fit comfortably in the clearance spaces in bearings, seals, dampers, etc., and be capable of withstanding the shear forces produced by the relative velocities prevailing in these devices.

6.1.1 PARTICLE SIZE

For particles of simple geometric shapes, such as spheroids, cuboids, disks, or rods, their shape can be specified in simple terms. However, most particles in a powder assembly have far from simple shapes; moreover, they vary throughout the assembly in both size and shape. During operation, they often undergo a comminution process, changing from one form to another. Stochastic and mathematical approaches have not fully resolved this seemingly simple problem. Optical micrographs of a number of powder particles under various conditions of pressure and temperature are shown in Figures 6.1 to 6.7, which give some idea of the variety of configurations likely to be encountered, even in commonly used powder materials.

Among other effects, particle size is also related to the rate of wear in mating surfaces. A qualitative relationship between particle size and rate of wear is given in Figure 6.8. When the particles are large, their behavior is that of an in situ elastic body; when small, they tend to aggregate to form a solid body. In between, there is a range of particle sizes suitable for tribological applications in which the rate of wear is at a minimum, which is one of the primary functions of a proper lubricant.

FIGURE 6.1   Pure nickel powder prior to exposure (left) and nickel plus graphite mixture after exposure at 1500°F (right) (300×).

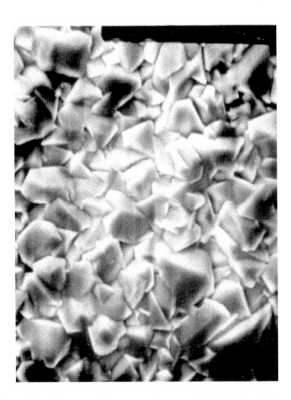

FIGURE 6.2   Scanning electron microscope photograph of nickel oxide on the surface of the heated powder at 5000×.

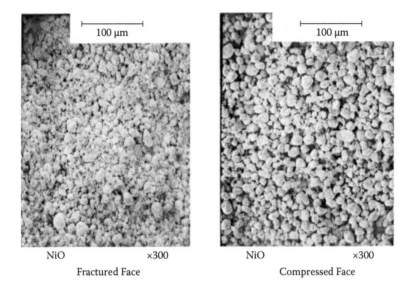

FIGURE 6.3   Light micrographs of NiO particles compact prepared by isostatically pressing at 300 psi: fractured face diagonal to the compressed face.

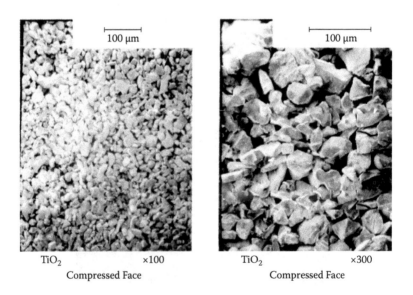

FIGURE 6.4   Light micrograph of TiO$_2$ rutile particles compact prepared by isostatically pressing at 10,000 psi.

FIGURE 6.5   Boron nitride powder before elevated temperature exposure (1000×).

FIGURE 6.6   Nickel oxide and graphite powder mixture after elevated temperature exposure (1000×).

FIGURE 6.7 Re_2O_7 and CuO powder mixture before (left) and after (right) exposure at 1500°F (3000×).

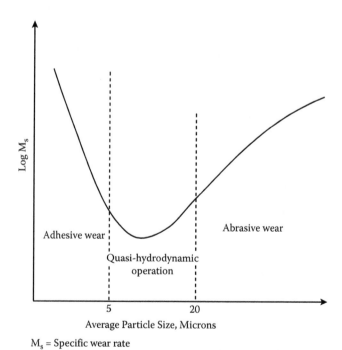

M_s = Specific wear rate

FIGURE 6.8 Qualitative curve for relation between particle size and rate of wear in mating surfaces.

6.1.2   PARTICLE SHAPE

To deal correctly with particle shape is an even more speculative task than tackling its dimension. One attempt was made in Chapter 5 when the Wadell index was defined as the ratio of (d_v/d_s): the ratio of the diameter of a sphere of volume equal to that of the irregular particle to the diameter of a sphere with equal surface area. Because a sphere has the smallest surface area for a given volume, the maximum value of the Wadell index is 1; for a cube it is 0.806, and for round sand its value ranges from 0.8 to 0. Of course, this index is not unique, and different particle shapes may have the same Wadell index number.

Another approach is the geometric signature waveform. This is applied to the projected area of a particle profile. After picking an appropriate centroid in the profile, a vector is drawn from it to the periphery. The vector is then rotated in equal angular increments, and its length recorded as sketched in Figure 6.9. The waveform can be subjected to a Fourier analysis, and the harmonic spectrum used to define its shape. But even here, a certain regularity of shape is required. For example, when the vector intersects the profile periphery at several points, uncertainties may arise, as would be clear from the particle shapes shown in Figure 6.10. It must also be borne in mind that, should this method prove adequate for a planar profile, a particle is a three-dimensional body; a planar analysis is incapable of taking this into account.

The problems associated with particle size and shape are not only pertinent, per se, but they affect the value of mass. Though one may know the bulk density of the material, the mass and inertia of single or aggregate particles would depend on how the size and shape of these particles are defined.

6.1.3   TRIBOPARTICULATES

A subgroup of powders of the appropriate size and material properties has been termed triboparticulates, or candidates for use as lubricants in powder-lubricated devices. A number of such candidates are listed in Table 6.1. About ten of these were selected for detailed testing on the basis of the following capabilities:

- Tolerance of high pressures and shear strain
- Frictional and wear characteristics
- Past experience showing promise in tribological applications
- Tolerance of temperatures up to 2000°F
- Adequate oxidation resistance
- No deleterious chemical interaction with the mating surfaces

Powders A and B—rutile form of TiO_2. Titanium dioxide powder has a melting point of 2480–3000°C and a molecular weight and specific gravity of 80 and 4.26, respectively. They had a 50% cumulative particle size distribution of 2 microns for powder A, 6 microns for powder B, and were 99.5% pure with small amounts of Al, Ca, Cr, Cu, Fe, Mg, and Si. Figure 6.11 shows a photograph of powder A compressed under 50,000 psi. Scanning

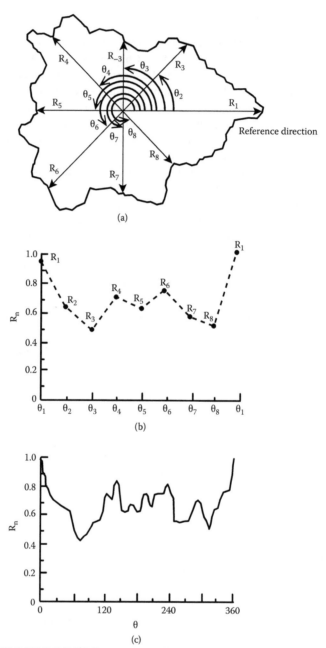

(a)

(b)

(c)

⊠ ⊠ ⊠ ⊠ ⊠ ⊠ ⊠ ⊠ ⊠ ⊠ ⊠ ⊠odm⊠tri⊠ signature waveform is governed by the number of vectors
used to unravel the boundary. (b) Low resolution eight point signature waveform using the
boundary values shown in (a). (c) High resolution-72 point geometric signature waveform of
profile shown in figure (a) R = vector at angle θ. All values of R normalized with respect to the
maximum value of R chosen to be the referenced direction at θ = 0.

FIGURE 6.9   Method of geometric signature waveform.

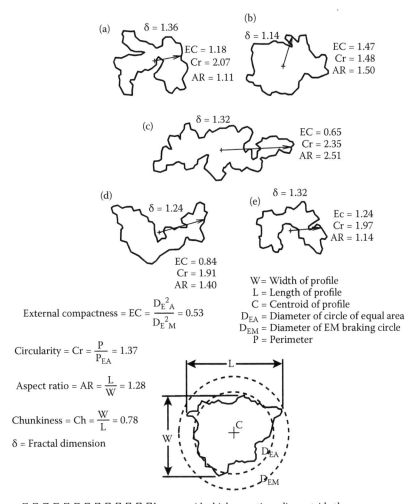

profile and the second difficulty is the vector R drawn from the centroid which, even when it lies within the profiles, crosses the boundary several times and it is difficult to know which value of R to use in setting up the geometric signature waveform of the profile.

FIGURE 6.10 Complications in using the geometric signature waveform for complex shapes.

electron photomicrographs (SEP) of pressed powders A and B are shown in Figures 6.12 and 6.13, respectively. In both of them, fractured faces resulted from the high loading.

Powder C—anatase form of TiO₂. This is a white crystalline powder with a boiling point 2500°C–3000°C, a melting point of 1830°C–1850°C, and a molecular weight and specific gravity of 80 and 3.84 g/cc. They had a 50% accumulative particle size distribution of 0.4 microns. Spectrographic

TABLE 6.1
Some Available Lubricant Powders

Sulfides	Selenides	Tellwides	Oxides	Chemical Composition of Lubricious Material
Ag_2S	—	AgTe	SiO_2	$35CaF_2$—$62BaF_2$
Bi_2S_3	Bi_2Se_3	Bi_2Te_3	Bi_2O_3	BiOCd
CdS	CdSe	CdTe	CdO	Cs_2MoS_3[a]
Ce_2S_3	—	—	K_2MoO_4	$CuCl_2$
HgS	—	—	NiO	BN
MoS_2	—	—	TiO_2	Graphite
PbS	—	—	PbP	PCH_2
ZnS	—	—	ZnO	$ZnMoO_2S_3$[a]
Cs_2SO_4	ZnSe	ZnTe	—	$ZnMoO_2$ 57 LiF–42 CaF_2

[a] These are from a group of complex oxides [Metal]$_x$ [Metal]$_x$ O$_x$ S$_x$ that are under development.

analysis showed it to be 99.9% pure, consisting of tiny amounts of Al, Co, Mg, and Si. Figure 6.14 shows a photograph of this powder compressed under 50,000 psi; its fractured face is given in Figure 6.15. In general, they have a spheroidal shape and a tendency to form agglomerates.

Powders D and E—lubricating-grade MoS2. This powder was obtained in two grades—suspension grade (D) and technical grade (E)—the two differing essentially only in particle size. Their 50% accumulative size distribution

Powder A; TiO_2 (Rutile Form); $\bar{P}_d < 2\ \mu m$; $\sigma_{yy} = 50$ ksi

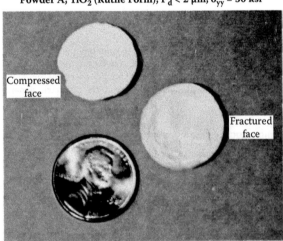

FIGURE 6.11 Powder A, TiO_2 rutile form compressed up to 50,000 lb; fractured in a plane diagonal to the *Fy* direction.

TiO_2 **(Rutile Form);** $\overline{P}_d$ = 2 µm; [A]

Compressed Face

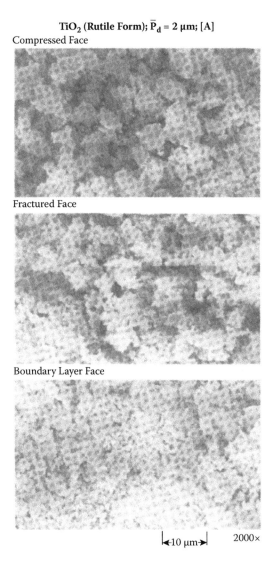

Fractured Face

Boundary Layer Face

|←10 µm→| 2000×

FIGURE 6.12   Scanning electron photomicrographs of powder A.

was in the range of 1–2 microns for powder D and 12 microns for powder E. The compressed samples are presented in Figures 6.16 and 6.17. Powder E, with its coarser particles, readily forms agglomerates. SEPs of pressed MoS_2 powders are shown in Figures 6.18 and 6.19. The fractured sites show the particles to be irregularly shaped platelets. There was no significant difference between the uncompressed and compressed samples, except for the reduction in porosity. The most significant observation was the generation of a distinct pattern in the sheared faces. Packed particles at these sites seem to form clusters in the shape of cauliflowers. Induced by shearing, this

TiO₂ (Rutile Form); $\bar{P}_d$ = 6 μm; [B]

Compressed Face

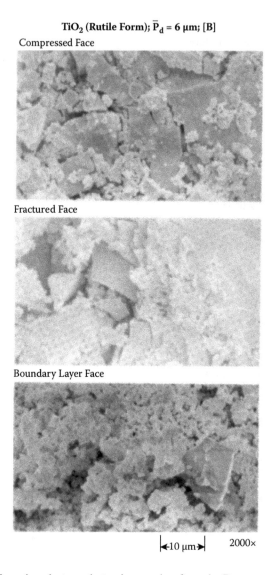

Fractured Face

Boundary Layer Face

|←10 μm→| 2000×

FIGURE 6.13 Scanning electron photomicrographs of powder B.

is a reproducible pattern independent of particle size. It is considered to be
a form of boundary layer between the shearing surface and the bulk of the
powder; hence, its known tribological merit.

Powder F—CeF₃. Cerium trifluoride is a good solid lubricant for high-
temperature applications, though it has a high coefficient of friction. The
powder has a hexagonal shape and is of an off-white to pinkish color. Its
specific gravity is 6, with a melting point at 2620°C. Its average size is
about 10 microns and, as shown in Figure 6.20, tends to form aggregates.

Powder C: TiO$_2$ (Anatase Form); P$_d$ < 0.4 μm; σ$_{yy}$ = 50 ksi

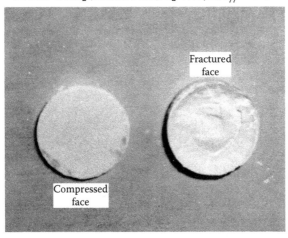

FIGURE 6.14   Powder C, TiO$_2$ (anatase form) $P_d = 0.4$ μm. The sample was pressed up to $\sigma_{yy} = 50$ ksi.

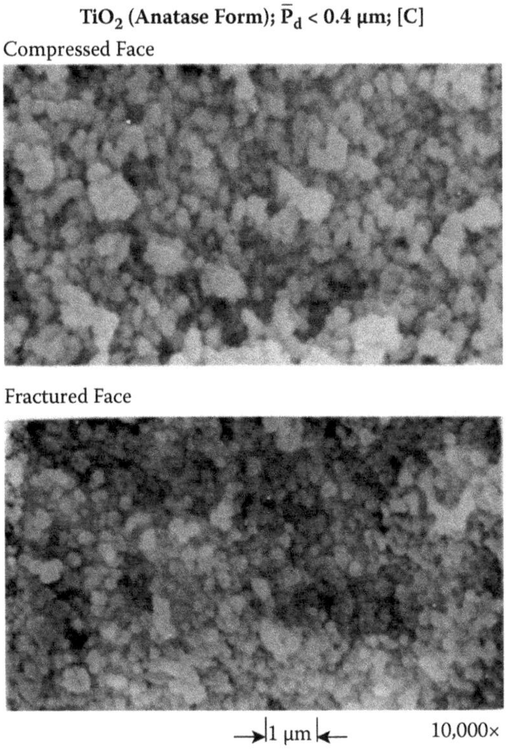

TiO$_2$ (Anatase Form); $\bar{P}_d$ < 0.4 μm; [C]

Compressed Face

Fractured Face

→|1 μm|←   10,000×

FIGURE 6.15   Scanning electron photomicrograph of powder C.

Powder D; MoS$_2$; $\bar{P}_d$ < 1–2 µm; σ_{yy} = 50 ksi

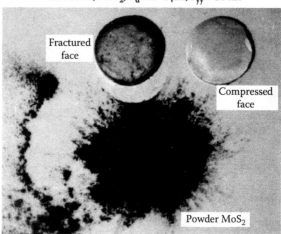

FIGURE 6.16   Powder D (1) bottom powder form, (2) top left fractured after test approximately perpendicular to F_y, (3) top right compressed face, σ_{yy} = 50 ksi, 20,000 lb.

SEPs of the compressed samples show fractured sites consisting mostly of spheroidally shaped particles.

Powder G—ZnMoO$_2$S$_2$. Zinc dithiomolybdate has a tendency to oxidate at relatively low temperatures, but the products constitute good lubricants. It has an orange color, a specific gravity of 3.17, and a molecular weight of

Powder E; MoS$_2$; $\bar{P}_d$ < 12 µm; σ_{yy} = 50 ksi

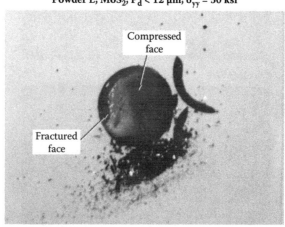

FIGURE 6.17a   Compressed powder E extruded from test cylinder, fractured in planes perpendicular to F_y.

Powder E; MoS$_2$; P$_d$ < 12 μm

FIGURE 6.17b Appearance of poured powder molybdenum disulfide.

474. Its average size is around 5 microns. A photograph of the compressed powder is shown in Figure 6.21, showing fractures normal to the applied load (50,000 psi). Figures 6.22 and 6.23 show all three sites for two compression loads of 20,000 and 5,000 psi, respectively. Sheared face particles appear as clusters similar to smeared chalk. Figure 6.24 demonstrates the three-dimensional deformation of the powder under normal stress, similar to that of liquids. When the applied load was increased, the powder yielded in all directions.

Powder H—carbon graphite. This familiar solid lubricant showed a coefficient of friction from 0.01 to 0.04. The powder obtained from this material had an average size of 2 microns and 99.9% purity. It is an amorphous powder that oxidizes at near 600°C, has a boiling point of 4827°C, and a melting point at 3652°C. Its molecular weight is 12 and specific gravity is 2.22–2.23. Figure 6.25 shows a picture of the compressed powder at 50,000 psi. It easily forms aggregates and also flows easily through narrow gaps. Its behavior is close to that of grease. Figure 6.26 shows the radial outflow of the powder under a compressive load. SEPs of the compressed powder are given in Figure 6.27. The fractured sites show staggered thin plates, which seem to flow in layers over each other.

Powder I—fumed SiO$_2$. Fumed silica is a submicron powder ranging from 70 to 500 Å. The present sample had an average size of 0.012 microns consisting of 99.5% silicon dioxide, free of any metallic oxides. Its specific gravity is 2.2. It is useful in high-temperature applications, particularly as an

MoS$_2$ (Finer Sized Particles); $\bar{P}_d$ < 2 µm; [D]

Compressed Face

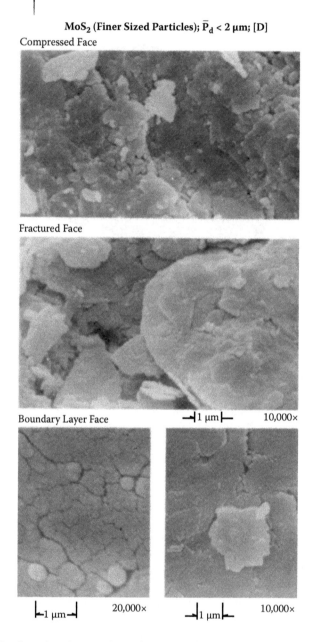

Fractured Face

Boundary Layer Face   ⊢1 µm⊣   10,000×

⊢1 µm⊣   20,000×   ⊢1 µm⊣   10,000×

FIGURE 6.18   Scanning electron photomicrograph of powder D.

additive to other lubricating materials. Figure 6.28 shows samples of this powder after being extruded from the test cylinder. Under 50,000 psi, the compressed sample tended to agglomerate. Figure 6.29 shows SEP pictures of the fractured faces, the particle shape being spheroidal, regardless of the level of compression.

MoS$_2$ (Moderate Size); $\bar{P}_d$~12 μm; [E]

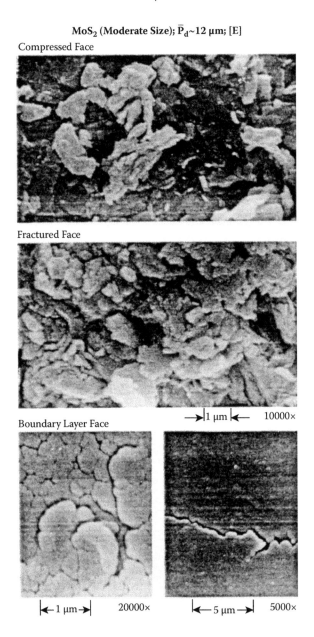

FIGURE 6.19   Scanning electron photomicrographs of powder E.

Powder J—NiO. Nickel oxide powders are suitable for high-temperature applications. It has a melting point of 1984°C and a molecular weight and specific gravity of 75 and 6.7, respectively. It showed a 58% accumulative particle size distribution of 5 microns and was 99.5% pure. Figure 6.30 shows a compressed sample of the powder under 50,000 psi with no tendency to

Powder F; CeF$_3$; $\bar{P}_d$ < 10 µm; σ_{yy} = 50 ksi

FIGURE 6.20   Powder F: top: powder form, bottom: after compression test up to 20,000 lb and extruded from load cylinder, platelets, and flakes.

agglomerate, thus assuring high flow ability. Figure 6.31 shows SEPs of three sides of the extruded powder. The sheared face appears to consist of much denser and finer particles than the other surfaces.

Tables 6.2 and 6.3 summarize some of the foregoing data and provide additional information on the size and shape aspects of the powders described earlier.

Powder G; ZnMoO$_2$S$_2$; $\bar{P}_d$ < 5 µm; σ_{yy} = 50 ksi

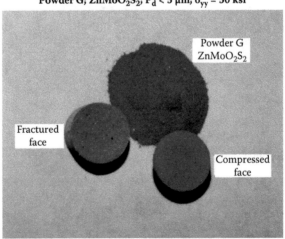

FIGURE 6.21   Sample was pressed up to σ_{yy} = 50,000 psi.

Powder G; $ZnMoO_2$; $\bar{P}_d < 5$ μm; $\sigma_{yy} = 50$ ksi

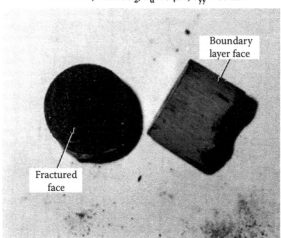

FIGURE 6.22   Powder G after test subject to 20,000 lb of compression load.

6.2   THERMODYNAMIC PROPERTIES

The familiar basic thermodynamic quantities are pressure, temperature, and density. While these are of fundamental relevance to the study of gases (and, to a lesser degree, liquids), when it comes to powders, the crucial quantity is density. The reason for it is that, while in fluids, pressure and temperature are sufficient to fix their density, this is far from being the case with powder assemblies. The reason for it is that a powder can exist in a wide and variable spectrum of concentration,

Powder G; $ZnMoO_2S_2$; $\bar{P}_d < 5$ μm; $\sigma_{yy} = 12.5$ ksi

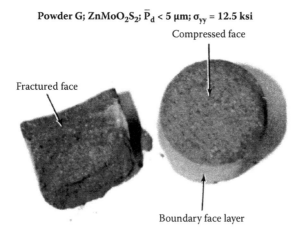

FIGURE 6.23   Powder G after 5000 lb of compression load fractured to compressed faces.

$ZnMoO_2S_2$, $\overline{P}_d = 5\ \mu m$; [G]

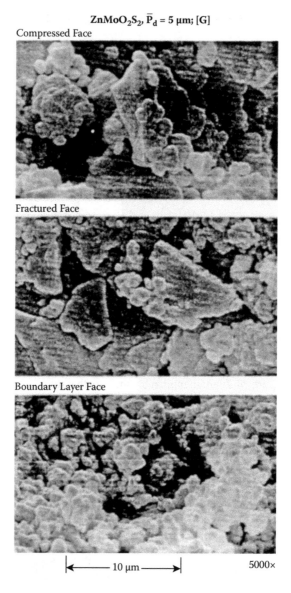

FIGURE 6.24   Scanning electron photomicrograph of powder G.

porosity, voids, and compaction. These can vary not only in bulk but also in the space it occupies and with its use (i.e., in time, too). It should, thus, not be surprising that most of the studies invested in a thermodynamic description of powders are devoted to density. Next in level of importance comes compaction, not so much because of the resulting pressure but because compaction vitally affects the density.

Powder H; Carbon Graphite; $\bar{P}_d$ < 2 µm; σ_{yy} = 50 ksi

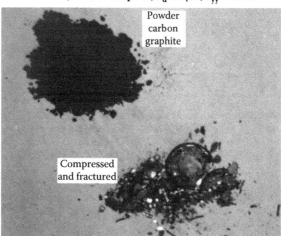

FIGURE 6.25 Powder H: top: powder graphite; bottom: compressed powder under 20,000 lb and extruded from test cylinder (fractured in the form of needles and plates).

6.2.1 DENSITY

The density of a powder is a function of the degree of packing. The characteristics of the individual particles are important as well because they determine the nature of interparticle behavior, as well as the interaction with the carrier fluid—be it a liquid or a gas. Particle ensembles have to be categorized in terms of mass, space, and

Powder H; Carbon Graphite; $\bar{P}_d$ < 2 µm; σ_{yy} = 50 ksi

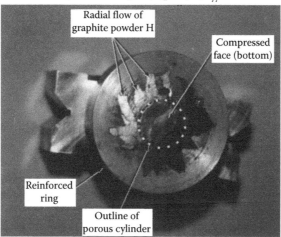

FIGURE 6.26 Powder H, carbon graphite, showing squeezed powder from the bottom surface contact area under 20,000 lb (50 ksi).

Carbon Graphite, $\bar{P}_d$ = 2 μm; [H]

Compressed Face

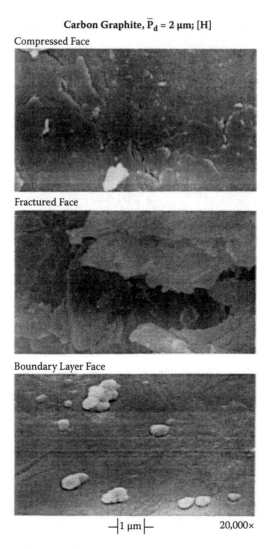

Fractured Face

Boundary Layer Face

$\dashv$1 μm$\vdash$ 20,000×

FIGURE 6.27   Scanning electron photomicrograph of powder H.

time. This relationship between mass, sketched in Table 6.4, affects the ability of the powder to flow and, hence, to act as a lubricant. To this, one must add the effects of rheological properties. Conversely, density affects the yield value, effective viscosity, and heat transfer properties of the medium. Figure 6.32 shows how some of the foregoing elements affect the overall performance of a powder aggregate. This maze of interlocking relations has to be systematically sorted out in order to arrive at some sort of hierarchical order of importance of the parameters involved. Furthermore, the particle density alone is a complex material property due to the exclusion or inclusion of open and pore spaces, as shown in Fig 6.33.

Powder I; SiO; $\overline{P}_d < 0.012$ μm; $\sigma_{yy} = 50$ ksi

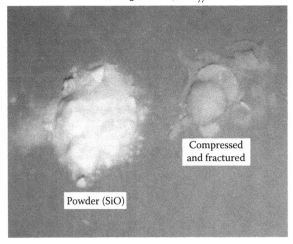

FIGURE 6.28 Powder I extruded from test cylinder; sample was fractured into platelets (almost transparent) and thin plates.

SiO_2 (Fire-Dry Fumed Silica); $\overline{P}_d = 12$ nm; [I]

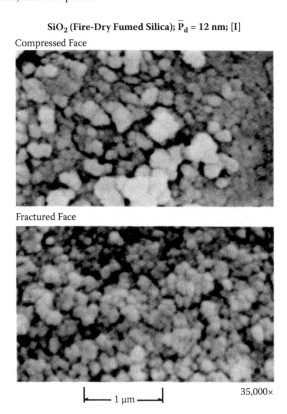

FIGURE 6.29 Scanning electron photomicrograph of powder I.

Powder J; NiO; $\overline{P}_d < 5\ \mu m$; $\sigma_{yy} = 50$ ksi

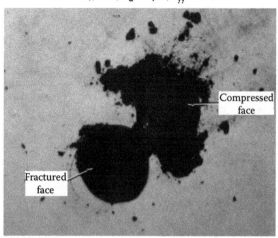

FIGURE 6.30 Powder J compressed up to 20,000 lb and extruded from test cylinder, showing fractured surface perpendicular to F_y.

Because powders consist of solid particles in a fluid, the bulk density lies somewhere between the density of the fluid and of the solid particles. Bulk density is not a single specific value but a complex property depending on the particles' shape, size, degree of compaction, concentration, and the operating conditions. Furthermore, particle density itself is complex because it can exist in an aerated, poured, tapped, or compacted condition, each yielding a different value for bulk density. In one series of tests (8a), the variation of density was investigated using nickel oxide with a size of 5 microns and the other titanium oxide with an average size of 5.6 microns. The samples were compressed up to 10,000 psi, which, upon examination, were found to have fractured faces. Photographs showed that the compressed faces had a profusion of interparticle porosities. On the fractured faces, the particles appeared to be finer and closer together, indicating a higher density. These differences were more pronounced with the nickel oxide, probably due to its finer size. Proceeding from the center of the fractured toward the compressed face, the state of the powder slowly changed to that of the compressed face. It appeared that, under compression, the particles reorient themselves so as to have the highest concentration in the center of the bulk material.

A schematic interpretation of the foregoing phenomenon in a powder lubricant is given in Figure 6.34. The concentration C (solid fraction of powder) is a function in both x and y with the variation in the longitudinal direction given by

$$C(x) = C(p) + C(0)$$

Based on the experimental results, an empirical relation for the preceding function can be written as

$$C(x) = C(0)[1 - p/K_c ....] \tag{6.1}$$

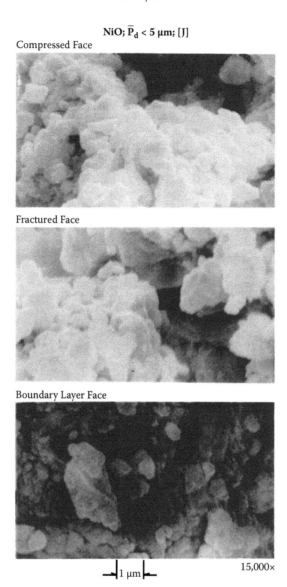

FIGURE 6.31   Scanning electron photomicrograph of powder J.

where $C(0)$ is the reference solid fraction, p the local pressure, and K_c a solid-fraction coefficient. The variation across the film is, likewise, formulated as being given by

$$C(y) = \left[\frac{(C - C_w)}{(C_\infty - C_w)} \right] = 1 - \mathrm{Cos}h \frac{\left[\beta h \left(\frac{y}{h} - \frac{1}{2} \right) \right]}{\left(\frac{\beta h}{2} \right)} \tag{6.2}$$

TABLE 6.2
Physical Properties of Selected Triboparticulates

ID	Formula	Powder Name and Form	d_s (µm)	ρ_s (gm/cc)
A.	TiO_2	Titanium dioxide (rutile form)	1–2	4.26
B.	TiO_2	Titanium dioxide (rutile form)	5–6	4.26
C.	TiO_2	Titanium dioxide (anatase form)	0.4	3.84
D.	MoS_2	Molybdenum disulfide (finer-sized particles)	1–2	4.80
E.	MoS_2	Molybdenum disulfide (moderate-sized particles)	12	4.80
F.	CeF_3	Cerium trifluoride (cerous fluoride)	10	6.16
G.	$ZnMoO_2S_2$	Zinc oxythiomolybdate	5	3.17
H.	Carbon	Carbon graphite (finer-sized particles)		2.25
I.	SiO_2	Fire-dry fumed silica	0.012	2.20
J.	NiO	Nickel oxide, green powder	5	6.67

TABLE 6.3
Qualitative Image of the Individual Particles of TiO_2–Anatase

Diameter (X-FERET)	Compressed Face	Fractured Face	Sheared Face
Sample size	164	187	164
Mean value	0.33	0.36	0.34
Variance	0.010	0.008	0.008
Standard deviation	0.101	0.092	0.089
Limits	0.167–0.78	0.128–0.735	0.164–0.59

Perimeter	Compressed Face	Fractured Face	Sheared Face
Sample size	164	187	164
Mean value	1.22	1.33	1.17
Variance	0.235	0.103	0.083
Standard deviation	0.485	0.322	0.289
Limits	0.527–3.85	0.575–2.59	0.571–1.90

Areas	Compressed Face	Fractured Face	Sheared Face
Sample size	164	187	164
Mean value	0.084	0.095	0.090
Variance	2.72×10^{-3}	2.16×10^{-3}	2.04×10^{-3}
Standard deviation	0.052	0.046	0.045
Limits	0.018–0.298	0.015–0.390	0.018–0.246

Shape Factor	Compressed Face	Fractured Face	Sheared Face
Sample size	164	187	164
Mean value	0.71	0.66	0.79
Variance	0.027	0.016	0.009
Standard deviation	0.164	0.126	0.096
Limits	0.028–0.999	0.098–0.976	0.528–0.987

TABLE 6.4

Categories of Particle Assemblies in Terms of Mass, Length, and Time

Dimensional Groups	Bulk Powder Properties	
Volume:Volume	Packing	Densification
Mass:volume	Single particle	Porous
		Nonporous
	Ensemble of particles	Compressed
		Sheared
		Aerated
		Vibrated
		Poured
Mass:mass (interaction)	Particle–particle	Cohesiveness/tensile strength
		Densification
		Expansion
		Fluidization
	Particle–fluid	Pneumatic transport
		Fluidization
		Catalytic
		Agglomeration
Mass:time	Particle–particle	Flowability
		Floodability
Volume:time (transportation)	Particle–fluid	Gaseous
		Liquid

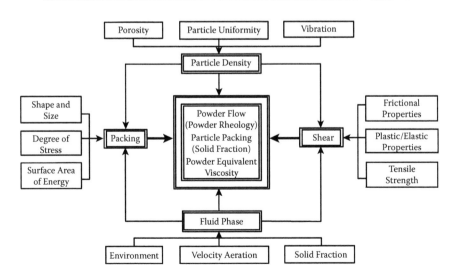

FIGURE 6.32 Variables affecting powder flow behavior and powder properties.

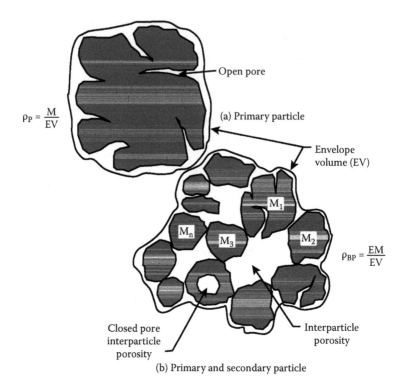

FIGURE 6.33 Definition of particle density.

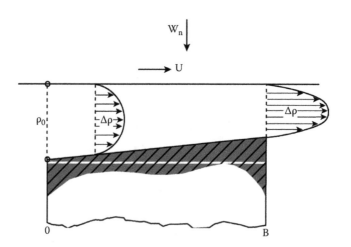

FIGURE 6.34 Transverse and longitudinal variation in density with powder lubrication.

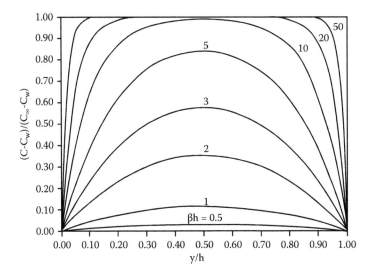

FIGURE 6.35   Mathematical model of solid fraction variation across film.

Here, C_w is the solid fraction near the wall, C_∞ the maximum bulk solid fraction (near one), h the film height, and β a constant strongly dependent on the local pressure and material properties that is determined empirically.

While a completely free assembly of particles can be considered to be in a state of aerated density, the most common usage of such an aggregate results in what is called tapped density. The ratio of these two densities is a useful factor in evaluating many aspects of shear-driven powder films. It is also used in defining the compressibility of powders

$$\text{Compressibility} = 100\left[\left(\frac{\rho_t}{\rho_a}\right)-1\right] \tag{6.3}$$

where subscript t refers to the tapped and a to the aerated density. Powders with a compressibility value in the range of 5%–15% are considered to have good flow characteristics, whereas those in excess of 20% have poor flow capabilities. The compressibility index is also a good indicator of uniformity of particle size, shape, and other relevant material properties.

In the tests conducted in Reference 8a, an apparatus was constructed to obtain powder compaction and measure its properties as a function of degree of compression. This consolidometer, shown in Figure 6.36, has a container with a piston made of porous nickel to allow the carrier fluid to escape during compression. The dimensions of the cylinder, with the piston having a loose clearance to prevent metal-to-metal contact, were ID = 18.14 mm and height 38 mm. This was installed in the test rig shown in Figure 6.37 for the application of a range of pressures on the piston. Each test was initiated by pouring and tapping the powder sample until it was flat

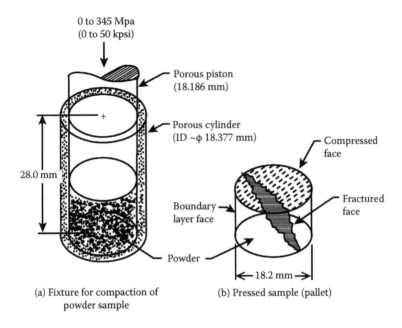

0 to 345 Mpa
(0 to 50 kpsi)

Porous piston
(18.186 mm)

Porous cylinder
(ID ~φ 18.377 mm)

28.0 mm

Boundary
layer face

Powder

Compressed
face

Fractured
face

18.2 mm

(a) Fixture for compaction of
powder sample

(b) Pressed sample (pallet)

FIGURE 6.36   Preparation of packed powder (pallet) and determination of compacted density.

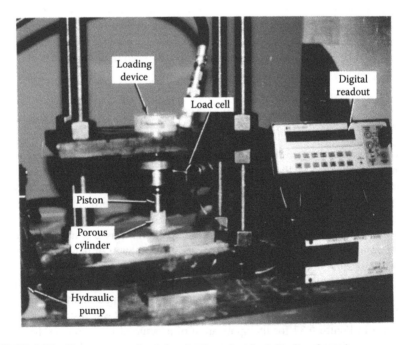

Loading
device

Load cell

Digital
readout

Piston

Porous
cylinder

Hydraulic
pump

FIGURE 6.37   Test apparatus for determination of pressed density of powders.

with the top of the cylinder. The ratio of powder weight to cylinder volume was considered its tapped or reference density. A steady load was then applied on the piston and, following each load increment while maintaining the load constant, the change in powder volume was measured. This change gave the variation of density of loading.

The results for bulk density P_p, as well as its concentration as a function of pressure, are given in detail in Reference 8a. Here, the results are summarized in several basic plots of pressure versus P_p. Figures 6.38 and 6.39 show composite plots for two groups of powders. For powders A, B, C, G, H, and J, the maximum density achieved was below 3 g/cc; for the group consisting of powders D, E, F, and J, the maximum density exceeded 5 g/cc. As is clear from the test data, compacted powders exhibit two regimes in response to compaction. At low values of p—up to some 35 MPa—the bulk density is strongly affected by the rise of pressure. A transition occurs in the range of 26 to 50 MPa. Above 50 MPa, the packed powder exhibits a high stiffness, yielding only a small rise in density with an increase in loading. The solid fraction C—ratio of powder density to the solid phase—is plotted in Figures 6.40 and 6.41. In these plots, the data are extrapolated linearly to the limit of C = 1. Here, too, at pressures higher than 50 MPa, the curves level off. Beyond C = 1, the substance can no longer be considered a powder but rather a material characteristic of powder metallurgy.

Plots of the TiO_2 group of powders (A, B, and C) are presented in Figure 6.42. This behavior seems an anomaly for materials so different in particulate shape, size, and crystal structure. A similar statement can be made when grouping powders B and G, or J, D, and E, particularly when their slope (dp_p/d_p), rather than the absolute values of p_p, is taken up for comparison.

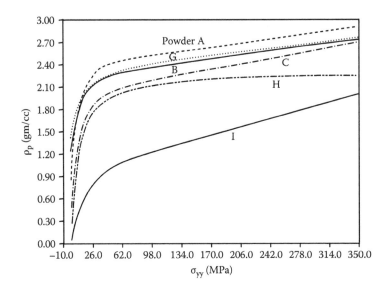

FIGURE 6.38 Powder density p_p as a function of compression for powders TiO_2 (A,B,C), $ZnMoO_2$ (G), graphite (H), and SiO_2 (I).

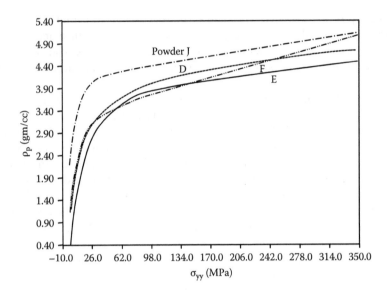

FIGURE 6.39 Powder density as a function of compression for powders MoS$_2$ (D,E), CeF$_3$ (F), and NiO (J).

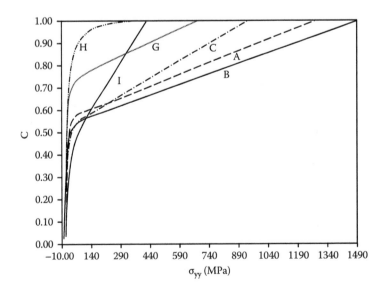

FIGURE 6.40 Solid fraction versus compression for powders TiO$_2$ (A,B,C), ZnMoO$_2$ (G), graphite (H), and SiO$_2$ (I); extrapolated for $\sigma_{yy} > 350$ MPa.

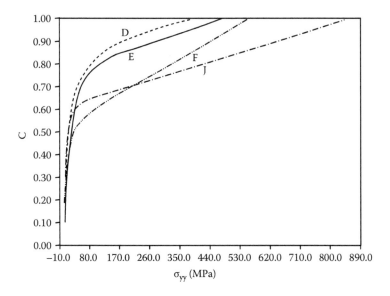

FIGURE 6.41 Solid fraction versus compression for powders MoS_2 (D,E), CeF_3 (F), and NiO (J); extrapolated for $\sigma_{yy} > 350$ MPa.

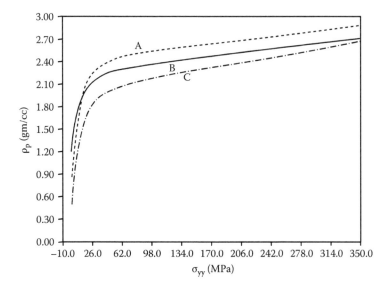

FIGURE 6.42 Bulk density of TiO_2 powders versus compression.

6.3 RHEOLOGICAL CHARACTERISTICS

The rheological nature of powders is the main determinant of their lubrication qualities. Being a bridge between fluids and solids, it is no surprise that their rheological
properties most relevant to powder flow are the equivalent viscosity, as is the case
with liquid lubricants and their stress–strain relationship—the determining property
of stressed solid bodies. These are the two primary parameters in investigating a
powder's suitability to act as a lubricant.

6.3.1 EFFECTIVE VISCOSITY

The most common instruments used to determine viscosity are the flow- and the
shear-type viscometers. The first measurements are based on the time it takes for a
given volume to flow out of a container through a standard constrictor. Calibrated
against a substance of a known viscosity, that of an unknown substance can be
determined. In shear-type devices, the viscosity is determined by measuring the
force required to overcome the substance's resistance. Using either a calibration
standard or an analytical formula containing a viscosity term, the unknown viscosity can be determined.

Figure 6.43 shows one type of instrument that can be used for powders: an arrangement similar to the familiar Redwood or Saybolt viscometers. In it, the powder is
allowed to flow out of a hopper through a restrictor, and the time required for a given

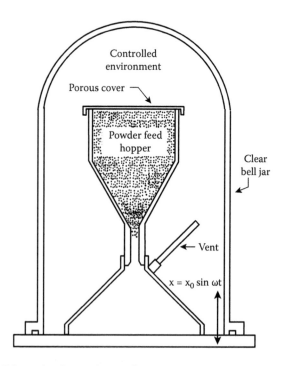

FIGURE 6.43 Schematic of venturi-type viscometer.

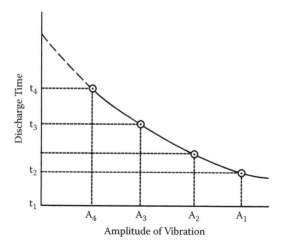

FIGURE 6.44 Expected discharge time versus amplitude of vibration utilizing scheme shown in Figure 6.43.

volume to be discharged provides the kinematic viscosity. Liquids of known viscosity are used as a standard to obtain the powder's effective viscosity. Some powders may not flow under gravity alone and require agitation to make them flow. This is provided by mounting the viscometer on an electromagnetic shaker to induce vertical vibration in the device. The discharge time is then measured as a function of amplitude at a given frequency. The time–amplitude data is then sketched in the manner shown in Figure 6.44 and the data extrapolated to yield the time at zero amplitude. The zero amplitude time provides the effective kinematic viscosity of the powder tested.

The shear-type viscometer used in the following experiments (8a) is depicted in Figure 6.45; a photograph of it is shown in Figure 6.46. A stationary disk is prevented from rotating by an instrumented torque arm while the rotating element, of an elliptical shape, is set upon a rolling element bearing attached to the base. Provisions are made for choosing a predetermined clearance between the mating parts. The conical bore of the mating collar is slightly eccentric with the outer diameter, thus, introducing nutation in addition to the rotation. A controlled amount of powder is supplied to the center of the rotating collar and, hence, to the clearance space where it is sheared and the corresponding torque recorded. Usually, a 50 cc powder sample was poured into the center of the opening and dry nitrogen gas admitted to the top to ensure the existence of a full powder film in the clearance.

The transmitted shear force through the lubricant film to the mating collar was measured and converted to units of viscosity. The following assumptions and calculations for data reduction purposed were considered:

$$R_1 = \frac{(R_{min} + R_{maj})}{2} = 2.58 \text{ in.}$$

$$R_2 = R_1 + \Delta R = 2.78 \text{ in}$$

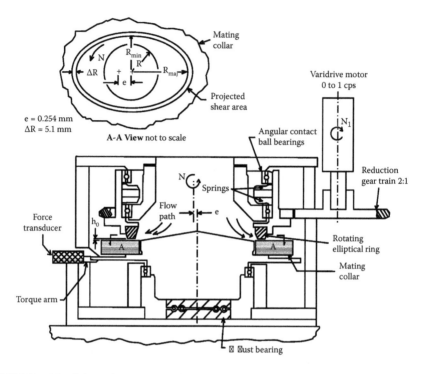

e = 0.254 mm
ΔR = 5.1 mm

A-A View not to scale

FIGURE 6.45   Schematic of powder viscometer.

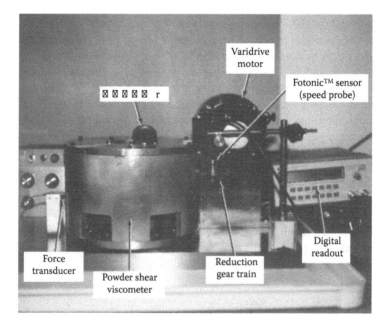

FIGURE 6.46   Test apparatus of powder shear viscometer.

$$e = \frac{1}{2}\Delta R$$

$$\Delta R = 0.2 \text{ in}$$

$$R_T = 4.285 \text{ in. } (R_T = \text{torque arm radius})$$

Because ΔR and e are small relative to the mean radius of the rotating collar ($(\Delta R/R) = 0.07$), the viscous shear force $\overline{F}_r$ developed at the mating ring may be deduced from the following equation:

$$F_T R_T = \overline{F}_r \overline{R}$$

$$F_T = \frac{(2\pi\mu_o\omega)}{(R_T h_o)} \int_{R_1}^{R_2} r^3 dr \qquad (6.4)$$

$$F_T = \left(\frac{\mu_o N \pi^2}{60 R_T h} \right) \left(R_2^4 - R_1^4 \right) \qquad (6.5)$$

Thus, by substitution of numerical value in Equation 6.5, we have:

$$\mu_o = (3.75965 \times 10^{-3}) \frac{h_o F_T}{N} \qquad (6.5a)$$

where
  F_T = Measured force at torque arm (in grams)
  R_T = Torque arm radius
  F_r = Viscous shear force
  $R = \frac{(R_1+R_2)}{2}$
  ω = Angular speed (rad/s)
  N = Rotational frequency (rpm)
  h_o = Average film thickness
  μ_o = Viscosity (lb-s/in.^{2})

As a preliminary to the powder tests, SAE 20, 30, and 50 oils at 70°F were tested to verify the reliability of the apparatus. The results are plotted in Figures 6.47 and 6.48. As can be seen, the agreement between the known viscosities of these oils and those calculated from the analytical expressions is fairly good.

The first series of viscosity measurements on powders was conducted with the rutile form of powder A in a size range of 1–2 microns. The results charted by the recorder are given in Figure 6.49. Values of the viscous shear forces were obtained from the following relations:

$$F_a = \frac{(F_1 + F_2)}{2}; \qquad F_m = \frac{(F_2 + F_3)}{2}; \qquad F_{avg} = \frac{(F_a + F_m)}{2}$$

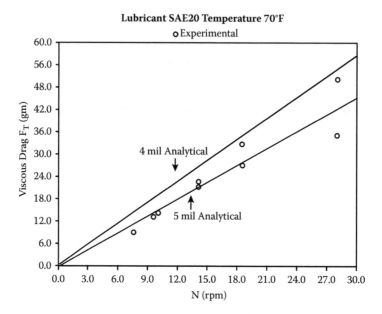

FIGURE 6.47   Measured and calculated viscous drag versus shear viscometer speed; lubricant SAE 20 @ 70°F, $\mu_0 = 28\ \mu$ Reyn.

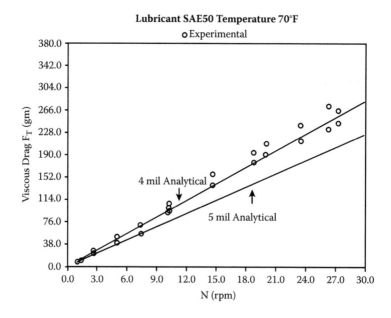

FIGURE 6.48   Measured and calculated viscous drag versus shear viscometer speed; lubricant SAE 50 @ 70°F, $\mu_0 = 116\ \mu$ Reyn.

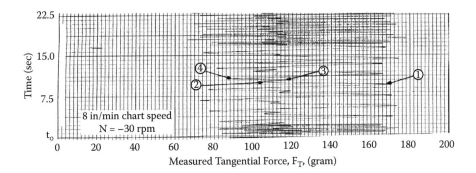

FIGURE 6.49 Typical experimental data powder viscous shear force transducer output recorded against test time, N = 30 rpm.

where subscripts 1 to 4 are the apparent peak values of the signature in Figure 6.49. The corresponding values of the tangential forces F_a, F_m, and F_{avg} were converted to effective viscosities using Equation 6.2. Overall results with powder A are given in Table 6.6, while Figure 6.50 depicts the effective powder viscosity as a function of the viscometer speed.

Tests were also run with powder B at the two clearances of 3.5 and 7.5 mil. The results are given in Tables 6.6 and 6.7. Figure 6.51 provides the arithmetical averages of the test data given in the tables. The effective viscosity of the two powders was found to be 55 micro-Reyn for powder A and 47 micro-Reyn for powder B. These viscosities place both powders somewhere between an SAE 30 and an SAE 40 oil.

TABLE 6.5
Bulk Density of Powder and Particles Terms and Definitions

	Powder
Absolute powder density	Mass of powder percent absolute volume
Absolute powder volume	Space occupied by a powder excluding pores or interstices
Apparent powder density	Mass of a powder divided by the volume occupied by the powder
Pressed density or green density	Apparent density of a compact
Aerated density (ρ_a)	Fluidization is used to measure bulk density of the powder
Tap density (ρ_t)	Apparent powder density of a powder bed formed in a container of stated dimensions when a stated amount of powder is vibrated or tapped under stated conditions
	Particle
Apparent particle density	Mass of particle divided by the volume of the particle excluding open pores but including closed pores
Effective particle density	Mass of a particle divided by the volume of the particle including open and closed pores
True density	Mass of a particle divided by the volume of the particle excluding open and closed pores

TABLE 6.6

Equivalent Viscosity of Powder A Experimental Results Using Shear Viscometer

Test No.	N (rpm)	F_a (g)	F_m (g)	F_m (g)	μ_a (μ Reyn)	μ_m	μ_{avg}	Remarks
1	10	54.1	44.5	49.3	71.2	58.6	65.4	N_2 gas was introduced.
2	10	35.0	30.0	33.0	47.1	39.5	43.4	No gas flow. Insufficient powder in the gap.
3	10	51.5	39.8	45.7	67.8	52.3	60.1	Resume gas flow.
4	20	50.0	41.0	46.0	—	—	—	No gas flow. Insufficient powder in the gap.
5	20	95.2	73.3	85.2	62.7	48.2	56.1	N_2 gas was introduced.
6	25	63.6	51.0	57.3	—	—	—	No gas flow. Insufficient powder in the gap.
7	25	92.2	86.0	89.0	48.6	45.3	46.9	N_2 gas was introduced.
8	25	115.0	99.0	107.0	60.6	52.2	56.4	N_2 gas was introduced.

Note: Clearance gap 0.0035 in., $T_a = 70°F$.

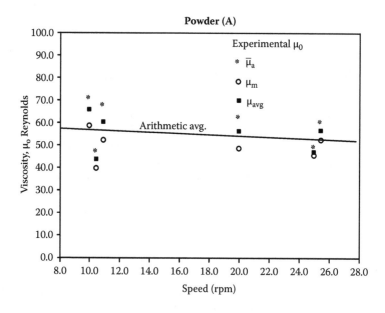

FIGURE 6.50 Equivalent powder viscosity as a function of shear viscometer speed; $T_a = 70°F$, powder A.

TABLE 6.7

Equivalent Viscosity of Powder B Experimental Results Using Shear Viscometer

Test No.	N (rpm)	F_a (g)	F_m (µ Reyn)	F_m	μ_a	μ_m	μ_{avg}	Remarks
1	10	15.0	13.0	14.0	42.3	36.5	39.4	N_2 gas was introduced 7 h 8 mil.
2	10	37.0	42.4	37.0	48.7	55.8	52.3	N_2 gas was introduced 3 h 8 mil.
3	29.7	128.0	104.0	116.0	56.7	45.6	51.4	N_2 gas was introduced 2 10 s. At the start of the run $t = 0$ min.
4	30	113.0	103.0	108.0	49.6	45.2	47.4	Data taken at $t = 3$ min run.
5	30	116.0	104.0	110.0	50.9	45.6	48.3	Data taken at $t = 3$ min run.
6	30	109.0	105.0	107.0	47.8	46.1	46.9	Data taken at $t = 3$ min run.
7	30	112.6	104.1	108.3	49.0	46.0	47.5	Data taken at $t = 3$ min run.

Note: Clearance gap 0.0035 in. except for Test 1 ($h_0 = 0.0075$ in.); $T_a = 70°F$.

6.3.2 Stress–Strain Relationship

For the following series of tests (8a), a powder shear apparatus (shown in Figure 6.52) was constructed. Its components consisted of three vertically stacked cylinders with the middle one separated from the others by a gap of 229 microns (9 mil). The powder to be tested occupied the center container, and a normal load was applied on it via a hydraulic piston having a radial clearance of 2 mil to ensure that the entire load

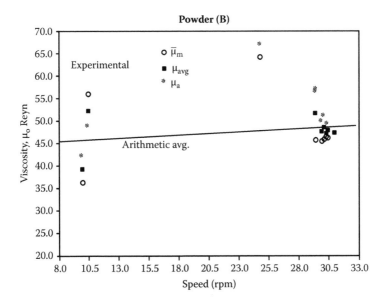

FIGURE 6.51 Equivalent powder viscosity as a function of shear viscometer speed; $T_a = 70°F$, powder B.

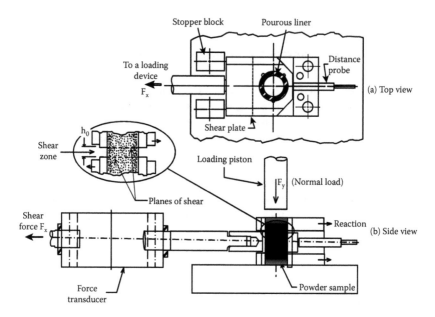

FIGURE 6.52 Schematic showing the principle of a shear cell tester.

was exerted on the tested sample. The central cylinder was attached to the loading device via a connecting rod and a force transducer, as depicted in the photographs of Figures 6.53 and 6.54. A shear force was generated by a high-pressure nitrogen gas. The maximum possible travel, delimited by slider blocks, was 1.98 mm (78 mil) and was recorded by a capacitance gauge located behind the shear plate within a

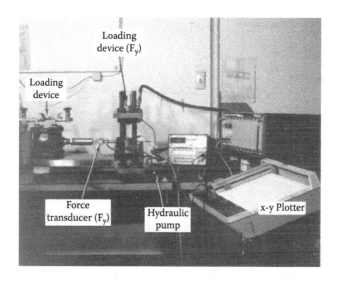

FIGURE 6.53 Experimental setup of powder shear cell.

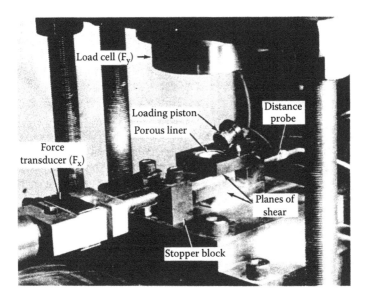

FIGURE 6.54 Close-up of powder shear cell test rig.

range that ensured linearity of the measured travel. Sample plots of the shear force and strain, as functions of time, shown in Figure 6.55, indicate that the shear force increased up to some limiting value after which there was a leveling off in its value. The subsequent increase was merely due to the shearing plate contacting the stopper blocks.

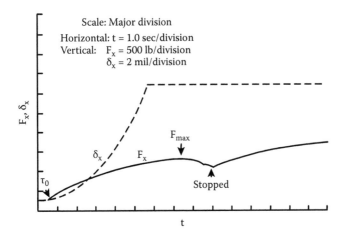

FIGURE 6.55 A sample plot of shear cell test data for powder D (MoS_2) shear force and strain versus time: $W_n = 3000$ lb.

A prodigious quantity of data was generated in the course of testing powders A to J (see Table 6.9). Only a compendium of these data will be presented, sufficient to establish the basic shape of the stress–strain curves of powder layers. A generic sample (for powder C) is given in Figures 6.56a and 6.56b. In order to extract additional relationships from the measurements of F_x and δ_x with respect to time, the following assumptions were necessary:

- There was no time delay between the applied shear force, reactive force, and instrumentation; likewise, for the measured travel.
- All the applied and shear forces were transmitted solely to the powder.
- The motion of the powder film and the shear force were in the same plane with no particulate migration in the vertical direction.
- There was no slip at the boundaries.

With the foregoing conditions fulfilled, one can obtain the following relevant quantities for powder behavior:

1. Shear stress (sheared area $A = 0.4051$ in.^{2}, of which there are two planes)

$$\tau_x = \frac{F_x}{2A}$$

2. Coefficient of friction, where $P = (W_n/A)$ is the applied normal pressure

$$f = \frac{\tau_x}{p}$$

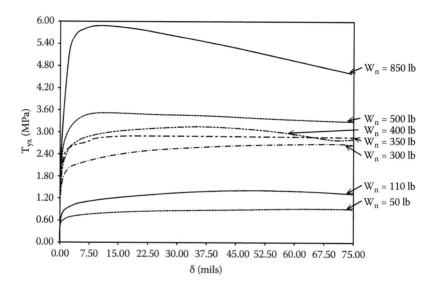

FIGURE 6.56a   Shear stress versus strain for powder C (TiO$_2$ anatase form $P_d = 0.4$ µm); $W_n = 50$ to 850 lb.

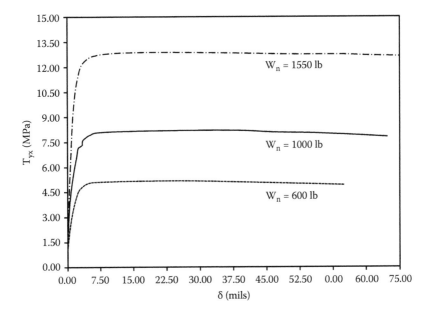

FIGURE 6.56b   Shear stress versus strain for powder C (TiO_2 anatase form $P_d = 0.4$ μm); $W_n = 600$ to 1550 lb.

3. Shear rate

$$\overset{\circ}{\tau}_{yx} = \frac{\Delta\tau_{yx}}{\Delta t}$$

4. Strain rate

$$U_x = \frac{\Delta\delta_x}{\Delta t} = \overset{\circ}{\delta}_x$$

Two normalized parameters used in presenting the shear stress–strain data are

$$\varepsilon = \frac{\delta x}{h_o} \quad \text{and} \quad \varepsilon = \frac{\delta x}{h_o} \quad \text{with} \quad h_o = 229 \text{ μm (9 mils)}$$

6.3.2.1   Shear Stress versus Shear Strain

A sample of the tabulated data corresponding to plots of Figures 6.56 and 6.57 is presented in Table 6.8. The detailed data for shear stress and coefficient of friction are given in the work of Reference 13, for all the powders tested. Some pertinent comments for the individual powders are as follows:

• No results are given for the SiO_2 powder because its particle size was so fine as not to conform to the general behavior of the other powders, exhibiting, in fact, the characteristics of a liquid.

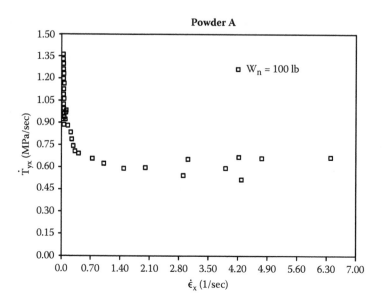

FIGURE 6.57   Shear stress rate versus strain rate for powder A (TiO$_2$); $W_n = 100$ lb ($p = 247$ psi).

- Powder C. In the friction coefficient plots, the maximum values of f decrease as the normal load is increased to 850 lb; after 1500 lb, it increases again up to 3000 lb; beyond that, it again starts to decrease.
- Powder D. Here, too, the values of f decrease with an increase in W_n up to 4000 lb, after which it stays constant.
- Powder G. Very high levels of friction were encountered so that the normal load had to be restricted 3500 lb. The variation of f with strain is, here, opposite to the other powders.
- Powder H. Beyond a normal load of 1000 lb, it was not possible to retain the graphite in the shear cylinder, producing appreciable leakage via the 9 mil gap.
- Powder J. Here, too, the friction versus normal load showed peculiar behavior, with an increase in W_n.

6.3.2.2   Shear Stress Rate versus Strain Rate

Powders A and B were selected to depict the shear stress rate relation to strain rate. Figures 6.57 through 6.61 show plots for the TiO$_2$ powder for a range of normal loads. As is evident from these plots, above $\varepsilon = 1.5$ the shear stress rate is independent of the strain rate. This trend is more pronounced at the higher loads, as shown in Figure 6.62. Figure 6.63 shows similar plots for the MoS$_2$ powder. Despite the fact that these are substantially different powders, their behavior is quite similar.

TABLE 6.8

Pressure-Dependent Properties of Triboparticulates

Powder ID	Yield Shear Stress (Mpa)			Limiting Shear Stress (Mpa)		
	ζ_{τ_o}	$\zeta_{o\ell}$	$\sigma_{yy}{}^{++}$	$\zeta_{o\ell}$	$\tau\ell^*$	$\sigma_{yy}{}^{**}$
A	0.0667	4.25	65.0	0.601	43.0	71.5
B	0.0467	1.80	40.0	0.542	42.5	78.3
C	0.0333	2.128	62.0	0.620	42.2	68.1
D	0.079	0.425	64.0	0.173	14.6	85.1
E	0.040	0.53	13.0	0.245	16.7	68.1
F	0.032	2.55	80.0	0.480	40.9[a]	85.1[a]
G	0.025	1.25	55.5	0.625	38.4[a]	59.6[a]
H	0.015	0.25	17.0	0.210	33.1	17.0
J	0.018	0.15	83.0	0.592	42.3	72.3

[a] Limiting values have not been achieved in the present test program; these data refer to maximum test.

$$\begin{cases} \tau_\ell = \tau_{\ell_{(o)}}\left(1+\left(\dfrac{\zeta_{\tau_\ell}}{\tau_{\ell_{(o)}}}\right)P\right) \text{ if } P \le \tau_\ell{}^* \\ \tau_\ell = \sigma_{yy}{}^{**} \text{ if } P > \tau_\ell{}^* \end{cases}$$

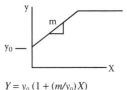

$$Y = y_0\,(1 + (m/y_0)X)$$

$$\begin{cases} \tau_o = \tau_{o_{(o)}}\left(1+\left(\dfrac{\zeta_{\tau_o}}{\tau_{o_{(o)}}}\right)P\right) \text{ if } P \le \tau_{o\ell} \\ \tau_o = \sigma_{yy}{}^{++} \text{ if } P > \tau_{o\ell} \end{cases}$$

Where $\zeta_{\tau_o} = \Delta\tau_o/\Delta P$; $\zeta_\tau = \Delta\tau_\ell/\Delta P$
(see also Equation 6.6)

6.3.2.3 The Yield and Limiting Shear Stresses (τ_o and τ_ℓ)

The stress at which a powder begins to flow, the yield shear stress τ_o, is a function of the morphology of the powder and degree of compaction. Values for this shear yield stress are given in Figures 6.64 through 6.72. The straight lines were obtained via an averaging of the data; thus, up to some limiting point, τ_o is a linear function of the normal load. This holds up to a point, called the limiting yield shear stress $(\tau_o)_\ell$ at some value of τ_{yy} denoted by $(\sigma_{yy})_{max}$. As can be seen from the plots of Figures 6.64–6.72, the relationship between τ_o and P may be expressed by the general form:

$$\tau_o \begin{cases} (\zeta_{\tau_o})P & \text{for } (\sigma_{yy})_{max} \le a \\ (\tau_o)_\ell & \text{for } (\sigma_{yy})_{max} \ge a \end{cases} \qquad (6.6)$$

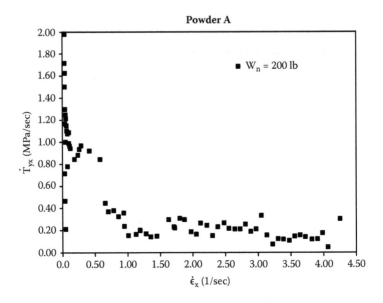

FIGURE 6.58   Shear stress rate versus strain rate for powder A (TiO$_2$); $W_n = 200$ lb ($p = 494$ psi).

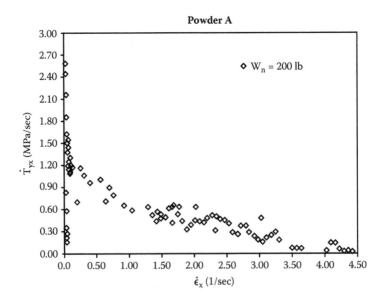

FIGURE 6.59   Shear stress rate versus strain rate for powder A (TiO$_2$); $W_n = 300$ lb ($p = 740$ psi).

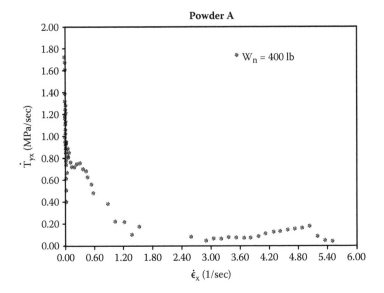

FIGURE 6.60 Shear stress rate versus strain rate for powder A (TiO_2); $W_n = 400$ lb ($p = 987$ psi).

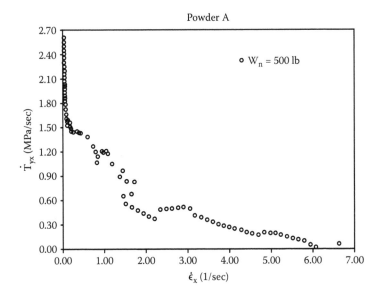

FIGURE 6.61 Shear stress rate versus strain rate for powder A (TiO_2); $W_n = 500$ lb ($p = 1234$ psi).

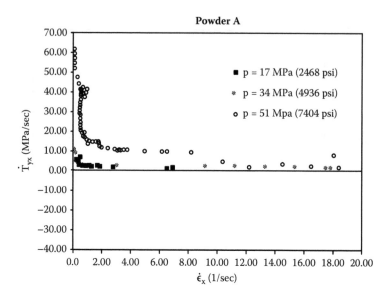

FIGURE 6.62 Shear stress rate versus strain rate for powder A (TiO$_2$, rutile form); $\tilde{P}_d < 2$ μm.

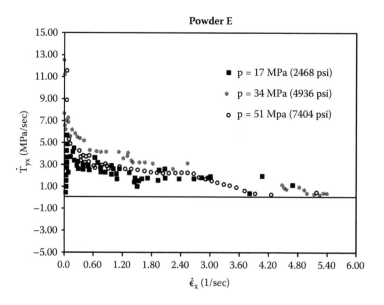

FIGURE 6.63 Shear stress rate versus strain rate for powder E (MoS$_2$, rutile form); $\tilde{P}_d < 2$ μm.

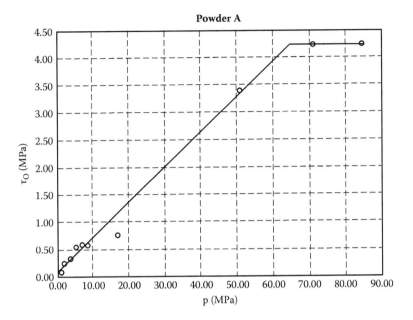

FIGURE 6.64   Yield shear stress versus normal stress for powder A.

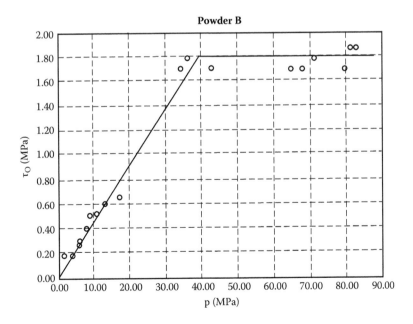

FIGURE 6.65   Yield shear stress versus normal stress for powder B.

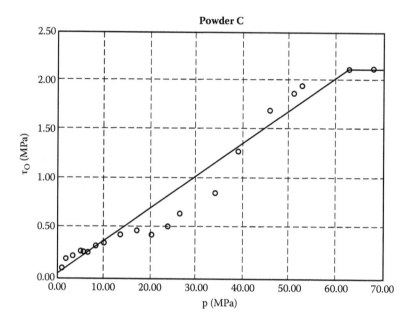

FIGURE 6.66   Yield shear stress versus normal stress for powder C.

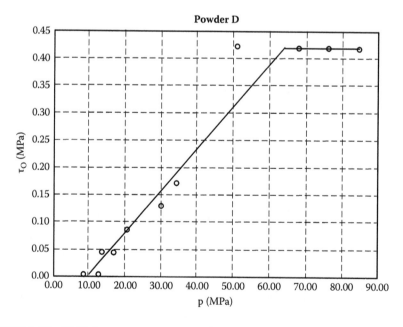

FIGURE 6.67   Yield shear stress versus normal stress for powder D.

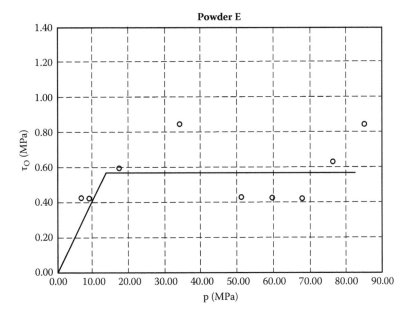

FIGURE 6.68   Yield shear stress versus normal stress for powder E.

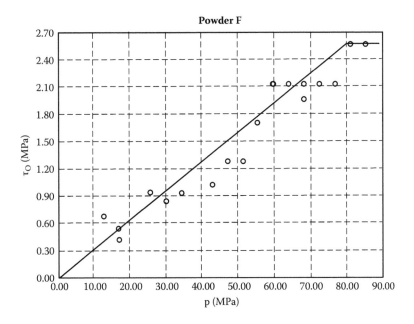

FIGURE 6.69   Yield shear stress versus normal stress for powder F.

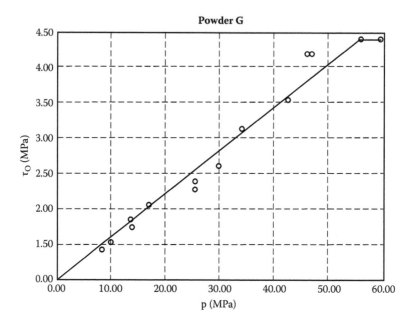

FIGURE 6.70   Yield shear stress versus normal stress for powder G.

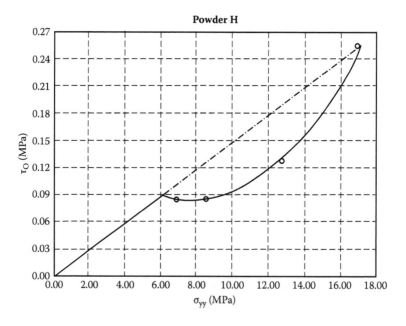

FIGURE 6.71   Yield shear stress versus normal stress for powder H.

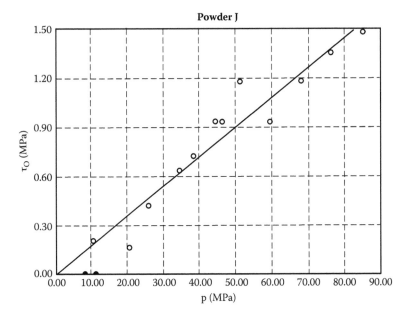

FIGURE 6.72 Yield shear stress versus normal stress for powder J.

where

$$\xi_{\tau_o} = \frac{\Delta\tau_o}{\Delta P} \tag{6.6a}$$

For example, considering powder A, we have:

$$\tau_o \begin{cases} \left(\dfrac{1}{15}\right)P & \text{for } (\sigma_{yy})_{max} \le 65 \text{ MPa} \\ 4.25 \text{ MPa} & \text{for } (\sigma_{yy})_{max} > 65 \text{ MPa} \end{cases} \tag{6.7}$$

Constants of the general linear function of τ_o are empirically determined and are given in Table 6.8.

In addition to τ_o, a powder film has a limiting shear strength τ_ℓ, which is the maximum stress the powder can support—a function of both the powder and the tribomaterial characteristics, as shown on Figures 6.73 to 6.75. Values of the limiting shear stress were extracted by computing the slope of τ_x with respect to δ_x. When the slopes approached zero, the corresponding τ_x was considered to represent the sought-after τ_ℓ.

There is a special interest in the data of Figures 6.42 and 6.75, where the powders showed a similar variation of density with normal load. Also, the slopes of the limiting shear stress were similar: 0.54 and 0.62. A linear correlation was performed, shown in Figure 6.75, in which the slope is about 0.59. One,

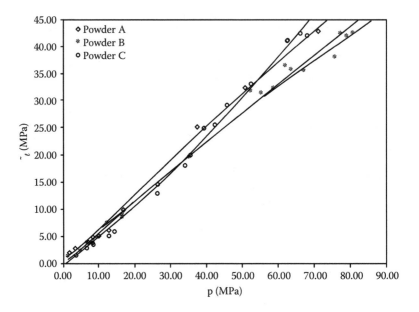

FIGURE 6.73 Limiting shear stress versus normal stress for powders A, B, and C.

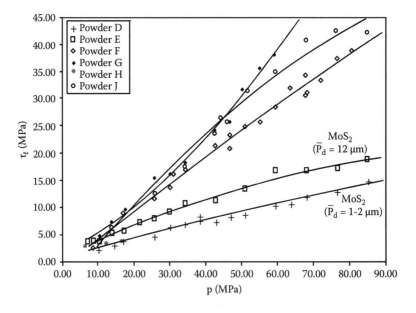

FIGURE 6.74 Limiting shear stress versus normal stress for powders D, E, F, G, H, and J.

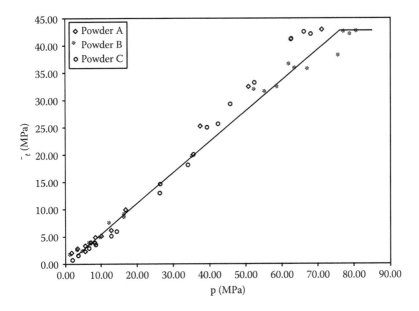

FIGURE 6.75 Limiting shear stress versus normal stress for powders A, B, and C showing linear approximation for data correlation.

therefore, can postulate that for powders A, B, and C, τ_ℓ is a linear function of the pressure or

$$\tau_\ell \begin{cases} \dfrac{p}{1.7} & \text{for } P \le 75 \text{ MPa} \\[2mm] 42.5 \text{ MPa} & \text{for } P > 75 \text{ MPa} \end{cases} \tag{6.8}$$

Approximate values for the constants τ_o and τ_ℓ are given in Table 6.8.

In summary, one can state the following:

- The shear stress and the coefficient of friction first increase sharply with displacement then level off to some limiting values.
- τ_{yx} increases with an increase in normal load; at high levels of normal load, it is only a weak function of this variable.
- f is a strong function of p, particularly at low levels of normal load (100 to 1000 psi); at moderate normal loads, it increases slightly; it then decreases again at high levels of normal load, reaching a limiting value (exceptions being powders G and J).
- Shear stress rate first increases dramatically with the strain rate (as shown in Figure 6.76), then decreases exponentially up to an ε_x of 1–1.5 (1/s).
- The yield shear stress is a strong function of the normal load. Initially, it increases linearly with normal load up to some point, and then levels off at a limiting value.

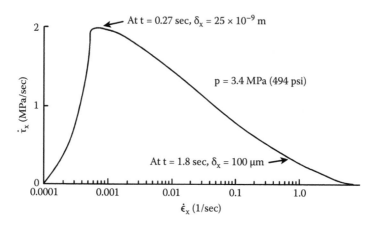

FIGURE 6.76 Initial behavior of a typical shear stress rate as a function of strain rate, Powder TiO_2, Rutile from; $\dot{P}_d < 2\ \mu m$.

- The limiting shear stress τ_ℓ is also a strong function of the normal load. In some ways, its behavior is akin to that of the friction coefficient.
- In a qualitative way, it can be said that the limiting shear stress and the friction coefficient parallel the behavior of density variation with pressure.

6.3.3 TRACTION

Along with the stress–strain relations, in the previous section we have dealt with the frictional characteristics of powders. Data for the sliding friction coefficient are given in the previous section. What is of related interest here—such as in rolling element bearings or gears—is the friction induced under a combination of sliding and rolling motion. The following experiment presents the basic features of powder traction, where they are also compared with traction in dry contacts and liquid lubricants.

An advanced rolling/sliding tribometer capable of handling both powder and liquid films was designed (10–16,18) and is illustrated in Figures 6.77 and 6.78. The mating materials used were silicon nitride and silicon carbide, lubricated with TiO_2 and MoS_2 powders. The lower cylindrical disk had a radius of 36 mm; the upper crowned disk had a radius of 14.4 mm. The lower disk was driven by a variable-speed electric motor; the upper by an integral air turbine. Both spindles are mounted on a hydrostatic bearing to keep them aligned and to eliminate all resistance to the rotational motions. The loading was achieved via a pneumatic cylinder that pulled down on the upper disk via a load cell. For high-temperature testing, the two disks were enclosed by an insulated oven.

The powders ranged in size from 0.1 to 10 microns and had a 99.99% purity, yielding a 50% cumulative average particle size of 5.95 microns. The TiO_2 powder was in rutile form; the zinc oxythiomolybdate had particles of TiO_2 and $ZnMoO_2S_2$. A conventional airbrush was used to deliver powder to the interface. A tube from

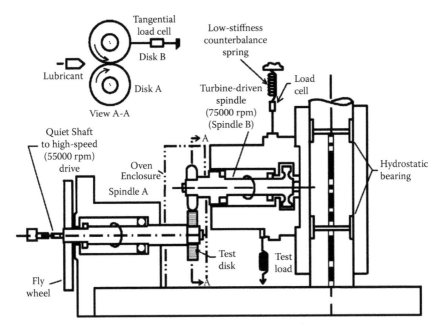

FIGURE 6.77   Schematic of powder traction apparatus.

FIGURE 6.78   The two-disk assembly.

the airbrush deposited the air–powder mixture some 25 mm ahead of the contact zone at a rate of 2 cc/min and a supply pressure of 35 psi. Calculations showed that, for the given film thickness and speed, a maximum of 0.23 cc/min could be accommodated; thus, a tenfold supply of lubricant was maintained. The speed of the lower disk was kept constant while the upper one was given higher speeds. The disks were then loaded and the air turbine shut down, making the upper disk coast down to rest with the traction forces recorded at proper intervals throughout the coastdown.

The test conditions were as follows:

$$W_n = 5 - 15 \text{ lbs}; \ w_1 = 150 \text{ cps}; \ w_2 = 150 - 300 \text{ cps};$$

$$T = 22°C \text{ to } 650°C, \text{ or room temperature to } 1200°F$$

The traction coefficient f_T is defined here as the ratio of the tractive force W_T to the normal load W_n and slip as the difference between the two surface velocities, $\Delta U = (U_2 - U_1)$. The traction data are presented mostly in the form of f_T versus $\Delta U = (U_2 - U_1)/U_1$ in %. Figure 6.79 presents a useful plot for comparing the traction data. As can be seen, traction increases with slip until η_{τ_ℓ} reaches a limiting shear, after which it gradually decreases. This is believed to be due to the effect of increasing temperature. The slopes at maximum traction and at the stable portion of the curve are designated α_1 and α_2, respectively. The smaller the magnitude of α_1, the higher the stable region; the greater α_2, the higher the ratio of tractive force to slip. The data that follow are given for three powders—NiO, TiO$_2$, and ZnMoO$_2$S$_2$—as well as for dry contacts and surfaces lubricated with a synthetic traction fluid.

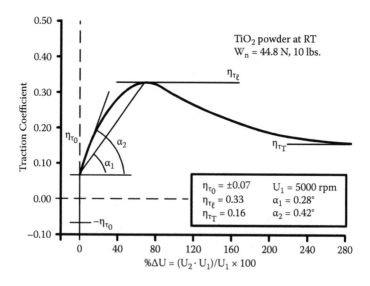

FIGURE 6.79   Characterization of traction curve parameters.

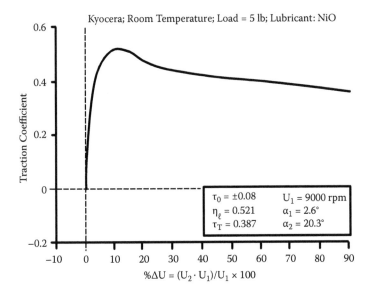

FIGURE 6.80   Traction coefficient versus slip for NiO powder at room temperature.

6.3.3.1   Results for the NiO Powder

The results for the NiO powder are given in Figures 6.80 to 6.83. The values of τ_o and τ_ℓ decreased with normal load and increased with a rise in temperature. Also, at higher temperatures, the traction curve reached its maximum at lower slip ratios. Beyond τ_ℓ, the tractive force decreased as slip increased. As seen from the figures, the value of τ_ℓ dropped from 0.387 to 0.23 at a slip ratio of 90%.

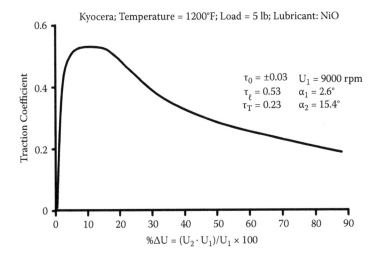

FIGURE 6.81   Traction coefficient versus slip for NiO powder at 600°C.

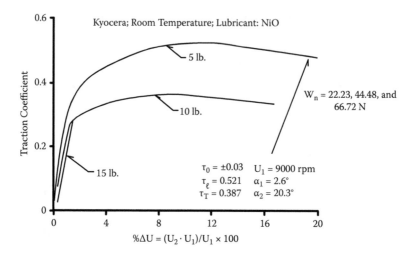

FIGURE 6.82   Traction coefficient for NiO powder at various loads and room temperature.

6.3.3.2   Results for the TiO$_2$

The data for this powder are given in Figures 6.84 through 6.87. At room temperature, the values of τ_o and τ_L were lower than at higher temperatures; however, at 1200°F and higher slip, the traction curve approaches a maximum. The depth of the wear track was insignificant, though the surface roughness increased fourfold. The width of the wear track was 70% larger than the calculated Hertz half-width of the major axis.

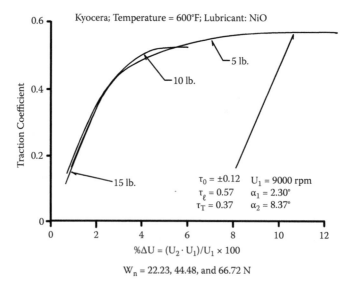

FIGURE 6.83   Traction coefficient for NiO powder at various loads and 316°C.

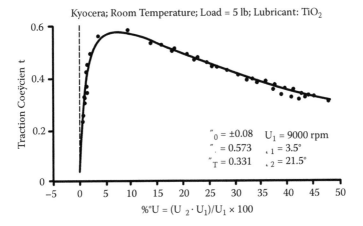

FIGURE 6.84   Traction coefficient versus slip for TiO_2 powder at room temperature.

6.3.3.3   Tests with $ZnMoO_2S_2$ Powder

For this powder, the results are shown in Figures 6.88 and 6.90. Here, τ_o was higher at low than at high temperatures but not τ_ℓ. At room temperature, the traction approaches its peak at low values of slip. There is a progressive decrease in the value of τ_ℓ with load. In all cases, there was no apparent yield stress for this powder, and the limiting stress was 50% lower than for the other powders.

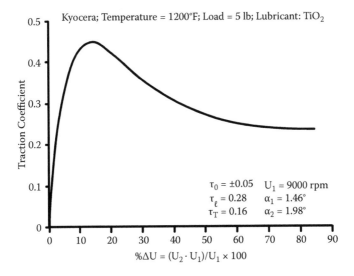

FIGURE 6.85   Traction coefficient versus slip for TiO_2 powder at 650°C.

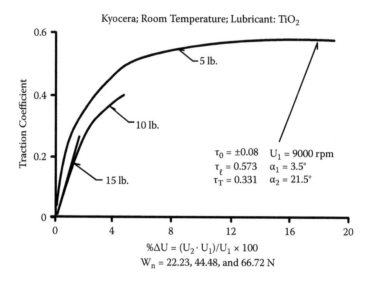

FIGURE 6.86   Traction data for TIO_2 powder at various loads and room temperature.

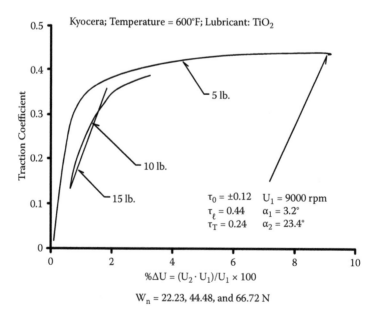

FIGURE 6.87   Traction data for TiO_2 powder at various loads and 316°C.

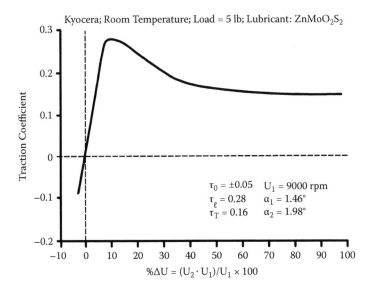

FIGURE 6.88 Traction data versus slip for $ZnMoO_2S_2$ powder at room temperature.

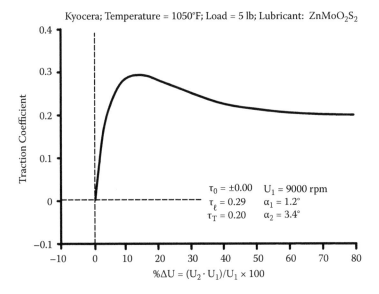

FIGURE 6.89 Traction data for $ZnMoO_2S_2$ powder at 566°C.

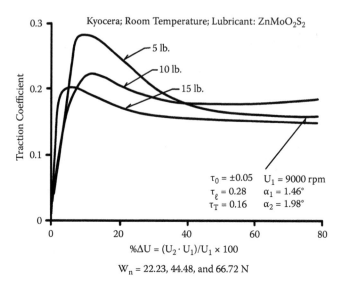

FIGURE 6.90   Traction data for $ZnMoO_2S_2$ powder at various loads and room temperature.

6.3.3.4   Dry Contact and Lubricated Tests

To assess the benefits of using powders in tractive drives, three tests were run with three types of Si_3N_4 materials. A sample result at 1200°F is shown in Figure 6.91. There was no traction dependence on either load or temperature. The figure also shows the comparative performance of the TiO_2 powder, showing the appreciable

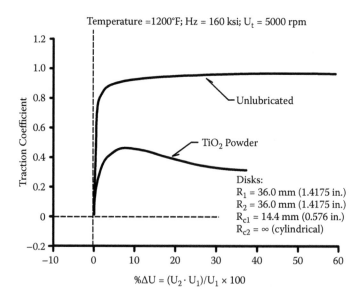

FIGURE 6.91   Comparison of dry contact and TiO_2 powder-lubricated traction performance.

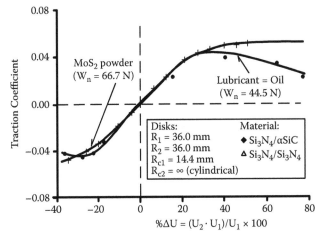

Traction data for dry MoS_2 powder (Hz = 1.11 Gpa) and synthetic
fluid equivalent to SAE 10 (Hz = 0.97 Gpa) at RT, U_t =188.5 m/s

FIGURE 6.92 Comparison of TiO_2 powder- and synthetic-fluid-lubricated-traction
performance.

advantage of using a powder-lubricated traction drive. Figure 6.92 compares the dry
contact results with a 10 weight synthetic traction fluid at room temperature. Past a
slip of 33%, the lubricated traction coefficient begins to decrease, while the curve for
the MoS_2 powder flattens out.

6.3.3.5 Summary of Results
The foregoing results can be summarized as follows:

- Powders change the shape of the traction curve. Compared with dry contacts, powders reduce the traction coefficient over a wide range of slip ratios;
  the contact shear stress τ_ℓ was reduced by some 50%. These results apply to
  the NiO and TiO_2 powders only.
- The behavior of the $ZnMoO_2S_2$ powder was quite different. It showed no
  yield stress and had low traction coefficients under all conditions. Overall,
  its behavior was close to that of liquid lubricants.
- The persistent presence of a powder buildup at the edges of the contact track
  suggests that there was side leakage from the powder lubricant, as occurs
  with a conventional lubricant.
- Until further tests with different materials are conducted, the foregoing
  conclusions must be restricted to the particular contact materials used in
  the present tests.

6.4 POWDER FLOW AND VELOCITY

The following experiments (22) supply evidence of a powder's ability to act as a lubricant: to flow under the action of a surface velocity, to simulate liquid behavior in the converging and diverging films, as well as to produce side leakage. The apparatus for these tests is shown schematically in Figure 6.97; its dimensions are given in Table 6.9. The top end of the vertically mounted journal was open, while its bottom was pressed against an O-ring to prevent escape of the powder. The powder, packed into the clearance space, consisted of three layers: the two outward ones contained the rutile form of TiO_2, 6 μm in size, while the middle contained MoS_2, 12 μm in size. This facilitated visualization of the flow pattern, which was recorded by two sets of cameras. As occurred during the determination of the effective viscosity, it was necessary to impart some vertical vibration to the test bed to obtain a tapped powder film in the clearance.

When viewed from the side, the powder flow in the circumferential direction could be followed. The following observations were made:

- The powder film assumes a higher velocity near h_{min}, as well as a higher slip rate.
- In the diverging portions, the powder flow becomes turbulent (for the high clearances used in the present tests).
- A form of cavitation sets in the diverging portions in the form of longitudinal cracks.
- Packing or tapping of the powder produced a higher slip at the boundaries than with loose powders.
- Powder was ejected from the edge of the bearing in the region ahead of h_{min}, as shown in Figure 6.93. Near the bearing edge, the velocity of the ejected powder was one third of the journal's linear surface velocity. This indicates the presence of an axial pressure gradient in the powder film (22).

TABLE 6.9
Dimensions of Flow Visualization Apparatus

	Bearing Dimensions and Test Conditions	
Item	Test With Dry Powder mm (in.)	Test With Liquid (Oil) mm (in.)
Journal diameter	51.11 mm (2.012 in.)	50.34 mm (1.982 in.)
Bearing diameter	64.39 mm (2.535 in.)	Same
Bearing length	12.7 mm (0.50 in.)	Same
Clearance	6.642 mm (0.2615 in.)	7.024 mm (0.277 in.)
Speed (cps)	0–15	Same
Min. film thickness	0.127 mm (0.005 in.)	0.508 mm (0.020 in.)
Eccentricity ratio, $\in$ (–)	0.981	0.928

FIGURE 6.93   Side leakage of powder.

Next, the powder film was viewed from the top, giving a transverse (y direction) view of the flow. To facilitate viewing the powder's shear and velocity distribution across the film, two powders were employed: TiO_2, which has a yellow-tan appearance, and NiO, which is dark green. These powders were packed into the clearance space in an alternating pattern. Then, the rotor was started up and stopped, and the changes in the positions of the powder segments noted. The results can be summarized as follows:

- Figures 6.94a, 6.94b, and 6.96 show the movement of the bright powder segments under the influence of journal rotation. This could be established by noting the position of the segments' radial lines before and after rotation. An exploded view of the relevant segments is shown in Figure 6.95. These radial lines are clearly skewed in the direction of the rotating journal.
- Figure 6.95 also indicates that, in the diverging film, the powder segments show circumferential cracks where the powder could not withstand the rate of dilatation and "cavitated." In the converging region wherever cracks appear, they are in the radial direction, which is an indication of compressive stress or pressurization of the powder film.
- Figures 6.96a and 6.96b show similar results following a larger number of journal rotations, 10 cycles compared to the previous five. Here, of course, the displacement of the powder segments was much higher.

Both series of tests show that the shear stress induced by rotation is transferred across the height of the film. In the converging section, the stresses are compressive, while tensile stresses are formed where the powder is forced into dilatation. The originally circumferential motion of the powder acquires an axial component as h_{min} is approached, producing side leakage analogous to a fluid film. These are all basic characteristics of a quasi-hydrodynamic behavior of triboparticulates.

Standstill: dark color-Nio; light color - TiO$_2$

(a) Before start of rotation

Line of radial displacement

Powder displacement after 5 cycles

(b) After rotation

FIGURE 6.94   Displacement of powder segments after five cycles.

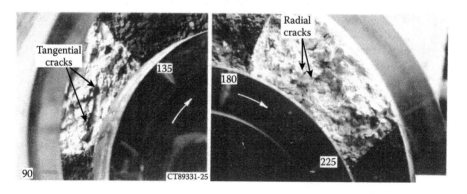

Tangential cracks

Radial cracks

FIGURE 6.95   Close-up of Figure 6.94b.

FIGURE 6.96a   Displacement of powder after 10 rpm.

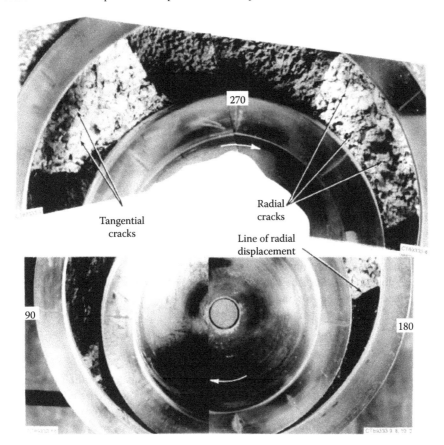

FIGURE 6.96b   Close-up view of Figure 6.96a.

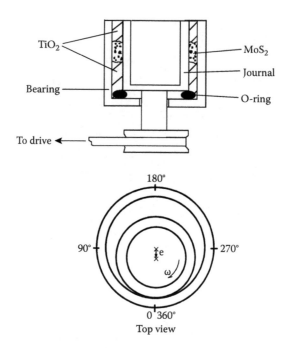

FIGURE 6.97 Apparatus for flow visualization.

6.5 TRIBOLOGICAL QUALITIES OF POWDERS

Two important tribological facets of powders will be considered here. One is that its shear and damping properties, as in addition to its role as lubricants, a powder layer can serve as an effective medium in attenuating vibrations in dampers, turbine blades, aircraft control surfaces, and other structures subject to high-frequency fatigue. The other outstanding issue is to secure a powder supply to the interface on a sustained basis. While, in most cases, gas lubricants are taken from the environment and liquid lubricants are delivered by a system of pumps, sumps, and appropriate piping, the question arises how to have powder fed to a bearing on a continuous basis, although conceivably, as discussed here, it would be feasible to have the powder pellets pressed against the rotating journal, thus supplying the interface with a steady injection of powder lubricant.

6.5.1 SHEAR AND DAMPING CHARACTERISTICS

A powder is likely to be a more suitable medium to attenuate vibration than conventional elastomers for a number of reasons. One is that powders are not subject to creep, which degrades viscoelastic materials, nor do they suffer from elevated temperature the way conventional elastomers do. It is, therefore, of interest to ascertain experimentally the frictional and damping behavior of powders under an imposed shear stress and elevated temperature. A test facility, shown in Figure 6.98, was

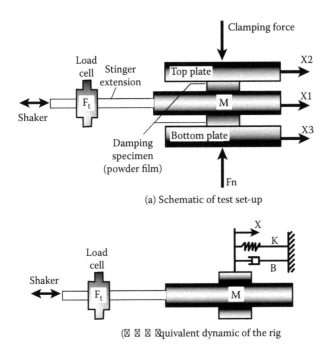

(a) Schematic of test set-up

(b) Equivalent dynamic of the rig

FIGURE 6.98   Test apparatus for measuring powder shear and damping.

constructed that could impose a range of frequencies, tangential forces, and high rates of shear and high temperatures on a thin layer of powder. The total motion of one of the moving components was on the order of 40–50 μin. at frequencies of 2000 Hz, corresponding to levels encountered in industrial equipment. The powder used was TiO_2 with an average particle diameter of half a micron, which was subjected to the following range of variables:

Tangential force, F_t: 111.2–233.6 N (25–75 lb)
Frequencies, f: 100–2000 Hz
Temperatures, T: 21°C–538°C (7–1000°F)
Film thickness, h: 178–311 microns (7–12 mil)

The measurements recorded were force F, stringer displacements X_1, and top and lower plate displacements X_2 and X_3, from which the following quantities were obtained:

- Stinger extension inertia force $F_m = M \omega^2 X_1$, where M is the stinger mass
- Average relative displacement $X = [(X_1 - X_2) + (X_1 - X_3)]$
- Fourier transforms were obtained for F, the input force F_m, and X, from which a transfer function K_d is obtained

$$K_d = [F(\omega) + F_m(\omega)]/X(\omega)$$

where $X(\omega)$ is the Fourier transform of the average displacement $X(t)$ at the excitation frequency and K_d represents the shear dynamic stiffness in the complex plane, namely,

$$K_d = K + j\omega B = D \tan \Phi, \quad \text{where } D = K^2 + \omega^2 B_e^2 \quad \text{and} \quad \Phi = \arctan(\omega B/K)$$

where K is the real part of the stiffness coefficient and B_e is the equivalent viscous damping coefficient.

At the load cell $F = -M\omega^2 X_1 - KX + j\omega B_e$; after eliminating the inertia term, this becomes

$$K_d = (F + M\omega^2 X_1)/X = K + j\omega B$$

• Assuming that the powder behaves like a viscoelastic material, we have

$$K_d = G(1 + j\eta) \, 2A/h = 2AG/h + j\eta \, 2AG/h$$

where A is the area of the sheared layer.
• Comparing the two expressions for K_d from the last two equations with $D \cos \Phi = 2AG/h$ and $D\eta \sin \Phi = 2 \, AG/h$, we obtain for the shear coefficient

$$G = K_d h/[2A \, (1 + \eta^2)^{\frac{1}{2}}] \tag{6.9}$$

and for the loss factor

$$\eta = \tan \Phi = \tan \, [\arctan \, (\omega \, B/K)] = \omega \, B/K \tag{6.10}$$

Figure 6.99 shows the experimentally obtained shear modulus at various combinations of tangential force, film thickness, and temperature. The results all fall within a narrow band from 69 to 138 MP (10,000–20,000 psi). The low-temperature results are near the low values of G. Figure 6.100 shows the loss factor for various values of F_t; here, too, the results show only a mild dependence on either tangential force or temperature. The main dependence of the loss factor is on frequency; this is of considerable interest, as high-frequency fatigue occurs, as the name implies, at high frequencies. No variation in the normal load F_n was used, as from a previous work (19,20,21), shown on Figure 6.101; it, too, had little effect. The increase of the loss factor with frequency and, to some degree with temperature, is in direct contrast to viscoelastic dampers. Figure 6.102 contains upper and lower brackets for the assembly of data of loss factor versus temperature. Most of the low-temperature points fall above the median curve, while the high-temperature points congregate below that line. The ratio of the lower to the upper bracketing line, as a function of frequency, shows the following ratios:

Frequency (Hz)	Loss Factor Limits Ratio
300	0.17
600	0.29
1000	0.41
2000	0.90

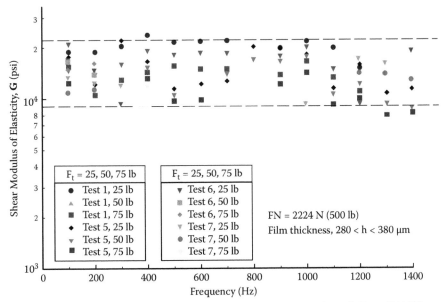

Composite plots of powder TiO_2; Test 1 (70°F), Test 5 (500°F), Test 6 (750°F), Test 7 (1000°F)

FIGURE 6.99 Stress modulus versus frequency of a TiO_2 powder.

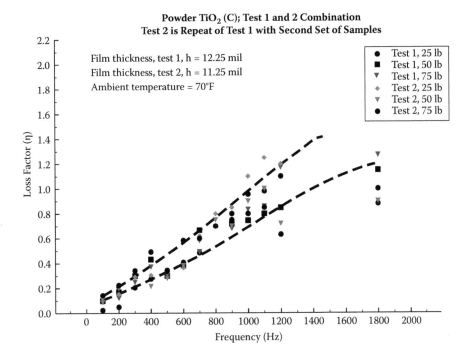

FIGURE 6.100 Loss factor versus frequency of a TiO_2 powder.

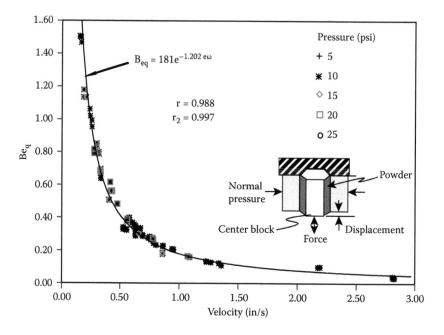

FIGURE 6.101   Powder damping versus velocity at various normal loadings.

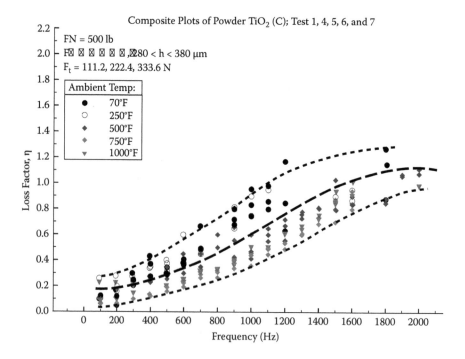

FIGURE 6.102   Loss factor versus frequency as a function of temperature.

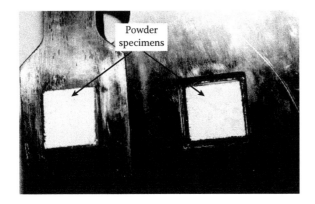

FIGURE 6.103a   Test rig—TiO$_2$ powder film test specimen before test.

Thus, the difference in the value of the loss factor at low and high temperatures decreases with a rise of frequency.

The tests show a number of significant attributes of powder films, namely,

- They are capable of producing high levels of damping.
- The shear modulus is constant over a wide range of frequencies.
- The loss factor increases steeply with a rise in frequency, which is of particular relevance to high-speed machinery.
- There is no noticeable effect on powder film performance with an increase in tangential or normal forces.

It is worth noting that the present tests, albeit in an indirect way, confirm some of the basic tenets of the quasi-hydrodynamic theory for powders. While Figure 6.103a shows the powder film test specimen before testing, Figure 6.103b gives evidence

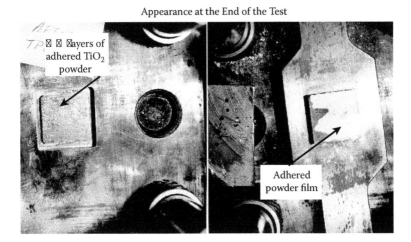

FIGURE 6.103b   Damping characterization of powder films.

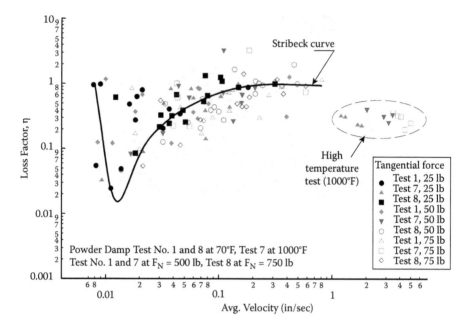

FIGURE 6.104 Loss factor of a TiO$_2$ powder versus velocity on a Stribeck curve.

that thin layers of TiO$_2$ powder had adhered to the test surfaces. Recall that, unlike in fluid films, portions of powder layers adjacent to the surfaces tend to stick to them with only a central core accommodating the relative velocity. The photographs confirm this basic fact. Another postulate for powders was that they can somehow be accommodated along the traditional Stribeck curve, perhaps to the right of the hydrodynamic regime. This is a consequence of the powder reaching a limiting shear stress, typical of non-Newtonian media. An attempt was, therefore, made to accommodate the present test points on the Stribeck curve. The results plotted on Figure 6.104 show considerable scatter. This is not surprising given that a powder film, which possesses a number of characteristics not present in fluids, could not automatically conform to the hydrodynamic grouping of ($\mu\omega/W'$) or the Sommerfeld number. An attempt was, therefore, made to provide a refinement on the traditional hydrodynamic group of variables by including a powder's shear stress characteristics. This modified dimensionless number is given by the following expression:

$$\Lambda' = f_p(\tau)\left(\frac{B}{h}\right)^2\left(\frac{U}{\tau_{max}}\right) \tag{6.11}$$

where $f_p(\tau)$ represents the powder's resistance to flow and was taken to be a constant equal to 0.75×10^{-6}. The other terms in Equation 6.11

$$\tau_{max}, psi = \frac{F_t}{A}$$

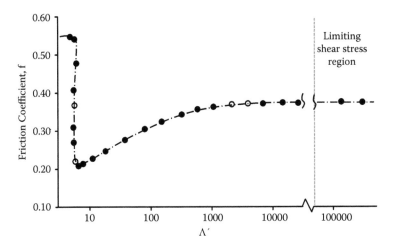

FIGURE 6.105   Friction coefficient of a TiO_2 powder versus Λ' on the Stribeck curve.

B, in. = Length of specimen in direction of motion
U, in. = Average velocity based on an average amplitude of motion
h, in. = Powder film height

In order to show the trend of friction coefficient f vs. Λ', the experimental setup was theoretically modeled. Figure 6.105 depicts a data plot of friction coefficient of a TiO_2 powder versus Λ', including the limiting shear stress region. It should be mentioned that a powder film's internal damping has a relationship with the powder f.

When the data for both theoretical friction and experimental loss factor are replotted against the new parameter Λ', Figures 6.105 and 6.106 are obtained, which are a much more compatible with the Stribeck curve.

6.5.2   PELLETIZED POWDER LUBRICANTS

Molybdenum disulfide was chosen to investigate the feasibility of using powder in the form of compressed pellets to provide a steady rate of lubricant supply. These pellets were run against a titanium carbide disk, and the resulting friction and rate of wear recorded under a range of operating conditions. Three different powders were used with variations in pellet makeup and powder compositions, as listed in Table 6.10. A special granular compaction system was developed for producing the pellets, and Figure 6.107 shows the basic parts, which consisted of top and bottom dies to house the powder during compaction. Figure 6.108 shows the familiar two regimes of compression response discussed previously. The tribometer for obtaining performance data, shown in Figure 6.109, consisted of a test disk mounted on a spindle driven by a variable-speed DC motor. The pellet was held in a vertical position in an L-shaped holder, mounted on a low-friction bearing, to enable the specimen to translate without constraint. Vertical pellet movement—that is, rate of wear—was measured with a linear-variable differential transformer, LVDT, having a resolution of 0.0025 mm.

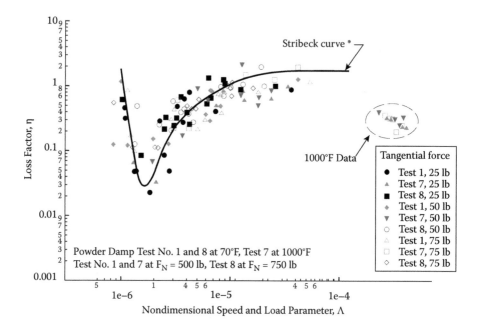

FIGURE 6.106   Loss factor of a TiO$_2$ powder versus Λ on a Stribeck curve.

Figures 6.110 and 6.111 show that both friction and wear rates were high initially, then decreased to a steady value; this was true of all three powders tested. Figure 6.112 shows the change in film thickness as speed was increased from standstill to 4.3 m/s; here, too, there was initially a noticeable rise in film thickness, which subsequently leveled off at higher speeds. The film thickness responded similarly when speed was decreased, exhibiting the definite presence of a hydrodynamic film. Figure 6.113 gives the overall results for all three powders, all at a load of 8.9 N, yielding coefficients of friction of approximately 0.1 to 0.2. Table 6.11 and Figure 6.114 show the test results at the higher speeds. At the highest test speed of 5000 rpm, the coefficient of friction and wear rate were both at their lowest; this, again, is related to the presence of an adequate lubricant film.

A common feature of all the tests is that, initially, the friction and wear rates were fairly high but then settled down to a steady level. This shows that after a transient initial period, a satisfactory lubricant layer was generated by the particles shed from the pellet, demonstrating the feasibility of this mode of lubrication. Of the three powders tested, sample A showed structural fractures and can be considered an inadequate candidate. The other two powders gave the following results:

Powder B at 4 lb load and 3000 rpm: wear factor = 7.17×10^{-9}, $f = 0.19$

Powder C at (a) 4 lb load and 2000 rpm: wear factor = 5.82×10^{-9}, $f = 0.17$

(b) 4 lb load and 3000 rpm: wear factor = 9.26×10^{-9}, $f = 0.18$

TABLE 6.10

MoS$_2$ Pellets Test Results at 500 rpm

Powder Samples of MoS2	Average Particle Diameter Size, P_d (μm)	Compaction Pressures, σ_{yy} (MPa)	Speed (rpm)	Pellet Load 8.9 N		Pellet Load 17.8 N	
				Pellet Wear Factor, φ (cm³/cm-kg)	Pellet Coefficient of Friction, f	Pellet Wear Factor, φ (cm³/cm-kg)	Pellet Coefficient of Friction, f
B	4–6	34.5	500	5.05×10^{-8}	0.2	3.63×10^{-8}	0.14
C	1–2	34.5	500	6.89×10^{-7}	0.2	3.94×10^{-8}	0.16
A	10–31	51.7	500	7.85×10^{-6}	0.19	—	—
A	10–31	69	500	5.47×10^{-7}	0.16	4.97×10^{-8}	0.15
B	4–6	69	500	—	—	1.33×10^{-8}	0.19
C	1–2	69	500	1.84×10^{-8}	0.1	—	—

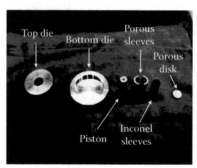
Fabrication Fixture

Powder Pellet

FIGURE 6.107   Fixture for the formation of powder pellets.

The wear factor given in the preceding was defined as

$$\Phi = \xi\, A/(L_T \times 0.719 \times W_n),\ (\text{cm}^3/\text{cm-kg}) \tag{6.12}$$

where ξ is the displacement, A the cross-sectional pellet area, L_T the distance traveled, and W_n the load on the pellet. In this manner, the volume of wear is made independent of the wear distance traveled or pellet loading.

Of the individual factors affecting friction and wear in pelletized powders, the following results were observed:

- Particle size and pellet integrity: Powder A, which consisted of particle sizes in the range of 10–31 microns and compressed at 34.5 MPa, had poor particle cohesion and showed, at the end of testing, structural fractures. Powders B and C, which had particle sizes in the range 4–6 microns and

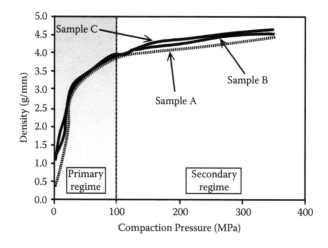

FIGURE 6.108   Density of powders tested as a function of compaction pressure.

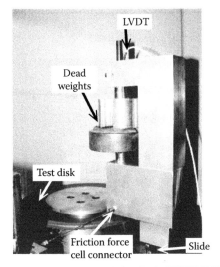

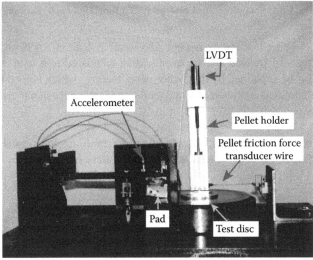

FIGURE 6.109   Tribometer for pelletized powder testing.

1–2 microns, respectively, exhibited no such deficiencies. This, too, conforms to the general postulate of the quasi-hydrodynamic theory, namely, that triboparticulates had to be within a relatively small range of sizes to act effectively as a lubricant. If larger sizes are to be used at all, they may require substantially higher levels of compaction.

- Powder compaction and wear rate: It was hypothesized that the wear factor would decrease with increasing compaction, because this would impose greater resistance to particle detachment. The tests fully confirm this, as the wear rate decreased when compaction pressures of 69 MPa were used in place of the standard 34.5 MPa.

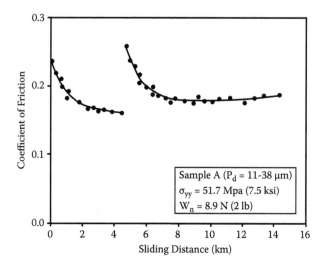

FIGURE 6.110 Friction coefficient of powder A as a function of test duration.

- Effect of pellet loading on friction and wear: A load of 8.9 N gave higher coefficients of friction and wear rate when compared to a load of 17.8 N. This may be explained by the higher pellet loading, which helped to form a more continuous lubricant film and facilitated better adhesion to the moving surface.

One qualification must be sounded with regard to the foregoing results concerning wear. Given that the pellets were small—their total volume being less than 15 cm³—the

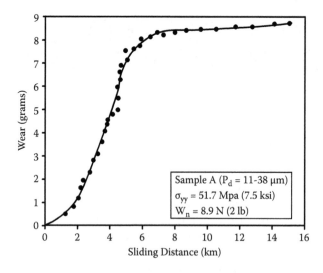

FIGURE 6.111 Amount of wear for powder A as a function of test duration.

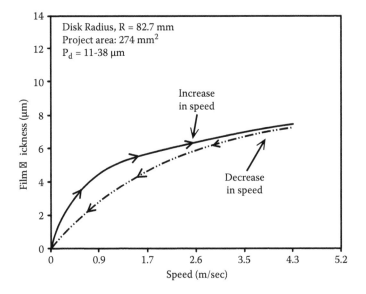

FIGURE 6.112 Powder lubrication film thickness as a function of speed.

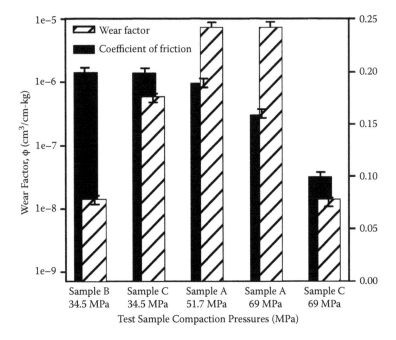

FIGURE 6.113 Comparison of friction coefficients and wear factors for three powders.

TABLE 6.11
MoS$_2$ Pellets Test Results at High Speeds

Powder Samples of MoS$_2$	Compaction Pressures, σ_{yr} (MPa)	Speed (rpm)	Load Applied on Pellet, 8.9 N		Load Applied on Pellet, 17.8 N	
			Pellet Wear Factor, φ (cm^3/cm-kg)	Pellet Coefficient of Friction, f	Pellet Wear Factor, φ (cm^3/cm-kg)	Pellet Coefficient of Friction, f
B	34.5	1000	1.23×10^{-5}	0.15	2.56×10^{-8}	0.24
B	34.5	2000	1.09×10^{-8}	0.2	3.31×10^{-9}	0.18
B	34.5	3000	1.44×10^{-9}	0.2	1.37×10^{-8}	0.2
B	34.5	5000	1.55×10^{-7}	0.3	1.37×10^{-6}	0.25
C	34.5	1000	2.36×10^{-8}	0.55	7.31×10^{-8}	0.19
C	34.5	2000	1.69×10^{-7}	0.4	1.87×10^{-8}	0.2
C	34.5	3000	1.22×10^{-8}	0.38	1.23×10^{-8}	0.15
C	34.5	5000	2.96×10^{-8}	0.08	1.36×10^{-8}	0.2

TABLE 6.12

MoS₂ Pellets Test Results at High Speeds

Powder Samples of MoS$_2$	Compaction Pressures, σ_{yy} (MPa)	Speed (rpm)	Load Applied on Pellet, 89 N		Load Applied on Pellet 17.8 N	
			Pellet Wear Factor, φ (cm³/cm-kg)	Pellet Coefficient of Friction, f	Pellet Wear Factor, φ (cm³/cm-kg)	Pellet Coefficient of Friction, f
B	34.5	1000	1.23×10^{-5}	0.15	2.56×10^{-8}	0.24
B	34.5	2000	1.09×10^{-8}	0.2	3.31×10^{-9}	0.18
B	34.5	3000	1.44×10^{-9}	0.2	1.37×10^{-8}	0.2
B	34.5	5000	1.55×10^{-7}	0.3	1.37×10^{-6}	0.25
C	34.5	1000	2.36×10^{-8}	0.55	7.31×10^{-8}	0.19
C	34.5	2000	1.69×10^{-7}	0.4	1.87×10^{-8}	0.2
C	34.5	3000	1.22×10^{-8}	0.38	1.23×10^{-8}	0.15
C	34.5	5000	2.69×10^{-8}	0.08	1.36×10^{-8}	0.2

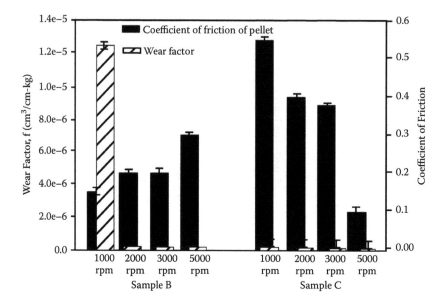

FIGURE 6.114   Friction coefficients and wear factors as a function of speed for powders B and C.

duration of each individual test was, likewise, short. Thus, the rates of wear measured reflect relatively brief operational periods. To be able to generalize more confidently about the nature of wear when using this mode of powder lubrication, longer periods of testing may be called for.

REFERENCES

1. Berthier, Y., Godet, M., and Brende, M., Velocity accommodation in friction, *STLE Trans.* Vol. 32, No 4, 1989.
2. Berthier, Y, Vincent, L., and Godet M., Velocity accommodation sites and modes in tribology, *Eur. J. Mech. A/Solids*, Vol. 11, No. 1, 1992.
3. Coulomb, C.A., Essai sur une Application des Regles de Maximis et Minimis a Quelques Problemes de Statique, Rel. a l'Architecture, *Memoires de Mathematiques et de Physique*, pp. 343–382, 1773.
4. Dareing, D.W. and Dayton, D.R., Non-Newtonian behavior of powder lubricants mixed with ethylene glycol, *STLE Trans.*, Vol. 35, No 1, 1992.
5. Godet, M., Third bodies in tribology, *Wear*, Vol. 136, 1990.
6. Gray, W.A., *The Packing of Solid Particles*, Chapman and Hall, London, 1968.
7. Heshmat, H., Pinkus, O., and Godet, M., On a common tribological mechanism between interacting surfaces, *STLE Trans.* Vol. 32, pp 32–41, 1989.
8a. Heshmat, H., Development of rheological model for powder lubrication, Vol. I, Tech. Report NASA Contractor Report CR-189043, October 1991.
8b. Heshmat, H., The rheology and hydrodynamics of dry powder lubrication, *STLE Trans.* Vol. 34 No 3, 1991.
9. Heshmat, H., High temperature solid lubricated bearing development dry powder lubricated traction testing, *AIAA J. Propulsion and Power*, Vol. 7, No 5, 1991.

10. Heshmat, H., High temperature solid lubricated bearing development—dry powder lubricated traction testing, *AIAA/SAE/ASME 26th Joint Propulsion Conference Proceedings*, Paper No. 90–2047 (July 1990), *AIAA J. Propulsion and Power,* Vol. 7, No 5, September–October (1991), pp. 814–820.

11. Heshmat, H. and Dill, J.F., Traction characteristics of high-temperature, powder-lubricated ceramics (Si$_3$N$_4$/SiC). ASME/STLE Tribology Conference 1990, *STLE Trans.*, Vol. 35, No 2 (1992) pp. 230–366.

12. Heshmat, H. and Brewe, D.E., On some experimental rheological aspects of triboparticulates, *Proc. 18th Leeds-Lyon Symp. on Wear Particles: From the Cradle to the Grave,* Lyon, September 3–6, 1991, Elsevier Science Publishers, *Tribology Series* 18, 1992.

13. Heshmat, H., Development of rheological model for powder lubrication, Volume II. Final Technical Report MTI 91TR31, prepared for U.S. Army AVSCOM, NASA contractor report CR-189043 (October 1991).

14. Dill, J. and Heshmat, H., Bearing technology for advanced propulsion systems, MTI 86TR56, Final Report prepared for DARPA MDA903-86-C-0050 (October 1991).

15. Heshmat, H., Rolling and sliding characteristics of powder lubricated ceramics at high-temperature and speed, presented at the *47th Annual Meeting of the STLE,* Philadelphia, PA, May 1992, *J. STLE, Lubrication Engineering*, Vol. 49, No 10 (1993), pp. 791–797.

16. Heshmat, H., Solid-lubrication roller bearing development: Static and dynamic characterization of high-temperature, Powder-Lubricated Materials, Tech. Report No. WL-TR-2-2050, for the period of July 1987 to June 1990, prepared under Contract No, F33615-87-C-2707, for Aero Propulsion and Power Directorate, Wright Lab., Air Force Systems Command, Wright-Patterson Air Force Base, Ohio, June 1992.

17. Heshmat, H. and Brewe, D.E., On the cognitive approach toward classification of dry triboparticulates, *Proc. 20th Leeds-Lyon Symp. on Dissipative Processes in Tribology*, Lyon, France (September 7–10, 1993), Dowson et al. (eds.), published by Elsevier Science B.V., *Tribology Series* 27 (1994), pp. 303–328.

18. Heshmat, H., Solid-lubrication roller bearing development: Static and dynamic characterization of high-temperature, powder-lubricated materials. Tech. Report No. WL-TR-2-2050, for the period of July 1987 to June 1990, prepared under Contract No. F33615-87-C-2707, for Aero Propulsion and Power Directorate, Wright Lab., Air Force Systems Command, Wright-Patterson Air Force Base, Ohio, June 1992.

19. Heshmat, H. and Walton, J.F., II, High temperature damper development. Final Report for 09-30-90—09/30/94, Aero Propulsion and Power Directorate, WLAFB, Ohio, Report No. WL-TR-94-2118, October 1994.

20. Heshmat, H., Experimental determination of powder film shear and damping characteristics. 2000-GT-365, submitted for publication and presentation at the ASME TURBO EXPO, IGTI 2000, Germany, May 2000.

21. Swanson, E.E., Xu, D.S., and Heshmat, H., On the application of powder shear damping to a structural element, presented at the *ASME TURBO EXPO*, IGTI 2000, Germany, May 2000, ASME Paper 2000-GT-0365.

22. Heshmat, H., The quasi-hydrodynamic mechanism of powder lubrication, Parts I and II, STLE Lubrication Engineering Vol. 48, Nos. 2 and 5, 1992.

23. Heshmat H. and Brewe D.E., On some experimental rheological aspects of triboparticulates, *Tribology Series* 18, Elsevier Science Publishers, 1991.

24. Kaneta, M., Nishsikawa, H., and Kameishi, K., Observations of wall slip in elastohydrodynamic lubrication, *ASME J. Tribology*, Vol. 112, pp. 447–452, 1990.

25. Nikas, G.K., Ioannides, E., and Sayles, R.S., Thermal modeling and effects from debris particles etc., *ASME J. Tribology*, Vol. 121, pp. 272–281, 1999.

26. Nikolakakis, I. and Pilel, N., Effects of particle shape and size on the tensile strength of powder, *J. Powder Technology*, 56, No 2, 1988.

27. Mansouri, M., Schmitt, M., and Paulmier, D., Third body effects on graphite XC48 steel magnetized sliding contact, *ASME J. Tribology*, Vol. 121, pp. 403–1999.

28. Scott, B. and Winer, W.O., A rheological model for elastohydrodynamic contacts based on primary laboratory data, *ASME J. Lubrication Technology*, Vol. 101, No 3, 1979.

29. Tevaarwek, J.L. and Johnson, K.L., The influence of fluid rheology on the performance of traction drives, *ASME J. Lubrication Technology*, Vol. 101, No 3, 1979.

7 Powder-Lubricated Devices

We have, in Chapter 6, elicited the nature of powders from the morphological, thermodynamic, and tribological standpoints. These properties appear to favor the idea that powders can serve as interface layers. The ultimate test, however, is to determine how powders will act when called upon to function as bona fide lubricants in tribological systems. We shall, therefore, in the present chapter, take a look at how powder films perform in such basic machine components as sliders, bearings, dampers, and piston rings, whose static and dynamic characteristics are vital for the satisfactory functioning of modern machinery.

7.1 SLIDERS

The work described in the following text was conducted on a plane slider, which, to a large extent, simulates a thrust bearing. The experiments offer the essential evidence that a hydrodynamic film is generated when powder is used as a lubricant. The slider was mounted on an eccentric pivot, which allowed it to assume any orientation compatible with the operating conditions. As shown in Figures 7.1 and 7.2, the pad rested on an annular ring, though the pad itself was square. The exact dimensions of ring and pad are listed in Figure 7.2. Both runner and slider were made of TiC cermet. The powder used was titanium dioxide, TiO_2, having a specific gravity of 4.26 with 50% accumulative particle size distribution of less than 5 microns compacted under a pressure of 67 MPa (10,000 psi). The powder lubricant was supplied by an airbrush placed 15 mm in front of the leading edge of the slider. The nozzle of the airbrush delivered dry air at pressures ranging from 172 to 414 kPa (25–60 psi). The rate of flow was controlled via adjustable orifices. Elaborate instrumentation was employed to measure the interface pressures, film thickness, levels of friction, and amount of powder entering the clearance space. The instrumentation used is described in some detail in the original published work (5). The system was checked out first by running a series of tests with conventional petroleum lubricant, which could be verified against known performance data.

7.1.1 DRY AND BOUNDARY-LUBRICATION RUNS

In preparation for full-scale powder tests, several runs were conducted on what could be termed dry and boundary-lubrication conditions. The test specimens were cleaned with acetone and the mating surfaces preoxidized in open air at 750°C. The slider was subjected to a load of 311.4 N (70 lb, or 64 psi) at 250 rpm for a period of 20 s. This removed most of the oxide layer formed during the heat treatment and

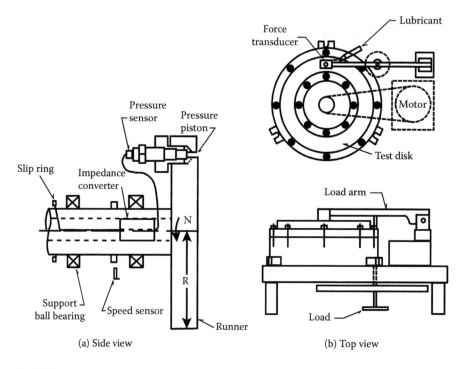

(a) Side view   (b) Top view

FIGURE 7.1 Apparatus for testing powder-lubricated bearings.

provided the pressure signature in the absence of either lubricant or oxide coating. As shown in Figure 7.3, no significant pressures were recorded under these conditions, marked t_o and t_1. A boundary lubrication regime was then generated by depositing powder on a dry paper tissue and periodically pressing it against the runner. The initially unloaded pad was, subsequently, subjected to a load of 156 N at 250 rpm. The pressure developed under these conditions, marked t_2, is shown in Figure 7.3. It is clear that a partial hydrodynamic pressure did develop under these conditions. This pressure signal could be made to either disappear or reappear by withholding or renewing the transfer of powder to the runner. In the initial stages of these runs, the pressure-based slider load deviated from the applied load by about 14%.

Of additional interest was the fact that the amplitude of the developed pressures was related to the amount of powder remaining on the mating surface after each procession. In addition, halfway through each run, the pressure profiles obtained under these boundary-lubrication conditions became more skewed toward the leading edge of the pad. As shown in Figure 7.4, the pressure profile, designated by t_2, was taken 7 s or 32 revolutions from the start, while t_3 was taken after 12 s or 55 revolutions. The shapes remain the same though the hydrodynamic film tends to decay with time due to starvation conditions produced by the diminution of powder. The pressure profile also has a strong similarity to that of the pressures obtained when a scraper is used with petroleum lubricants.

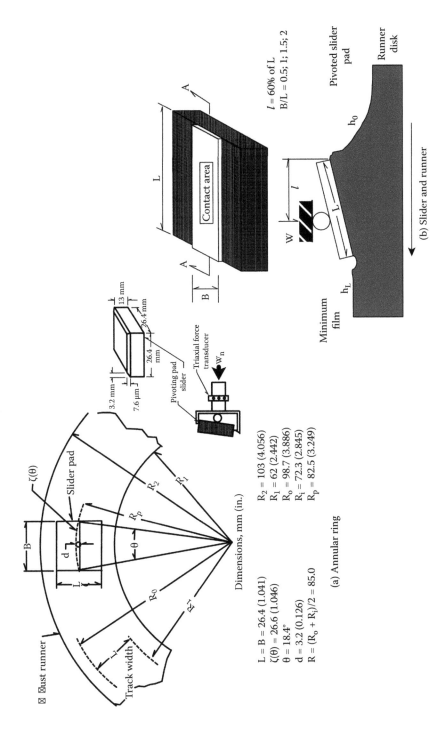

Dimensions, mm (in.)

$R_2 = 103$ (4.056)
$R_1 = 62$ (2.442)
$R_o = 98.7$ (3.886)
$R_i = 72.3$ (2.845)
$R_p = 82.5$ (3.249)

$L = B = 26.4$ (1.041)
$\zeta(\theta) = 26.6$ (1.046)
$\theta = 18.4°$
$d = 3.2$ (0.126)
$R = (R_o + R_i)/2 = 85.0$

(a) Annular ring

$l = 60\%$ of L
$B/L = 0.5; 1; 1.5; 2$

(b) Slider and runner

FIGURE 7.2 Arrangement for slider.

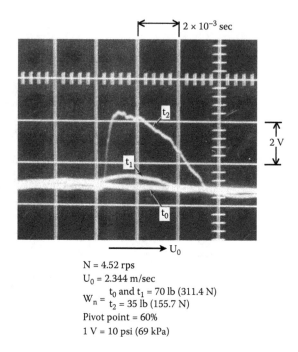

N = 4.52 rps
U_0 = 2.344 m/sec
$W_n = \begin{array}{l} t_0 \text{ and } t_1 = 70 \text{ lb (311.4 N)} \\ t_2 = 35 \text{ lb (155.7 N)} \end{array}$
Pivot point = 60%
1 V = 10 psi (69 kPa)

FIGURE 7.3   Pressure signatures for dry and boundary-lubrication runs.

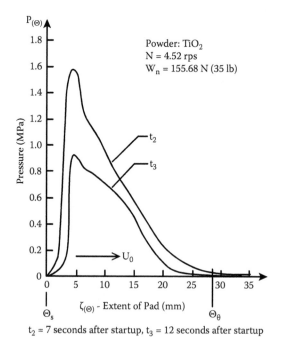

t_2 = 7 seconds after startup, t_3 = 12 seconds after startup

FIGURE 7.4   Digitized pressure profiles for powder boundary lubrication.

7.1.2 Full Film (Flooded) Powder Lubrication

The full film tests were all conducted at room temperature (22°C) and a runner speed of 532 rpm. The applied normal load was kept at 156 N (35 lb, or 32 psi). In actual testing it was found that, in addition to the steady load, there was a dynamic component from the test facility with a peak-to-peak force amplitude of 22.5 N. Under a dry air supply pressure of 414 kPa, the rate of powder supply was $9\,\frac{cm^3}{s}$. The slider pivot was kept 60% downstream from the leading edge. To ensure a full film upon start-up, the clearance space was filled with powder prior to loading. The runs were then continued for about 30 min.

Figure 7.5 shows the pressure oscilloscope traces taken 2 and 5 s after start-up. During the continuation of the run, the pressure profile oscillated between the profile designated t_o and t_1 in Figures 7.5a and 7.5b, mainly to the dynamic load component. The digitized pressures, replotted in Figure 7.6, show the apexes at 75% downstream from the slider, but the calculated centers of pressures for both curves yield a satisfactory 62% location. Using a parabolic pressure distribution in the axial direction, the force vector obtained from the integrated two-dimensional pressures field adds up to the externally imposed load.

Figure 7.7 depicts the frictional behavior of powder-lubricated sliders as a function of load, other conditions remaining constant. Each set of data points was recorded in three separate tests, their maxima and minima indicated on the curve. The calculated slopes of the friction–load curve are given in Table 7.1. In tests with a restricted supply of powder—that is, under starved conditions—the response of powder lubrication paralleled fluid film starvation, as reported in Reference 2. Finally, Figure 7.8 shows a comparison of the pressure profiles obtained with a petroleum lubricant with full powder lubrication. When plotted in nondimensional form, that is, with $p=(\frac{p}{p_{max}})$, the two profiles are, in all essential aspects, similar, the only noticeable difference being that, in powder lubrication, the pressure apex is skewed closer to the trailing edge; under starved conditions, the center of pressure shifts even further upstream.

7.1.3 Effect of (B/L) Ratio

The foregoing tests were extended to investigate the effects of several different (B/L) ratios, as well as using a different powder, MoS_2. The range and dimensions of the new slider geometries are given in Table 7.2—the (B/L) ratios ranging from 0.5 to 2. The new powder—MoS_2—had a 50% cumulative particle size distribution of 1–2 microns and a solid specific gravity of 4.8. For all tests, the pivot position was 60% from the leading edge and the unit pressure, $P=(\frac{W_n}{BL})$, was kept constant for all the different geometries. The variation of the coefficient of friction with load and (B/L) ratio is given in Figures 7.9 and 7.10 for two speeds. The first represents operation in the boundary-lubrication regime, while the latter provides a hydrodynamic film. As can be seen, full hydrodynamic operation shifts the (B/L) ratio from 1.5 to 1.0, which is also the optimum for fluid film lubrication.

Figure 7.11 shows some limited tests with the TiO_2 powder for different speeds at the constant load of 34.5 kPa (5 psi). The trend here is similar to the results with

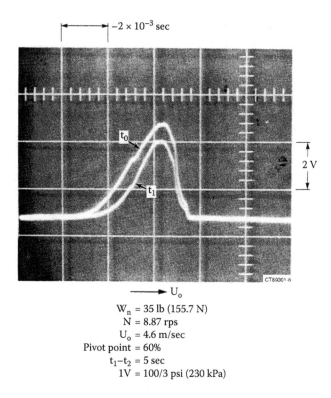

W_n = 35 lb (155.7 N)
N = 8.87 rps
U_o = 4.6 m/sec
Pivot point = 60%
t_1-t_2 = 5 sec
1V = 100/3 psi (230 kPa)

FIGURE 7.5a Photograph of Oscilloscope Traces Showing Sweep of pressure profiles for powder-lubricated Pivoted Pad Slider, t_0 = 2 sec. After startup, $t_1 = t_0 + 5$ sec.

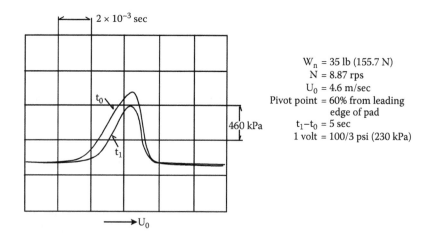

W_n = 35 lb (155.7 N)
N = 8.87 rps
U_0 = 4.6 m/sec
Pivot point = 60% from leading
edge of pad
t_1-t_0 = 5 sec
1 volt = 100/3 psi (230 kPa)

FIGURE 7.5b Hydrodynamic pressure signatures in powder-lubricated sliders.

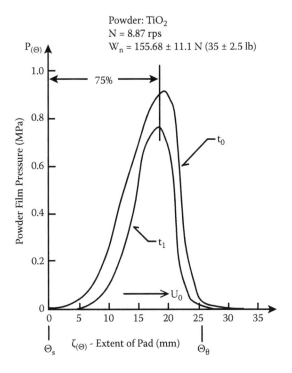

FIGURE 7.6   Digitized pressure profiles for powder lubrication sliders.

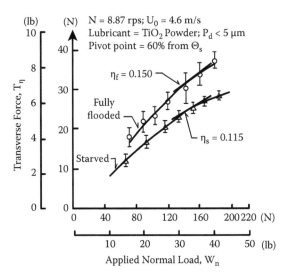

FIGURE 7.7   Frictional force in sliders under full and starved powder lubrication.

TABLE 7.1
Friction Coefficient in a Powder-Lubricated Slider

	Normal load, N (lb)			
	66.7 (15)	111.2 (25)	155.7 (35)	200.2 (45)
η_f Flooded	0.27	0.163	0.150	0.113
η_s Starved	0.182	0.148	0.115	0.093

Note: $U = 4.5$ m/s.

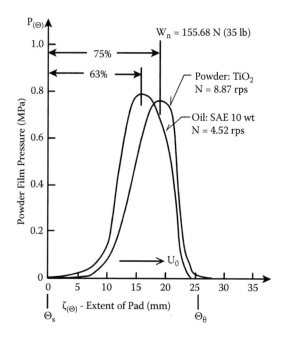

FIGURE 7.8　Comparison of pressure profiles in oil- and powder-lubricated sliders.

TABLE 7.2
Configurations of Tested Sliders

Aspect Ratio B/L	Width, B, mm (in.)	Length, L, mm (in.)	Area, mm² (in.²)
0.5	13.2 (0.52)	26.4 (1.04)	384.5 (0.54)
1	26.4 (1.04)	26.4 (1.04)	697 (1.08)
1.5	26.4 (1.04)	17.6 (0.693)	464.6 (0.72)
2	26.4 (1.04)	13.2 (0.52)	348.5 (0.54)

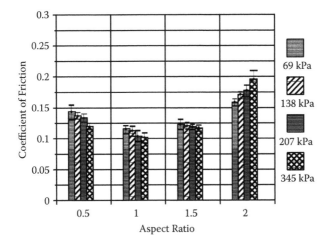

FIGURE 7.9   Effect of (B/L) ratio on coefficient of friction at $U = 0.27$ m/s.

MoS$_2$ powder, the coefficient of friction being high in the boundary lubrication regime and dropping as speed is increased. Over the entire range of speeds, a (B/L) value of 1 gives the lowest coefficient of friction (6,8,9,14–17,22).

7.2   JOURNAL BEARINGS

The test rig constructed for investigating the performance of powder-lubricated journal bearings is shown in Figure 7.12. It was equipped to impose high speeds and high temperatures, the former up to 2 million DN or 60,000 rpm (Ref. 7.4–7.6, 7.8, 7.9, 7.15, 7.16 and 7.21). A special sprayer, operated on the Venturi principle and shown in

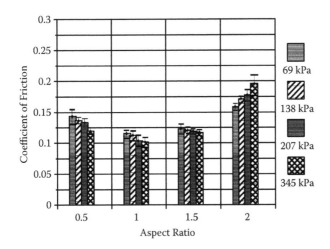

FIGURE 7.10   Effect of (B/L) ratio on coefficient of friction at $U = 4.5$ m/s.

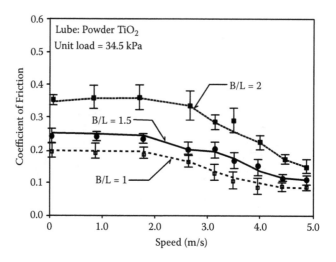

FIGURE 7.11   Effect of (*B/L*) ratio using TiO$_2$ powder.

Figure 7.13, was designed for powder delivery. The low pressure induced by the inner tube drew the powder from the hopper toward the discharge end. In optimization trials, it was found that efficient flow was dependent on the relative positions of the inner and outer channel tips and their distance from the inlet zone. The nozzle used an inlet air pressure of 60 psi, which yielded a powder flow of 11 mg/s. The powder

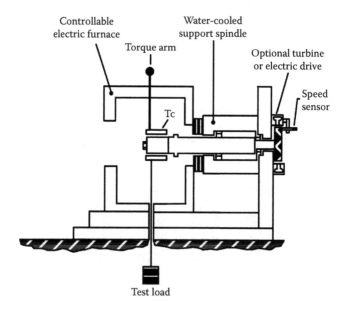

FIGURE 7.12a   Apparatus for testing powder-lubricated journal bearings.

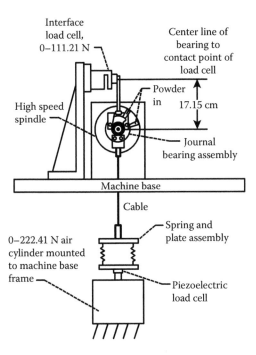

FIGURE 7.12b Apparatus for testing powder-lubrication journal bearings.

was fed to the sides (i.e., axially into the individual pads). The discharged powder was collected in a chamber enclosing the test bearing, then filtrated and returned to the cyclone.

7.2.1 THREE-PAD BEARING WITH MoS₂ POWDER

The three-pad pivoted bearing, shown in Figure 7.14, was specially constructed to facilitate powder lubrication. Its L × D dimensions were 2 × 3.4 cm (0.79 × 1.34 in.), yielding a projected area of 6.82 cm² (1.06 in.²); radial clearances of 0.05 and 0.1 mm (2 and 4 mil) were used. The clearances were built in by placing shims between the compliant mount and the bearing cartridge. The pads were pivoted 60% from the leading edge. The tests were mainly concerned with the thermal behavior of the

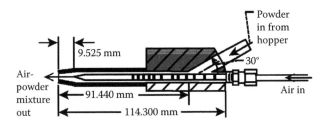

FIGURE 7.13 Device for feeding powder lubricant.

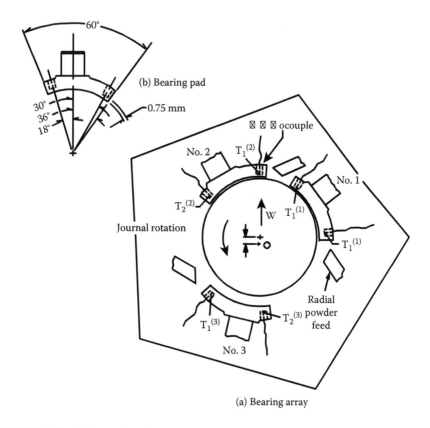

FIGURE 7.14 Configuration of a three-pad bearing.

bearing; thus, it was properly instrumented with several thermocouples, as shown in Figure 7.14. Each pad had thermocouples at the inlet and outlet boundaries, inserted along the centerline. The MoS_2 powder had the same properties as in the tests described in the previous section. Additional features of the bearing and individual pads are given in Figure 7.14.

Figure 7.15 shows sample plots of the coefficient of friction and pad surface temperature as a function of speed and test duration. At low speeds, f is at its maximum, with the onset of hydrodynamic conditions. The friction data, as well as the thermal behavior, corresponds closely with that prevailing with fluid films; there is even the effect of groove mixing temperature present here, as described in References 6, 9, 14, and 15. Figure 7.16 shows the temperature rise along the loaded part—pad # 1—while Figure 7.17 shows the overall bearing average temperature. Initially, there is a temperature decrease due to the appearance of an adequate hydrodynamic film; with higher speeds, both the average and maximum temperatures begin to rise. After a test duration of over 8 h—under various load and speed conditions—both the bearing and journal were inspected and found to be in excellent condition. Figure 7.18 shows the appearance of the pads with a thin layer of powder still attached to their

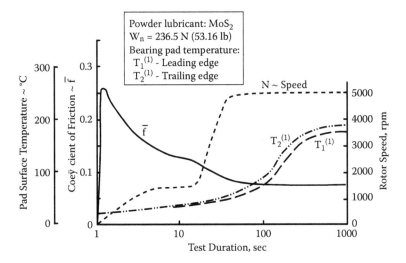

FIGURE 7.15   Performance of powder-lubricated three-pad bearing.

surfaces. On the other hand, when deliberate powder starvation was introduced, both the pads and the journal were damaged.

7.2.2   THREE-PAD BEARINGS WITH WS$_2$ POWDER

For comparative purposes, the same bearing was subsequently run using a WS$_2$ powder. In previous investigations (8,16), it was found that, up to 400°C, tungsten disulfide had about the same coefficient of friction as MoS$_2$; whereas MoS$_2$ deteriorated after that temperature, WS$_2$ performed well up to 600°C. Also, the oxidation

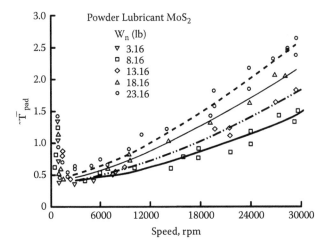

FIGURE 7.16   Temperature rise in loaded pad.

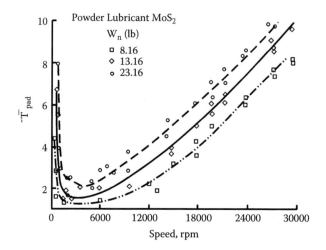

FIGURE 7.17   Average temperatures in a three-pad bearing.

Pad 3   Pad 2   Pad 1

FIGURE 7.18a   Condition of pads after testing with adheared film.

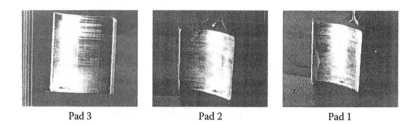

Pad 3   Pad 2   Pad 1

FIGURE 7.18b   Condition of pads after testing after cleaning.

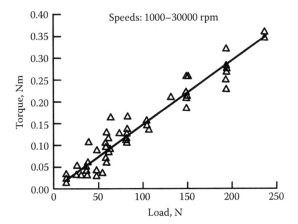

FIGURE 7.19   Friction versus load with WS$_2$ powder.

products of tungsten disulfide proved beneficial in reducing wear. It was, therefore, chosen for a comparison against molybdenum disulfide. The specific WS$_2$ used had particles ranging from 1 to 2 microns and a specific gravity of 7.5.

Tests were carried out with loads up to 236 N and speeds of 30,000 rpm. Figure 7.19 shows torque as a function of load and speed, which proves to be independent of speed and proportional to load. This converts to a constant coefficient of friction of 0.09. When plotting the coefficient of friction against the normalized load (i.e., the Sommerfeld number, as is done in Figure 7.20), one obtains a constant line, as compared with a similar plot for MoS$_2$, which yields the expected knee-shaped curve. Using the foregoing coefficient of friction, Figure 7.21 provides the power loss as a function of the load. Figure 7.22 shows the differences in leading and trailing edge temperatures. While in the unloaded pads it is minimal, there is a differential of 40°C in the loaded pad, as would be the case with a petroleum lubricant.

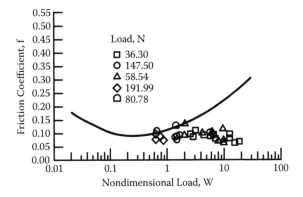

FIGURE 7.20   Friction coefficient in MoS$_2$ and WS$_2$ powders.

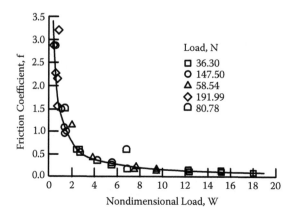

FIGURE 7.21 Power loss with WS$_2$ powders.

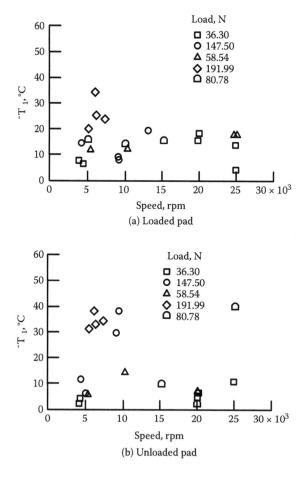

FIGURE 7.22 Temperature rise with WS$_2$ powders.

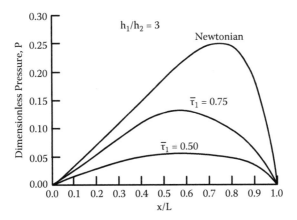

FIGURE 7.23 Pressure profiles for different levels of powder limiting shear strength.

The independence of friction of speed and the constancy of the coefficient of friction over a range of operating conditions brings out one of the essential features of powder lubrication: the presence of a limiting shear strength. In the present case with the WS_2 powder, this limiting shear strength was reached at very low sliding velocities. While yielding lower levels of friction, this has an adverse effect on the hydrodynamic pressures, leading to a lower load capacity. Figure 7.23 shows the pressure profiles as a function of a limiting shear stress on the performance of a powder film. In a Newtonian lubricant, no limiting shear stress exists at all. For non-Newtonian media, such as powders with two different upper limits of shear stress, there is a corresponding decease in the levels of the generated pressure profiles. The lower friction in the WS_2 film naturally leads to lower temperatures. With a load of 36.3 N and a speed of 25,000 rpm, the average pad temperature for the MoS_2 powder was 240°C; for the WS^2 powder, it was 136°C. Thus, depending on the design objective—high load capacity or low temperature levels—one can select a suitable lubricant from the many available powders.

7.2.3 FIVE-PAD BEARING WITH MoS_2

This journal bearing, shown in Figure 7.24, was similar to the previous one except that it had five instead of three pads. Its dimensions were $D \times L = 3.41 \times 2$ cm (1.34×0.79 in.), each pad having a projected area of 6.82 cm² (1.06 in.²). The two radial clearances used were 0.05 and 0.1 mm (2 and 4 mil). The powder lubricant was a molybdenum disulfide with the same properties as used previously in the tests on the three-pad bearing. The test apparatus was the same as described in the previous section. For the tests with the smaller clearance, Figure 7.25 shows the power losses as a function of load and speed. Here, too, qualitatively, there is a striking similarity to the power loss behavior with Newtonian oils. For the tests with the larger clearance, the load and powder flow rates were kept the same. Contrary to expectation, doubling the clearance had only a small effect on the levels of power loss.

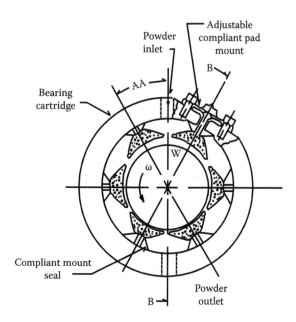

FIGURE 7.24 The five-pad journal bearing.

Special tests were run to investigate the effect of powder flow rates. Two rates were used: 15 and 30 cc/min (0.28 and 0.56 g/s). Figures 7.26a and 7.26b show the effect of changing the powder flow rates on the power loss along with the effect of external loading. Figure 7.27 shows the friction as a function of the normalized load for the parameters of clearance and powder flow rate. Here, too, the behavior of the

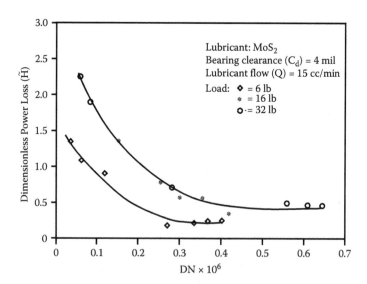

FIGURE 7.25 Power loss as a function of speed and load in a five-pad journal bearing.

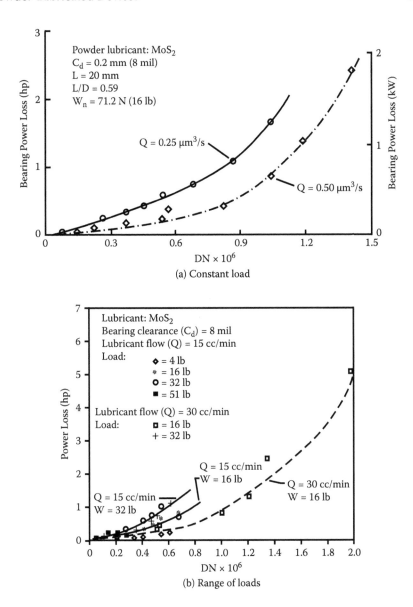

FIGURE 7.26 Power loss in five-pad journal bearing as a function of powder flow rate.

upper and lower frictional curves parallels the familiar Stribeck curve of hydrody-namic lubrication. A direct comparison of hydrodynamic and powder lubrication is shown in Figure 7.28. A summary of the major results of the test series with the five-pad bearing is given in Table 7.3. Following a total test time of three-quarters of an hour under a range of loads and speeds, the journal and bearing were inspected and found to be in excellent condition, as shown in Figure 7.29. A close examination of

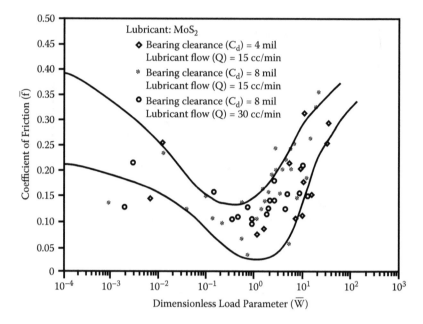

FIGURE 7.27   Locus of friction coefficient in five-pad journal bearings.

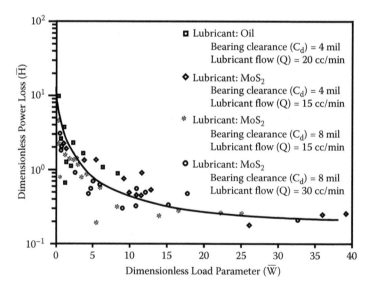

FIGURE 7.28   Power loss in powder- and oil-lubricated five-pad journal bearings.

TABLE 7.3

Summary of MoS$_2$-Lubricated Journal Bearing Experiments

W (lb)	Average DN (mm × rpm) × 10^6	Average Time (s)	$F \times D$ (lb-in.) × 10^6	Maximum Power Loss (hp)	Maximum Pad Temperature (°F)
4	0.6	870	1.37	0.24	146
6	0.4	210	0.31	0.23	188
16	1.3 to 2.0	1020	15.70	2.5	250 to 300
32	0.6	450	3.34	1.0	275 to 350
51	0.27	180	0.15	0.16	200 to 250
Total	—	2730	20.87[a]	—	—

Note: M2 tool steel, pad dimensions: L = 20, B = 17.8, D = 34 mm; Total number of pads = 5.
[a] Total work done on bearing = 1.75×10^6 ft × lb $\cong 0.655$ kWh.

the surfaces after testing revealed that a layer of powder 2.5 to 7.5 microns (0.1–0.3 mil) thick had adhered to the journal. As pointed out on numerous occasions, this is further evidence of the existence of an adhered boundary layer and slip near the boundaries. It is, in a way, an automatic form of protection of the mating surfaces when using powder as a lubricant.

7.3 DAMPERS

In gas turbine engines, dampers are required to maintain seal clearances consistent with engine dynamics and maneuver loadings. The high temperatures in these engines—of the order of 1000°F to 1500°F—make the employment of conventional dampers risky, and one possible solution is the use of powder-lubricated dampers. Figure 7.30 shows one such possible damper design. It supports the bearing outer race and is secured by an antirotation device to the static frame of the machine. To investigate the viability of such a damper, a series of tests (4,23) was conducted on square pads, as well as circumferential dampers, as shown in Figures 7.30 and 7.31. The pads 1 × 1 in extent had a half a mil crown machined in the circumferential direction and a chamber in the axial direction to facilitate the inflow of a powder lubricant. The damper elements and pads were made of a titanium carbide cermet because of its good wear and temperature properties. A rutile form of TiO$_2$ powder with a specific gravity of 4.26 was delivered by the spray method described previously. The controlled orbit test rig consisted of an eccentric shaft capable of running at 20,000 rpm. The forces transmitted during the imposed orbits were recorded for a range of loads, temperatures, and speeds. For the high-temperature tests, the pads were heated by air at a temperature of 1300°F.

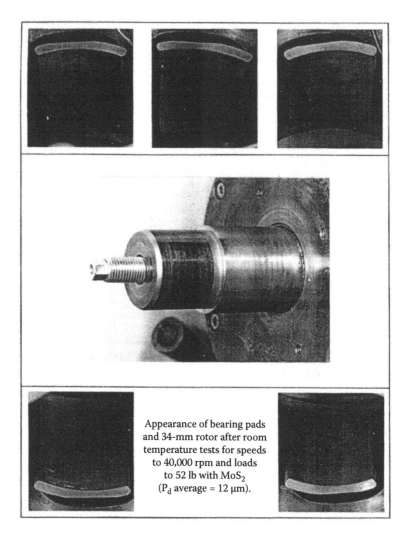

Appearance of bearing pads
and 34-mm rotor after room
temperature tests for speeds
to 40,000 rpm and loads
to 52 lb with MoS_2
(P_d average = 12 μm).

FIGURE 7.29   Posttest condition of journal bearing showing thin adhered layers of MoS_2 film.

Figures 7.32 through 7.36 show the results of the experiments highlighting the following characteristics of powder-lubricated dampers:

* Figure 7.32: Damping decreases linearly with speed but at a rate lower than that which occurs in conventional friction dampers. In a friction damper, the equivalent damping over the plotted speed range would be expected to drop by a factor of 3; here it decreased only by some 30%. Damping increased with load, so that a tripling of the normal load increased damping by a factor of 2.5.

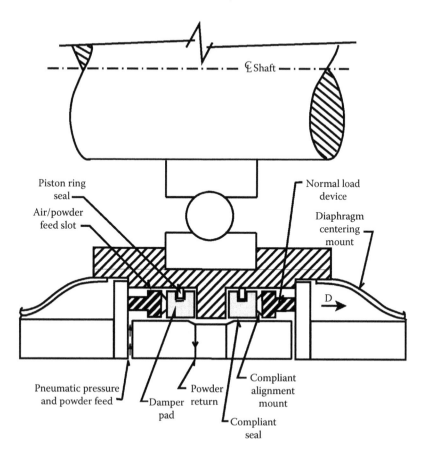

FIGURE 7.30 Location of a powder-lubricated damper in a gas turbine engine.

- Figure 7.33: Going from room temperature to 800°F, the damping as a function of speed first decreases and then rises again. The deviation between room and elevated temperatures decreases at higher speeds. This tendency to yield nearly the same damping levels over the speed range indicates that an energy dissipation mechanism other than pure friction is present in a powder-lubricated damper. This may, indeed, be due to the formation of a quasi-hydrodynamic film, which provides additional forces to accommodate the orbital motion.
- Figure 7.34: This presents the Dampers temperature rise as a function of speed. At the highest test speed of 10,000 rpm, it was less than 50°F, indicating appreciable heat convection by the flowing powder.
- Figures 7.35 and 7.36: These compare the results of powder-lubricated and unlubricated dampers. Whereas with powder lubrication, damping is nearly constant with speed, the damping of the unlubricated damper decreases rapidly with speed. The results in Figure 7.35 are most likely influenced by

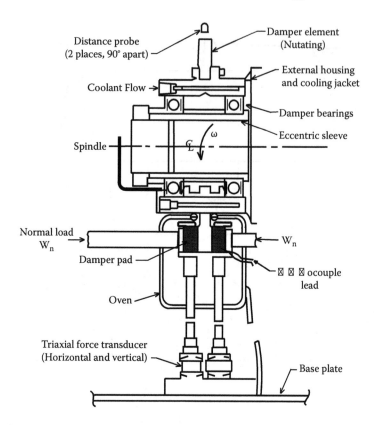

FIGURE 7.31 Test arrangement for powder-lubricated damper.

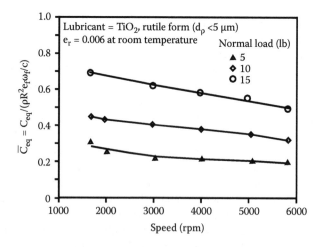

FIGURE 7.32 Equivalent damping as a function of speed in a powder-lubricated damper.

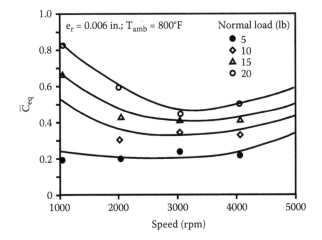

FIGURE 7.33 Equivalent damping at elevated temperature.

the formation of debris in the unlubricated contact, which would effectively equal the action of the powder-lubricated damper. The frictional behavior in the powder-lubricated damper in Figure 7.36 is similar to a hydrodynamic device, whereas the unlubricated surfaces produced a constant shear force for much of the speed range.

One general conclusion that emerges is that, whereas powder-lubricated damping may initially be lower than in a friction damper, the nearly linear equivalent damping and increased damper life are significant advantages for the powder-lubricated damper.

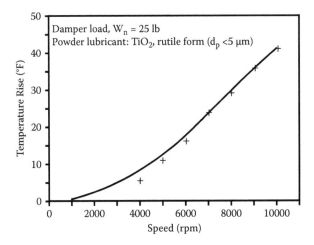

FIGURE 7.34 Temperature rise in powder-lubricated damper.

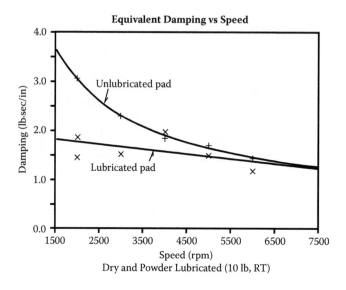

FIGURE 7.35 Damping in dry and powder-lubricated damper.

7.4 PISTON RINGS

In systems that use coal slurries either as fuels or as a process fluid, wear of the engine components is extreme and can be almost 100 times higher than in conventional engines. The problem could be alleviated if that very particulate mixture

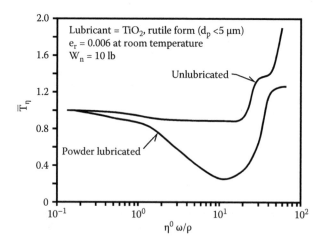

FIGURE 7.36 Shear forces in dry and powder-lubricated damper.

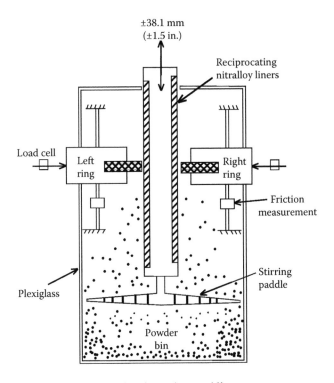

±38.1 mm
(±1.5 in.)

Reciprocating
nitralloy liners

Load cell

Left
ring

Right
ring

Friction
measurement

Stirring
paddle

Plexiglass

Powder
bin

FIGURE 7.37   Experimental setup for piston rings and liners.

could be utilized as a lubricant, particularly for particle sizes in the range of 1 to 10 microns. References 11–13 offer a series of tests aimed at utilizing coal ash, either alone or in conjunction with an externally supplied powder, as a means of alleviating this wear problem, particularly as it occurs in the piston rings and liners of such engines.

The test setup for these experiments is shown in Figure 7.37. A vertical block, capable of reciprocating at a frequency of 2000 cpm, with a full stroke of 7.62 cm, was lined on opposite sides with nitralloy plates against which two piston rings were positioned. A normal load was applied on the rings and the frictional forces measured. The lower end of the reciprocating block was equipped with a perforated stirrer or paddle to agitate and supply powder lubricant to the test rings. These features are shown in Figure 7.38, while the shape and dimensions of the test specimens are given in Figures 7.39a and 7.39b. The effects of geometry, materials, and powder lubrication were ascertained via a series of tests with oil, dry surfaces, and powder lubrication, recording the corresponding wear. With regard to geometry, three surface profiles were investigated: crowned, trapezoidal, and flat. A preliminary theoretical calculation showed the trapezoidal geometry to be superior, which was later confirmed in the course of testing. Two particulates were used: fly ash from a bituminous coal with a 55 Hardgrave index and carbon graphite with an average particulate size

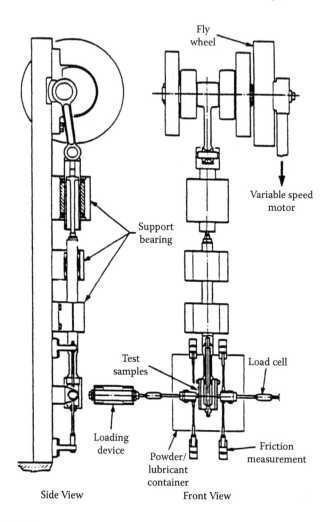

FIGURE 7.38 Test components arrangement.

of 2 microns. The various combinations of powders and materials used are shown in
Table 7.4; the matrix of performed tests is given in Table 7.5.

Figure 7.40 shows the method used in calculating the resulting wear. The abscissa
represents the cumulative energy term given by summing all the incremental ener-
gies $\Sigma(\Delta WL)$. The ordinate is made up of points representing the average wear function Φ given by

$$\Phi = \Sigma\left(\frac{\delta_n}{\Sigma WL}\right) = \frac{(Total\,Wear)}{\Sigma\Delta(WL)_n}$$

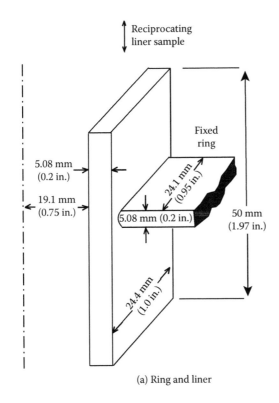

Reciprocating liner sample

Fixed ring

5.08 mm (0.2 in.)

19.1 mm (0.75 in.)

24.1 mm (0.95 in.)

5.08 mm (0.2 in.)

50 mm (1.97 in.)

24.4 mm (1.0 in.)

(a) Ring and liner

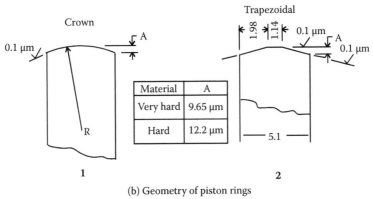

Crown

Trapezoidal

0.1 μm

1.98

1.14

0.1 μm

A

0.1 μm

A

R

Material	A
Very hard	9.65 μm
Hard	12.2 μm

5.1

1   2

(b) Geometry of piston rings

FIGURE 7.39 Dimensions of piston rings and liners.

TABLE 7.4
Specifications of Test Specimens

	Materials		
Ring Geometry	**Ring**	**Liner**	**Powder Lubricant**
Flat crowned	Ductile iron	Nitralloy	Graphite ×%[a] graphite + (1.×)% coal ash
Trapezoidal	Tungsten carbide	Tungsten carbide	Coal ash

[a] Percentage by weight.

The wear data given in Table 7.5, together with the visual observation and inspection of the surfaces following completion of testing, leads to the following general conclusions:

- The only measurable wear in test series 4–8 was when using a crowned geometry. In tests 6–8 (which used coal ash as a lubricant), no wear occurred at all, as was also the case when graphite was used—all of which were run with trapezoidal rings. A trapezoidal geometry, thus, seems the best configuration to use when triboparticulate lubricants are contemplated.
- As in previous experiments with powders, film starvation sets in with the passage of testing time. As starvation rises, the friction coefficient

TABLE 7.5
Matrix of Tests of Piston Rings

	Mating Surfaces				
	Type 100-7-03 Ductile Iron Ring Versus 135 Nitralloy Liner			Tungsten Carbide (GE Carboloy 44A,WC+Co) for Both	
Lubricant	**Flat**	**Trapezoidal**	**Crowned**	**Trapezoidal**	**Crowned**
Oil	P[a]				
Grease	P[a]				
Graphite	—	Test 2	Test 3	Test 4	Test 5
50% graphite + 50% ash[b]	—	—	—	Test 6	—
25% graphite + 75% ash	—	—	—	Test 7	—
100% ash	Test 1	—	—	Test 8	—

[a] Preliminary runs.
[b] Percentage by weight.

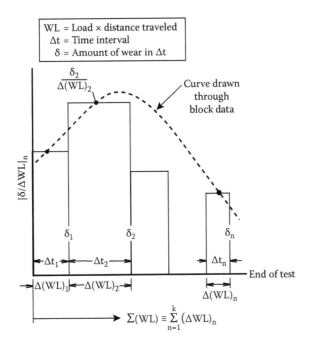

FIGURE 7.40 Method of compilation of wear data.

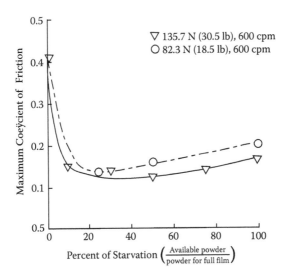

FIGURE 7.41 Effect of starvation on friction coefficient.

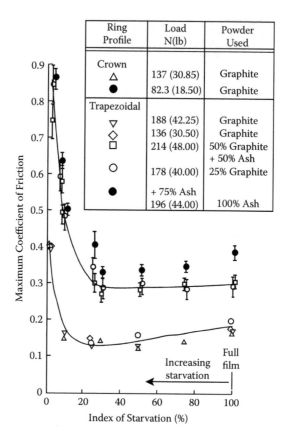

Ring Profile	Load N(lb)	Powder Used
Crown		
△	137 (30.85)	Graphite
●	82.3 (18.50)	Graphite
Trapezoidal		
▽	188 (42.25)	Graphite
◇	136 (30.50)	Graphite
□	214 (48.00)	50% Graphite + 50% Ash
○	178 (40.00)	25% Graphite + 75% Ash
●	196 (44.00)	100% Ash

FIGURE 7.42 General plot of friction coefficient.

initially decreases, but when starvation of the order of 80% is reached, the friction begins to rise and approaches levels of dry contact. This behavior is portrayed in Figure 7.41. The general behavior of friction for all tests is shown in Figure 7.42. The variation of the coefficient of friction with degree of starvation is similar to its behavior in oil-lubricated devices (2).

- It was found that, in combining graphite with coal ash, the two media did not mix, nor was there a tendency for the coal ash to coat itself with the graphite. This corresponds to the case when two liquids of different physicochemical characteristics are employed as a common lubricant.

TABLE 7.6

Results of Wear Measurements

Test No.	Wear Factor, ϕ (cm³ cm⁻¹ kg⁻¹)				Weight Change per Unit Length								
	Ring		Liner		Both Rings	Both Liners	Ring/Liner Pair		Total Rings and Liners	Ring		Liner	
	Left	Right	Left	Right			Left	Right		Left	Right	Left	Right
	($\times 10^{-9}$)						($\times 10^{-12}$)			(mg cm⁻³)			
1	1.24	29.7	1.54	0.9	14800	1116	1305	14610	7956	9.4	264	11.7	7.9
2	0.04	0.02	-0.05	0.003	23	-17	-2	11.4	3.03	2.6	0.8	-3.6	0.1
3	-0.05	1.6	0.02	0.2	755	104	-15.2	875	430	-1.13	48.6	-0.45	2.8
	($\times 10^{-12}$)						($\times 10^{-14}$)			(µg cm⁻³)			
4	0.54	0.05	0.54	0.05	54.3	5.43	29.9	29.9	29.9	42	42	4.2	4.2
5	5.9	0.8	43.9	0.8	335	2233	2487	79.8	1284	290	42	2279	42
6	-0.023	-0.022	-0.16	-0.16	-2.3	-16	-9.4	-9.0	-9.2	-4.1	-4.1	-29	-29
7	0	0	0	0	0	0	0	0	0	0	0	0	0
8	-0.02	-0.02	-0.14	-0.139	-1.99	-13.9	-8.0	-8.0	-8.0	4.0	-4.0	-29.0	-29.0
9[a]	5	5	5	5	—	—	—	—	—	—	—	—	—

Note: Negative sign (−) designates gain.

[a] Data from pin/disk test obtained from *Wear Control Handbook* edited by M.B. Peterson and W.O. Winer, the ASM United Engineering Center, 1990.

REFERENCES

1   Anson, D. et al., Development plan for a coal-water slurry-fired engine for cogeneration applications, U.S. Department of Energy Report No. DOE/CE/40741-T1, July 1987.

2. Heshmat H., Starved bearings technology: theory and experiment, Ph.D. thesis, Mechanical Engineering Department, Renselaer Polytechnic Institute, Troy, New York, December 1988.

3. Heshmat, H. and Gorski, P.T., Mixing inlet temperature in starved journal bearings, *13th Leeds Lyon Tribology Symposium*, Leeds, U.K., September 1986, Paper III (iii), Tribology Series 11, Elsevier Publication edited by Dowson et al., 1989, pp. 73–79.

4. Heshmat, H. and Walton, J. F., High temperature powder lubricated dampers for gas turbine engines, *AIAA J. Propulsion and Power*, Vol. 8, No. 2, 1992, pp. 449–456.

5. Heshmat, H., The quasi-hydrodynamic mechanism of powder lubrication: Part II—Lubricant film pressure profile, *STLE J. Engineering*, Vol. 48, No. 5, 1992, pp. 373–383.

6. Heshmat, H., On the theory of quasi-hydrodynamic lubrication with dry powder: Application to development of high speed journal bearings for hostile environment, *Proc. 20th Leeds Lyon Symposium on Dissipative Processes in Tribology*, Lyon, France, September 7–10, 1993, Dowson et al. eds., published by Elsevier Science V., *Tribology Series* 27, 1994, pp. 45–64.

7. Heshmat, H., Wear reduction systems for coal-fueled diesel engines: Part II—the basics of powder lubrication, *Wear*, 162–164, pp. 508–517, 1993.

8. Heshmat, H. and Brewe, D.E., Performance of powder lubricated journal bearing with WS_2 powder: Experimental study, *ASME J. Tribology*, 1996.

9. Heshmat, H., Performance of powder-lubricated journal bearing with MoS_2 powder: Experimental study of thermal phenomena, *ASME J. Tribology*, 1995

10. Higgs, C.F., Heshmat, H., and Heshmat, C., Comparative evaluation of MoS_2 and WS_2 lubricants in high speed, multi-pad journal bearings, *ASME J. Tribology*, 1998.

11. Heshmat, H., Powder-lubricated piston ring development, MTI Report No. 91TR14, prepared for U.S. DOE-METC, performed under Contract No. AC21-88MC25252, No. DOE/MC/25252-3024 (DE91016655), June 1991.

12. Heshmat, H., Wear reduction systems for coal-fueled diesel engines—Part I: The basics of powder lubrication, *Wear Elsevier Sequoia*, 162–164 (1993), pp. 508–517.

13. Heshmat, H., Wear reduction systems for coal-fueled diesel engines—Part II: The experimental results and hydrodynamic model of powder lubrication. *Wear Elsevier Sequoia*, 162–164 (1993), pp. 518–528.

14. Heshmat, H. and Brewe, D.E., On the cognitive approach toward classification of dry triboparticulates, *Proc. 20th Leeds-Lyon Symposium on Dissipative Processes in Tribology*, Lyon, France, September 7–10, 1993, Dowson et al., eds., published by Elsevier Science B.V., *Tribology Series* 27 (1994), pp. 303–328.

15. Heshmat, H. and Brewe, D.E., Performance of powder-lubricated journal bearings with dry MoS_2 powder: experimental study of thermal phenomena. *ASME Trans., J. Tribology*, Vol. 117, July 1995, pp. 506–512.

16. Heshmat, H. and Brewe, D.E., Performance of a powder lubricated journal bearing with WS_2 powder: Experimental study. *ASME Trans., J. Tribology*, Vol. 118, July (1996), pp. 484–491.

17. Heshmat, H. The effect of slider geometry on the performance of a powder lubricated bearing—theoretical considerations, submitted for presentation at the *54th Annual Meeting of STLE*, May 1999, publication in *Tribology Transactions*.

18. Heshmat, H. and Heshmat, C.A., The effect of slider geometry on the performance of a powder lubricated bearing—experimental illustrations, presented at the *52nd Annual Meeting of STLE*, Kansas City, May 18–22, 1997, publication in *Tribology Transactions*.

19. Heshmat, H., Experimental determination of powder film shear and damping characteristics. 2000-GT-365, submitted for publication and presentation at the ASME TURBO EXPO, IGTI 2000, Germany, May 2000.
20. Swanson, E.E, Xu, D.S., and Heshmat, H., On the application of powder shear damping to a structural element, presented at the *ASME TURBO EXPO*, IGTI 2000, Germany, May 2000, ASME Paper 2000-GT-0365.
21. Kaur, R.G. and Heshmat, H., On the development of self-acting and self-contained powder lubricated auxiliary bearing, presented at the *2000 ASME/STLE Joint Conference*, Nashville, Tennessee.
22. Iordanoff, I., Berthier, Y., Descartes, S., and Heshmat, H., A review of recent approaches for modeling solid third bodies. *ASME J. Tribology*, October 2002, Vol. 124, No. 4, pp. 725–735.
23. Heshmat, H., Powder-lubricated damper with wavy damper pads, Patent #5,205,384, issued April 27, 1993.

8 Theory of Powder Lubrication

Much of the phenomenological and conceptual work on powder behavior described in the previous chapters can be utilized for calculating the performance of powder-lubricated tribosystems. New approaches are needed, both for the analytical formulation of the problem—such as the rheological behavior of the powder lubricant—as well as to generate numerical subroutines for solving the relevant equations. This, in addition to an altered Reynolds equation, involves expressions for the non-Newtonian behavior of the powder. While these make a solution more difficult, there are also compensations. Thermal effects are here of a lower order than with viscous fluids, and side leakage affects the results less than is the case with liquid lubricants, so that one-dimensional solutions are more applicable to powders than to conventional lubricants.

8.1 ANALYSIS

8.1.1 Basic Equations

Unlike in Newtonian fluids, whose stress–strain relationship is given by $(\frac{\partial u}{\partial y}) = (\frac{1}{\mu_o})\tau$, powders that are made to flow through a thin gap under the action of sliding or rolling exhibit modes of flow behavior that are non-Newtonian. The following class of constitutive rheological relationships can be realized:

$$\dot{\varepsilon} = \frac{1}{G}\tau + \frac{1}{\mu_o}\phi(\tau) \tag{8.1}$$

Here, this relationship is represented by a fifth-order equation of the form

$$\phi(\tau) = \frac{1}{\mu_o}\tau + \alpha\tau^3 + \gamma\tau^5 \tag{8.2a}$$

where α and γ are constants characteristic of a particular powder. Due to variations in degree of compactness, a powder's density and yield shear stress are pressure dependent; namely,

$$P(p) = \rho(o)\left[1 + \frac{P}{K_p} + \cdots\right] \tag{8.2b}$$

A powder's shear stress τ will rise until it reaches some maximum value τ_ℓ, after which it will remain constant, whereupon slip will occur at the surfaces. The shear stress in a powder film will therefore behave as follows:

$$\tau_\ell(p) = \tau_\ell(o)\left[1 + \frac{P}{k_{\tau_\ell}} + \cdots\right] \qquad (8.2c)$$

The foregoing implies that the powder will not shear until a value of τ_y is reached, and it will remain constant at τ_ℓ once this level of shear stress is obtained. This behavior is portrayed in Figure 8.1 where it is seen that the quintic representation of the shear stress adequately reflects these physical conditions. The general flow pattern of a powder film and the associated nomenclature are portrayed in Figure 8.2.

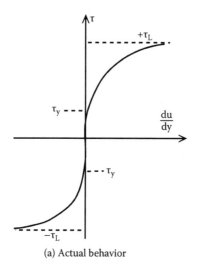

(a) Actual behavior

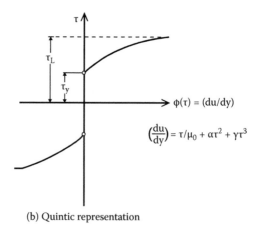

(b) Quintic representation

FIGURE 8.1 Analytical representation of shear rate and shear stress in a powder.

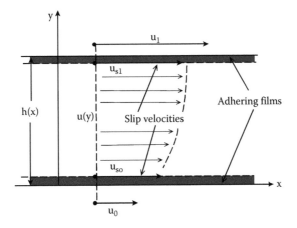

FIGURE 8.2 Velocities in a powder film.

The momentum, continuity, and the aforementioned constitutive equations provide a complete statement of the problem. For a given set of physical inputs, such as

$$U_1, U_o, B, h(x), \tau_\ell(p), \rho(p)$$

it is necessary to solve for u_{so} and u_{s1}—the film velocities at the surfaces (when they do not equal the surface velocities)—the local pressures $p(x)$, shear stress $\tau(x, y)$, and powder flow $Q(x)$. From a proper integration of these quantities, the overall performance of the system can be obtained.

The basic momentum equation when integrated with respect to y yields

$$\tau y = \left(\frac{dp}{dx}\right)y + \tau_o \tag{8.3}$$

From the constitutive equation we have

$$u = u_{so} + \int_0^y \phi[\tau(y')]dy'] \tag{8.4}$$

Applying the second boundary condition, $u = u_{s1}$, we have

$$\int_0^h \phi[\tau(y)]dy = u_{s1} - u_{so} \tag{8.5}$$

The continuity equation may be integrated by parts to yield

$$\int_0^h u\,dy = \left(u_{s_1}\right)h - \int_0^h y\left(\frac{\partial u}{\partial y}\right)dy = \frac{Q_o\rho o}{\rho} \tag{8.6}$$

Hence, from the $(\frac{\partial u}{\partial y}) = \phi(\tau)$ relationship

$$u_{s1}h - \int_0^y y\{\phi[\tau(y)]dy\} = \frac{Q_o\rho_o}{\rho} \qquad (8.7)$$

By integrating the pressure distribution, the resultant load is obtained, and integration of the shear stress yields the total friction. In addition, the energy E dissipated over the surface and an equivalent damping B_e can be obtained as follows:

$$E = B_e(x)xdt'$$

from which

$$B_e = \frac{E}{\omega\pi x_o^2}$$

per cycle, the energy dissipated and the damping then becomes

$$E = 4\pi x_o F_s \qquad (8.8)$$

$$B_e = \frac{4F_s}{\omega\pi x_o} \qquad (8.9)$$

where

$$F_s = \tau(x)dxdy \qquad (8.10)$$

To facilitate the computation and to obtain dimensionless results, the pertinent quantities are normalized as follows:

$$\bar{x} = \frac{x}{B}; \quad \bar{p} = \left(\frac{h_o^2}{\mu_o U_o B}\right)p; \quad \bar{y} = \frac{y}{h}; \quad \bar{\tau} = \left(\frac{h_o}{\mu_o U_o}\right)\tau$$

$$\bar{h} = \frac{h}{h_o}; \quad h* = \frac{2Q_o}{U_o h_o}; \quad \bar{u} = \frac{u}{U_o}; \quad \bar{\phi}(\bar{\tau}) = \left(\frac{h_o}{U_o}\right)\phi\left[\frac{\mu_o U_o}{h_o}\bar{\tau}\right]$$

with the quintic expression for the stress–strain relationship now becoming

$$\bar{\phi}(\bar{\tau}) = \bar{\tau} + \bar{\alpha}\bar{\tau}^3 + \bar{\gamma}\bar{\tau}^5$$

The pressure-dependent properties become

$$\bar{\rho}(\bar{p}) = \left[\frac{\rho(\bar{p})}{\rho(0)}\right]\{1 + \Lambda_\rho\bar{p} + \cdots\}$$

$$\bar{\tau}_\ell(\bar{p}) = \bar{\tau}_\ell(o)[1 + \Lambda_{\tau_\ell}\bar{p} + \cdots]$$

$$\bar{\tau}_y(o)[1+\Lambda_{\tau_y}\bar{p}+\cdots]$$

with

$$\Lambda = \frac{\mu_o U_o B}{h_o^2 K}$$

each Λ and K carrying the appropriate subscript relevant to either ρ, τ_y or τ_ℓ.

8.1.2 METHOD OF SOLUTION

The problem is solved using two primary levels of iteration. The inner level consists of solving Equations 8.3, 8.5, and 8.7, thereby determining the pressure and shear at each location. A Runge–Kutta integration procedure is used here to provide the values of p needed for evaluating the powder properties.

The outer level of iteration determines the value of Q needed to satisfy the pressure boundary conditions. A Mueller iteration procedure is used for the outer level. This requires two initial guesses to bracket the flow rate Q. In general, the flow is bracketed by Couette-type flows at the mean and minimum film thicknesses under no-slip conditions; namely,

$$\frac{\rho_o h_{min}(u_1+u_o)}{2\langle Q_o \rho_o \langle \rho_{max} h_{max} \left(\frac{u_1+u_o}{2}\right)}$$

The inner level, which is to determine $(\frac{dp}{dx})$ and τ_o to satisfy the integrals 8.5 and 8.7, is carried out by means of a Newton–Raphson iteration, the integrals evaluated by 10-point Gaussian quadratures. The internal span $0\langle x\langle B$ is subdivided into N equally spaced points, with the inner-level solution at each x taken from the preceding point. Starting values of τ_o and $(\frac{dp}{dx})$ at $x=0$ are borrowed from known Newtonian solutions with an equivalent viscosity defined as:

$$\frac{1}{\mu_o} = \left(\frac{d\phi}{d\tau}\right)_{\tau=\tau_y}$$

in the following manner:

$$\frac{dp}{dx} = 6\mu_o \frac{[(u_o+u_1)h(0)-2Q_o]}{h^3(0)}$$

$$\tau_o = \frac{-u_o[2u_1+4u_o)h(0)-6Q_o]}{h^2(0)-\tau_y}$$

with the value of τ_y subtracted from the Newtonian shear stress.

To start with, it is assumed that the limiting shear stress is not reached and the boundary velocities are thereby prescribed by the surface velocities. Because the shear stress varies linearly across the film, it is simple to determine the location of any point within the film where the yield stress is reached. Regions where the shear stress is below τ_o can then be excluded. In this manner, the inner-level solutions are obtained. If the limiting shear stress is exceeded, than noting that $\tau_1 = \tau_o + h(\frac{dp}{dx})$, one can consider Equations 8.5 and 8.7 to be functions of four variables, τ_o, τ_1, u_{so}, and u_{s1} of which two will be known. If τ_L is exceeded, a third solution must be obtained for the two film boundary velocities. The procedures involved in the foregoing two iterations are portrayed schematically in Figure 8.3.

Based on the foregoing equations, two sets of solutions will be derived. One deals with planar slider bearings operating under conditions of no-slip and

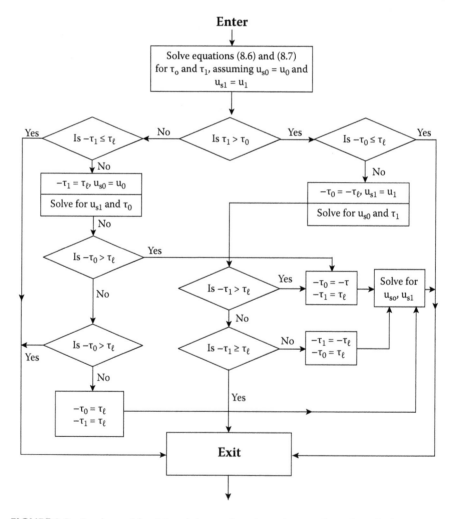

FIGURE 8.3   Logic used for determining surface shear stress and powder velocity.

boundary-slip velocities, the other pertaining to crowned surfaces as they apply to powder-lubricated dampers.

8.2 SLIDER BEARINGS

The sliders to be analyzed, shown in Figure 8.4, are lubricated with a powder having the properties of a titanium dioxide (TiO_2). The lubricant is thus assigned the following characteristic constants:

$$\alpha = 218; \quad \gamma = 2.14 \times 10^6; \quad \rho_O = 2.2\,\text{g/cc}; \quad \mu = 400\,\text{cps}; \quad \tau_y = 20\,\text{kPa}; \quad \tau_L = 724\,\text{kPa}$$

The film geometry h can range from slightly over 1—a nearly parallel surface—to $h = 10$, thus, covering the widest possible range of tapered surfaces. The velocity of the runner is taken to be $U_o = 4.6\,\text{m/s}$, with the slider at a standstill or $U_1 = 0$.

8.2.1 Solutions for No-Slip Conditions

Analytical results for the various slider performance quantities are given in Figures 8.5 to 8.11. Pressures levels are seen to rise with an increase of slider taper. However, as shown in Figure 8.5, one of the most interesting results—of crucial relevance to our concern with the operation of parallel surfaces—is that even for a practically zero taper ($h = 1.0001$), a measurable pressure profile and corresponding load capacity are generated in the bearing. From an optimization standpoint, load capacity levels off at a maximum near $h = 3$, whereas the coefficient of friction reaches a minimum at about the same value (Figure 8.6). Consequently, a favorable design for both load capacity and least power dissipation dictates an h value in the range of 3 to 4.

One of our concerns with powder lubrication is the rate of shear at the two surfaces, directly related to the velocity gradients at the surfaces. Figures 8.7 to 8.9 provide sample plots illustrating the effect of slider convergence on the velocity distribution. At low values of U, a Couette-type flow predominates, and no slip occurs.

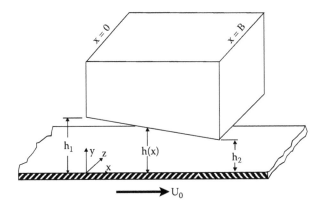

FIGURE 8.4 Geometry and nomenclature in a slurry film.

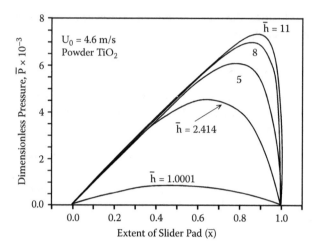

FIGURE 8.5   Effect of $\bar{h}$ on pressure profile.

As the convergence increases, the gradients, too, rise until, at $h = 10$, there is incipient slip at both edges of the slider. The corresponding levels of shear stress for the moving and stationary surfaces are given in Figures 8.10 and 8.11.

The additional points of interest that emerge from the foregoing plots are as follows:

- Shear increases with h but beyond $h = 8$, no further increase takes place.
- Maximum shear occurs at the trailing end, where, as noted, there is incipient slip.

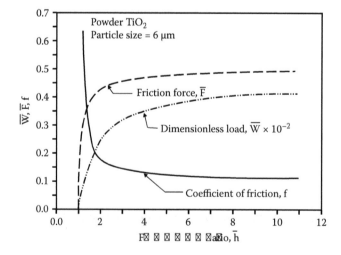

FIGURE 8.6   Effect of $\bar{h}$ on load and friction.

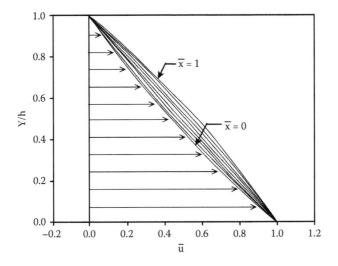

FIGURE 8.7 Velocity profile for $\bar{h} = 1.2$.

- For a parallel configuration, the shear is constant at both surfaces.
- The shear is slightly higher at the stationary than at the moving surfaces.

8.2.2 SOLUTIONS WITH SLIP AT THE BOUNDARIES

Two solutions were obtained for the case of $h = 10$ under conditions conducive to slip at the walls. This is brought about by reaching stress levels above the limiting shear

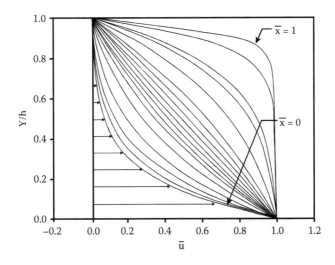

FIGURE 8.8 Velocity profile for $\bar{h} = 5$.

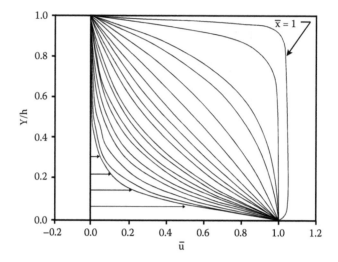

FIGURE 8.9   Velocity profile for $\bar{h} = 10$.

stress τ_L. The detailed results are given in Tables 8.1 and 8.2. The two runs presented in the tables were made for the same input parameters except that those in Table 8.2 used a higher yield stress, namely, $\tau_y = 0.414$ versus the $\tau_y = 0.276$ of Table 8.1. The results are plotted in Figures 8.12–8.14, which indicate the following:

- A higher stress yield point creates more favorable conditions for slip to occur.
- The pressure profiles assume an almost rectangular shape with nearly constant pressures throughout the region.

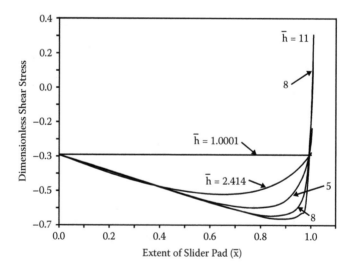

FIGURE 8.10   Effect of $\bar{h}$ on shear stress near moving wall.

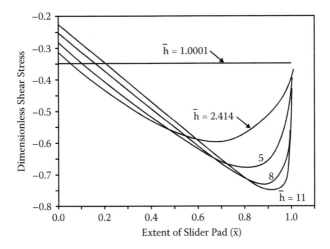

FIGURE 8.11 Effect of $\bar{h}$ on shear stress near stationary wall.

- The level of pressure is considerably below that for the case of a low-stress yield point due to the more efficacious hydrodynamic effects in the absence of slip.
- Taken as an average, the levels of stress are higher for the high value of τ_y, whereas for $\tau_y = 0.276$, half of the film was under a stress below the ultimate

TABLE 8.1
Solution for Slider Bearing with $\bar{h} = 10$ and $\bar{\tau}_y = 0.276$

x/L	$\bar{P}$	$\bar{Q}$	$\bar{\mu}_0$	$\bar{\mu}_0$	$\bar{\tau}_0$	$\bar{\tau}_1$
0.0	0	10.0	1.0	0	−0.329	−0.237
0.05	4.6×10^{-4}	9.12	1.0	0	−0.351	−0.267
0.10	9.2×10^{-4}	8.29	1.0	0	−0.373	−0.297
0.15	1.37×10^{-3}	7.50	1.0	0	−0.395	−0.326
0.20	1.83×10^{-3}	6.67	1.0	0	−0.417	−0.355
0.30	2.75×10^{-3}	5.41	0.972	0	−0.461	−0.413
0.40	3.11×10^{-3}	4.24	0.506	0	−0.461	−0.413
0.50	3.12×10^{-3}	3.25	0.525	0.087	−0.461	−0.461
0.60	3.12×10^{-3}	2.44	2.572	0.243	−0.461	−0.461
0.70	3.12×10^{-3}	1.81	0.672	0.427	−0.461	−0.461
0.80	3.12×10^{-3}	1.36	0.823	0.64	−0.461	−0.461
0.85	3.12×10^{-3}	1.20	0.908	0.746	−0.461	−0.461
0.90	3.12×10^{-3}	1.09	0.986	0.839	−0.461	−0.461
0.95	3.12×10^{-3}	1.02	1.04	0.903	−0.461	−0.461
1.00	$−1.75 \times 10^{-4}$	1.00	1.00	0	−0.324	−0.380

TABLE 8.2

Solution for Slider Bearing with $\bar{h}=10$ and $\bar{\tau}_y=0.414$

x/L	$\bar{P}$	$\bar{Q}$	$\bar{\mu}_0$	$\bar{\mu}_0$	$\bar{\tau}_0$	$\bar{\tau}_1$
0.0	0	10.0	0.735	0	−0.46	−0.40
0.05	2×10^{-4}	9.12	0.350	0	−0.46	−0.44
0.10	2.83×10^{-4}	8.29	0.284	0	−0.46	−0.45
0.15	3.13×10^{-4}	7.50	0.276	0	−0.46	−0.46
0.20	3.17×10^{-4}	6.67	0.283	0.013	−0.46	−0.46
0.30	3.17×10^{-4}	5.41	0.293	0.077	−0.46	−0.46
0.40	3.17×10^{-4}	4.24	0.320	0.150	−0.46	−0.46
0.50	3.17×10^{-4}	3.25	0.372	0.240	−0.46	−0.46
0.60	3.17×10^{-4}	2.44	0.458	0.360	−0.46	−0.46
0.70	3.17×10^{-4}	1.81	0.588	0.520	−0.46	−0.46
0.80	3.17×10^{-4}	1.36	0.762	0.710	−0.46	−0.46
0.85	3.17×10^{-4}	1.20	0.855	0.810	−0.46	−0.46
0.90	3.17×10^{-4}	1.09	0.938	0.890	−0.46	−0.46
0.95	3.17×10^{-4}	1.02	0.998	0.960	−0.46	−0.46
1.00	1.9×10^{-4}	1.00	1.06	0.940	−0.46	−0.46

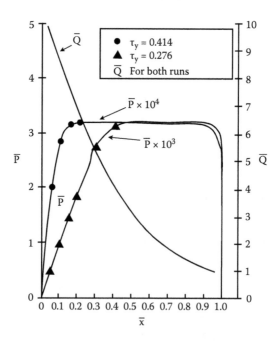

FIGURE 8.12   Pressure and flow with slip.

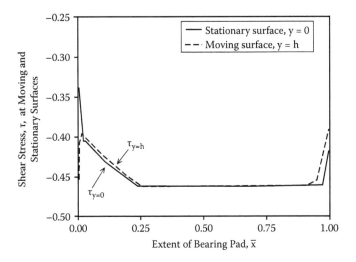

FIGURE 8.13 Levels of shear stress for $\bar{h} = 10$.

value of $\tau_o = 0.4$ For $\tau_y = 0.414$, both surfaces were at the maximum stress level of 0.46 throughout the film.

• The general pattern of flow is shown in Figure 8.14, with slip occurring at both moving and stationary surfaces.

A comparison was also made between powder- and petroleum-oil-lubricated sliders having the same viscosities of 14 cps. The operating conditions were the same

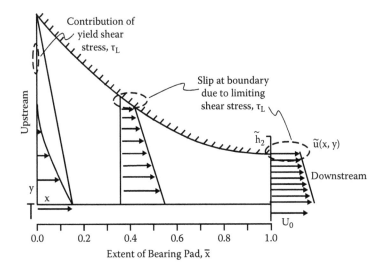

FIGURE 8.14 Schematic of slip flows at the boundaries.

TABLE 8.3

Performance of Tapered Land Slider with Powder and Liquid Lubricant as a Function of $h_1 = h_1/h_2$, $\tau_1 = 6.97$; $\tau_y = 6.13$; $U = 4.6$ ms^{-1}; $L \times B \times h_2 = 25.4 \times 25.4 \times 0.0254$ mm

		TiO$_2$ Powder $\eta = \mu_0$			Oil $\mu_0 = 13.79 \times 10^{-3}$ Pa s		
$\bar{X}$		P (kPa)	F_f (N)	Q (m^3 s^{-1}) $\times 10^{-6}$	P (kPa)	F_f (N)	Q (m^3 s^{-1}) $\times 10^{-6}$
0.01	1.05	0	11.68	1.558	46.2	4.90	1.557
0.02	1.5	111.7	11.68	1.901	280.6	4.20	2.147
0.023	1.75	189.5	11.45	4.064	339.7	3.94	2.491
0.027	2.0	216.8	11.41	4.261	375.8	3.94	2.819
0.04	3.0	278.1	11.30	5.113	422.0	3.13	4.130
0.053	4.0	281.3	11.25	5.293	417.2	2.78	5.441
0.08	6.0	280.2	11.19	5.473	379.2	2.38	8.095
0.107	8.0	279.4	11.16	5.555	335.8	2.14	10.65
0.13	10.0	275.8	11.16	5.572	304.1	1.99	13.27
0.2	15.0	268.9	11.12	2.819	232.4	1.73	19.83
0.53	40.0	268.0	11.12	2.851	102.1	1.28	52.44
1.0	75.0	262.01	11.10	2.850	53.4	1.07	98.81

as in the previous cases except for somewhat different rheological properties of the powder, namely,

$$\tau_y = 6.13 \text{ and } \tau_L = 6.97, \text{ where } \tau = \frac{h_2 \tau}{\mu_o U_o}$$

The resulting pressures—friction forces and flows—are listed in Table 8.3, where it is seen that though the oil produces the familiar bell-shaped pressure distribution, the powder yields almost constant values of pressure and friction throughout the region. Thus, while the maximum pressure for the oil was about 420 kPa, the powder's maximum hovered in the vicinity of 270–280 kPa. Consequently, as shown in Figure 8.15, the load capacity of a powder with an equivalent viscosity as that of a liquid lubricant produces lower load capacities over the entire region of slider tapers from $h = 1.5$ to 10. However, it should be realized that, due to its inherent properties and thermal effects during operation, it would difficult to maintain an oil at a high constant viscosity throughout the bearing region, whereas this would be inherent in a powder.

8.3 CROWNED SLIDERS

In some applications, the tribological surfaces, instead of being planar, have some sort of curvature. In most case, the shape s is either circular with a large radius of curvature or parabolic. One such application, shown in Figure 8.16, is a turbomachinery damper where the physical elements have curved surfaces separated by a

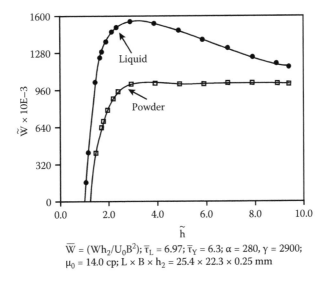

$$\overline{W} = (Wh_2/U_0B^2); \overline{\tau}_L = 6.97; \overline{\tau}_Y = 6.3; \alpha = 280, \gamma = 2900;$$
$$\mu_0 = 14.0 \text{ cp}; L \times B \times h_2 = 25.4 \times 22.3 \times 0.25 \text{ mm}$$

FIGURE 8.15 Comparison of slider performance with powder and oil lubrication.

lubricant—in our case, a powder—enabling relative motion so as to produce damping in the system.

The analyzed damper was operated with the previously mentioned titanium dioxide having the following properties:

$$\alpha = 280; \quad \gamma = 2900; \quad \mu_o = 2010^{-3} \text{ kPa}; \quad \text{and} \quad h = 3$$

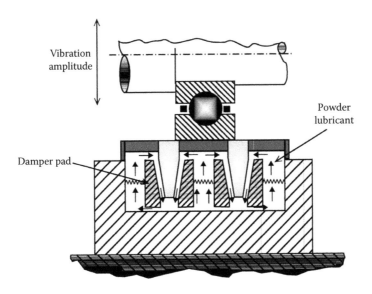

FIGURE 8.16 Powder-operated damper.

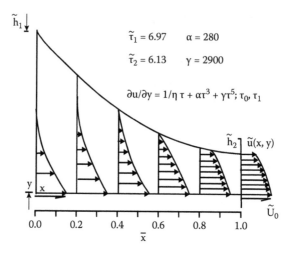

FIGURE 8.17 Velocity profiles in crowned slider.

Here, unlike in a slider bearing, the pressure boundary conditions at $p = 1$ at the inlet to the damper and $p = 0$ at the outlet.

The resulting velocity distribution is shown in Figure 8.17, where it is seen that there is a stagnant region at the inlet to the damper and a slip velocity, both at the moving and the stationary surfaces at the outlet. The effect of τ_y dominates at the leading edge, while the limiting shear stress τ_L dominates the flow at the exit. Figure 8.18 shows the resulting total force on the element, as well as the traction, both of which

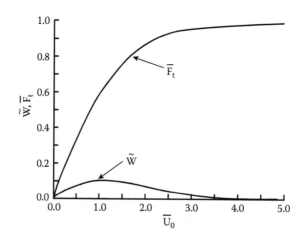

FIGURE 8.18 Load and traction in crowned slider as function of speed.

are given in normalized form by the expressions

$$W = \left(\frac{h_o}{\mu_o U_o B^2} \right) W \quad \text{and} \quad F_t = \left(\frac{h_o}{\mu_o U_o B^2} \right) F_t$$

From the latter plots, it seems that, unlike in conventional hydrodynamic films, an increase in U_o beyond 1.5 has a degrading effect on the load capacity. Due to the impact of the limiting shear stress, the traction approaches an asymptotic value with increasing speed. The consequence for damper performance is that a single element will have a decreasing damping ability with a further increase in speed. The solution lies in having a series of damping elements arranged in parallel, experiencing different speeds at various stages of the vibrational cycle.

REFERENCES

1. Heshmat, H., Development of rheological model for powder lubrication, NASA Lewis Research Center, CR 189043, October 1991.
2. Heshmat, H., The quasi-hydrodynamic mechanism of powder lubrication, Part II: Lubricant film pressure profile, *Proc. STLE,* Vol. 48, No 5, 1992.
3. Heshmat, H. and Walton, J.F., The basics of powder lubrication in high temperature powder lubricated dampers, *ASME J. Tribology*, Vol. 115, April 1993.

9 Granular Films

9.1 CONTINUUM APPROACH

Probably the most esoteric candidate for acting as an interface layer in tribological systems is an aggregate of discrete particles. Though not a continuum, such a layer, due to the interaction of each granule with its neighbors, generates tangential and normal stresses and is thus capable of supporting a load. While at one end, such a "lubricant" clearly abuts a powder film from which it differs essentially only in particle size, at the other extreme one is tempted to connect such a chain of particles with the balls or cylinders of rolling element bearings. We say "tempted" because, while physically there is good similarity between translating layers of granules and the procession of balls in a rolling element bearing, their mechanisms of operation are quite different. Still, when the nature of the tribological continuum is brought to full maturity, perhaps here, too, one would find a number of features common to these two regimes.

9.1.1 ANALYSIS

Some of the features assigned to the particles making up a granular film are that they are nearly spherical in shape, are free of intergranular cohesion, and the spacings between the particles are small compared to their diameter. While their separation is taken to be small, it is also postulated that it is never zero. The latter statement is to be understood in the sense that, while individual particles may contact each other, when a cell or a discrete volume of particles is considered, their average distance will never be zero; that is, they will not flocculate en masse. When the foregoing holds, the density of any part of the film—given by ρ $m/(s + d)^3$—can be written as ρ m/d^3, yielding a constant density. Given that in most cases the interstitial fluid is usually air, the equivalent viscosity assigned to the granules is so much larger than that of the intervening fluid it may be ignored.

The nomenclature employed in the subsequent equations is as follows:

a	Radius of particle
B	Extend of slider
B	Granular temperature at boundary
$\bar{B}$	$\dfrac{B}{U}$
c	Solids fraction
C_f	Coefficient of friction
d	Particle diameter

$\bar{d}$	$\dfrac{d}{h}$ in parallel plates
$\bar{d}$	$\left(\dfrac{d}{h_1}\right)$ in sliders
e	Coefficient of restitution
E	Energy fluctuation constant
fac	Dissipation constant
F	Intergranular constant
$F_{n,t}$	$\dfrac{F_{n,t}a}{mU_o^2}$
fr	See Equation 9.26
h	Film thickness
h_1	Inlet film thickness
h_2	Outlet film thickness
$\bar{h}$	$\left(\dfrac{h_2}{h_1}\right)$
hr	Particle hardness ratio
ΔH	Normal dilatation
I	Rate of energy loss
$k_{1,2\ldots}$	Constants
K	Thermal diffusivity
L	Length of granule
m	Mass
N, n	Number of granules
p	Pressure
$\bar{p}$	$\left(\dfrac{p}{\rho U^2}\right)$
P_o	Pressure at inlet
P_B	Pressure at outlet
Q	Volumetric flow rate
r	Separation of granule centers
s	Distance between granules
t	Time
u	Granule velocity
U_o	Surface velocity
U	Total velocity in granular motion
V	Volume
$\mho$	Linear velocity in granular motion
w	Fluctuation (thermal) velocity
$\bar{w}$	$\dfrac{w}{U}$
W	Load capacity

$\bar{W}$	$\dfrac{W}{BP_o}$
x	Coordinate in direction of motion
$\bar{x}$	$\left(\dfrac{x}{B}\right)$
x_{cp}	Location of pivot
y	Coordinate normal to motion
$\bar{y}$	$\left(\dfrac{y}{h_1}\right)$
μ	Film viscosity
σ	Shear stress
ρ	Density
τ	Collision time
ϕ	Surface roughness
ω	Particle rotational frequency

Subscripts

o	At lower surface
1	At upper surface
n, N	Normal
t, T	Tangential
i, j	Individual particles

One of the more comprehensive analyses of granular films is that given by Haff (4). As in ordinary fluid mechanics, the conservation of mass when the density is constant is given by the Navier–Stokes equations for Newtonian fluids if here, too, we assume that the velocity varies gradually from point to point and that the "viscosity"—still to be defined—arises from this very notion; we can then write

$$\frac{\partial}{\partial t}(\rho u_i) = -\frac{\partial}{\partial x_k}\left[p\delta_{ik} + \rho u_i u_k - \eta\left(\frac{\partial u_i}{\partial x_k} + \frac{\partial u_k}{\partial x_i}\right)\right] + \rho g_i \qquad (9.1)$$

where $p = p(x,t)$ is the pressure and g_i is the gravitational acceleration, the quantity η being considered the viscosity coefficient.

The conservation of energy, likewise, patterned on the hydrodynamic energy equation, is

$$\frac{\partial}{\partial t}\left(\frac{1}{2}\rho u^2 + \frac{1}{2}\rho v^{-2}\right) = -\frac{\partial}{\partial x_k}\left[\rho u_k\left(\frac{p}{\rho} + \frac{1}{2}u^2 + \frac{1}{2}v^{-2}\right) - u_i\eta\left(\frac{\partial u_i}{\partial x_k} + \frac{\partial u_k}{\partial x_i}\right)\right.$$

$$\left. - K\frac{\partial}{\partial x_k}\frac{1}{2}\rho v^{-2}\right] + \rho u_i g_i - I \qquad (9.2)$$

Here, the fluctuation velocity v is the analog of the thermal velocity in fluids and the coefficient K acts as thermal diffusivity. In the preceding equation consisting of a kinetic energy $(\frac{1}{2})\rho u^2$ and an internal energy $(\frac{1}{2})\rho v^2$, the second term corresponds roughly to the temperature of the system. The quantity I in the energy equation represents the amount of energy lost due to the inelastic collisions between the particles.

Before a solution can be attempted, expressions must be derived for the granular "viscosity" η, the thermal diffusivity coefficient K, and the energy sink I.

9.1.1.1 Granular Viscosity

The gradient of flow velocity u is taken to be in the y direction only. When grain collisions occur between particle layers, a net momentum $(m\Delta u)$ is transferred in the direction of flow (x). Because the collision rate is $(\frac{v}{s})$, the shear stress exerted by the upper on the lower layer in the x direction is

$$\sigma \sim \frac{m\Delta u}{d^2}\frac{\bar{v}}{s} \tag{9.3}$$

The velocity increment Δu is the change in u over a distance of the order d, so that $(\frac{\Delta u}{d})$ is proportional to $(\frac{du}{dy})$. The foregoing expression can thus be rewritten as

$$\sigma = qd^2\rho\frac{\bar{v}}{s}\frac{du}{dy} \tag{9.4}$$

where q is a constant. Comparing the foregoing equation with the viscous stress term in Equation 9.1 yields, for the viscosity coefficient,

$$\eta = qd^2\rho\frac{\bar{v}}{s} \tag{9.5}$$

This granular viscosity coefficient depends on the thermal velocity w contained in a solution of the energy equation, thereby coupling the momentum and energy equation via the foregoing viscosity term.

9.1.1.2 Thermal Diffusivity Coefficient

The energy flux due to grain–grain collisions is the average net energy transferred per collision multiplied by the collision rate divided by the area. The mean energy transfer $(m\bar{v}\Delta\bar{v})$ is, thus, the energy flux

$$Q \sim \frac{m\bar{v}\Delta\bar{v}}{d^2}\frac{\bar{v}}{s} \tag{9.6}$$

$$Q = -rd^2\frac{\bar{v}}{s}\frac{d}{dy}\left(\frac{1}{2}\rho\bar{v}^{-2}\right) \tag{9.7}$$

where r is a constant. The term in Equation 9.2 involving K is just the divergence of the internal energy flux or

$$K = rd^2 \frac{\bar{v}}{s} \tag{9.8}$$

Because in Equation 9.2 this term does not involve the velocity u, this heat term is present even in the absence of granular motion.

9.1.1.3 Collisional Energy Sink

With e, the coefficient of restitution and the relative collision velocity given by v, the loss of energy per collision is then

$$\Delta E \sim (1 - e^2) \left[\frac{1}{2} m \bar{v}^2 \right] \tag{9.9}$$

Multiplying the collision rate ($\frac{\bar{v}}{s}$) and the volume particle number n, yields the rate at which energy is lost per unit volume per unit time

$$I = \gamma \rho \left(\frac{\bar{v}^3}{s} \right) \tag{9.10}$$

where γ is a dimensionless factor proportional to $(1 - e^2)$.

The foregoing equations are now applied to flows between two parallel plates with the upper surface moving at a velocity U. Because the viscosity is now a function of only y, the y component of the momentum equation yields $\frac{dp}{dy} = 0$ and a constant pressure equal to

$$p = p_o = td\rho \frac{\bar{v}^2}{8} \tag{9.11}$$

where t is a constant. With the shear stress given by

$$\sigma = \sigma_o = \eta \frac{du}{dy} \tag{9.12}$$

the energy equation reduces to

$$0 = \frac{\partial}{\partial x_k} \left[u_i \eta \left(\frac{\partial u_i}{\partial x_k} + \frac{\partial u_k}{\partial x_i} \right) + K \frac{\partial}{\partial x_k} \frac{1}{2} \rho \bar{v}^2 \right] - I \tag{9.13}$$

The pressure expression can be used to eliminate powers in $\bar{v}^2$ in the energy equation, while Equation 9.12 eliminates its dependence on u. Thus, the energy equation becomes

$$\frac{d^2 \bar{v}}{dy^2} + \omega^2 \bar{v} = 0 \tag{9.14}$$

Because the frequency ω is an unknown, the preceding differential equation has to be treated as an eigenvalue problem.

The flow boundary conditions in the present solution are taken to have no-slip velocities, while the fluctuation velocities are taken to be zero at the boundaries; that is,

$$u(0) = 0, \quad u(h) = U \quad \text{and} \quad \bar{v}(0) = \bar{v}(h) = 0$$

With no fluctuation velocities at the boundaries (such as may be induced by plate
vibration), the only energy generated is by friction in the internal flow. For this energy
to be dissipated across the boundaries, a negative $\frac{d\bar{v}}{dy}$ is required at the surfaces. This,
in turn, demands that in Equation 9.14, ω^2 be positive.

The thermal velocity is given a solution in the form

$$\bar{v} = a \sin \omega y \tag{9.15}$$

where a is a constant and $\omega = \frac{\pi}{h}$.

The use of the foregoing boundary conditions and the appropriate integrations are
detailed in Haff (4), who obtains for the flow and thermal velocities the following
expressions:

$$u(y) = \frac{1}{2} U \left[1 - \cos\left(\frac{\pi y}{h} \right) \right] \tag{9.16}$$

$$\bar{v}(y) = \frac{1}{2} \left(\frac{q}{r} \right)^{1/2} \frac{1}{\left(\frac{1+h^2}{\pi^2 \lambda^2} \right)^{1/2}} U \sin \frac{\pi y}{h} \tag{9.17}$$

The preceding two velocity profiles are sketched in Figure 9.1.

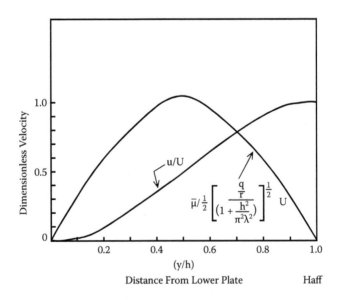

FIGURE 9.1   Velocity and "temperature" profiles in Couette flow.

9.1.2 Experimental Results

There is very little direct experimental data on the lubrication capacity of granular films, if for no other reason than the whole field is relatively new. One of the basic questions, and one that is central to the nature of granular lubrication, is whether such a layer can produce load-supporting pressures. The experiments conducted by Yu et al. (1) are aimed directly at providing such evidence.

The test rig used in the foregoing experiments is shown in Figure 9.2. The shear cell consisted of two concentric disks, the bottom one being mounted on a rotating shaft. The top disk was restrained by a torque arm to measure friction; it is also free to move vertically to accommodate any expansion or contraction of the granular film during operation. The dimensions of the circular test chan-nel are shown in Figure 9.2b. The granular beads were made of glass having a density of 2550 kg/m³ with diameters between 0.71 and 0.85 mm, and an aver-age $d = 0.787$ mm. The height of the film was $h = 4$ mm, yielding an $(\frac{h}{d})$ ratio of 5.1, which was retained throughout the tests. The solid fraction of 0.6 was, like-wise, kept constant throughout the tests. The rotational speed was in the range of 100–700 rpm, giving linear speeds between 0.4 and 2.8 m/s. Nominal shear rates corresponded to 100–700 (1/s). The forces applied to the top disk ranged from 17.8 to 44.5 N, producing average pressures in the range of 2250–5550 Pa. One of

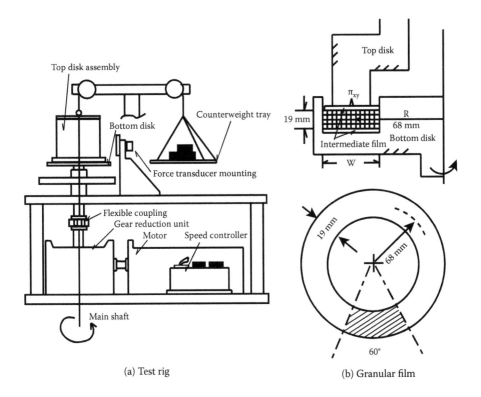

(a) Test rig   (b) Granular film

FIGURE 9.2   Test apparatus for investigating granular shear.

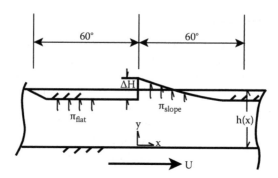

FIGURE 9.3 Geometries of the granular films.

the items discussed in the analysis by Haff is that a granular layer distinguishes between a smooth boundary and one composed of granules identical to the film itself. Consequently, the tests were conducted with both smooth surfaces and with the bottom surface lined with the beads making up the main layer. They, indeed, produced different results.

The study included both parallel and crowned sliders, as shown in Figure 9.3. Both geometries were mounted on the same annular ring. Each section—the parallel and crowned—extended over an arc of 60°. The shape of the crowned section was that of a cubic polynomial, with the trailing edge parallel to the runner. Three heights for the entry film were used—$a\Delta h$ of 1.29, 1.9, and 2.34 mm—giving $(\frac{\Delta h}{B})$ values of 0, 0.0162, 0.0239, and 0.0293.

The results of the experiments plotted in Figures 9.4 and 9.5 show the frictional stresses and the normal stresses, that is, the generated pressures. In general, the pressures are proportional to the shear rate squared. They are also proportional to the square of slider slope. This tallies with the theoretical predictions, as does the difference in the developed stress for smooth and rough surfaces. The coefficient of friction, based on the ratios of the stresses, is 0.3 to 0.4, which falls between the values of solid bodies and boundary lubrication. Again, it should be noted that parallel surfaces using granular films produce a load capacity not much different from that of a sloped surface.

9.1.3 COUETTE FLOW

Based on the conceptual work done by Haff, a complete analysis of Couette flow was undertaken by Mc Keague and Khonsari (6,7). This included slip velocities and the presence of granular "temperatures" and granular roughness at the boundaries of the channel. The mathematics involved are lengthy and still in the formative stage. Also, similar to Haff's work, the equations contain a number of unspecified constants. For these reasons, only some of the final results will be quoted here, while the focus will be on the performance characteristics of the system under discussion.

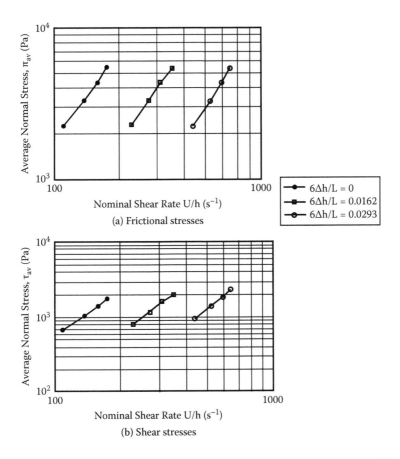

FIGURE 9.4 Stresses in Couette flow with smooth surfaces: experiment. $L = B$; $C = 0.6$; $(n/d) = 5.1$

It was shown earlier that the pseudo-energy equation for granular films has the form of

$$p = td\rho\frac{v^2}{s}; \quad \mu = qd^2\rho\frac{v}{s}; \quad Q_t = -rd^2\frac{v}{s}\frac{\partial}{\partial y}\left(\frac{1}{2}\rho v^2\right); \quad I = \gamma\rho\frac{v^3}{s}$$

where t, q, and r are dimensionless coefficients, $\gamma = 1 - e^2$, and P is the density of the grain particle.

Using, for the boundaries, the thermal conditions of

$$v(0) = B_o \, v(h) = B_h$$

the solution to the energy equation in dimensionless form is given by

$$v(x, y) = C_1(x) \cosh(\lambda y) + C_2(x) \sinh(\lambda y) \tag{9.18}$$

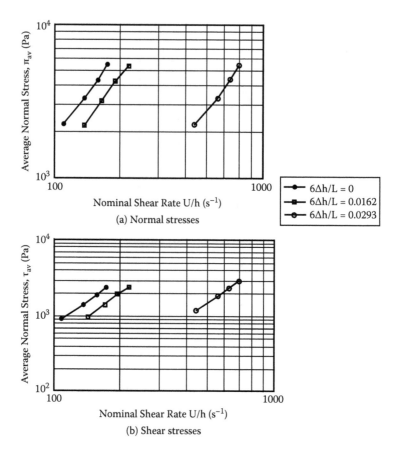

FIGURE 9.5 Stresses in Couette flow with rough stationary flow—$L = B$; $c = 0.6$; $(h/d) = 5.1$.

where

$$\lambda = \frac{1}{d}\sqrt{\frac{\gamma}{r}}, \quad C_1(x) = B_o(x), \quad C_2(x) = \frac{B_h(x) - B_o \cos h(\lambda h)}{\sinh(\lambda h)}$$

and $h = h(x)$ is the gap width between the top and bottom boundaries.

The solution for the granular velocity with slip and roughness at both boundaries yields

$$\bar{u}(\bar{y}) = \frac{1}{\lambda}[\sinh(\bar{\lambda}y) + \bar{G}_4 \cosh(\bar{\lambda}\bar{y})]\bar{C}_3 + \bar{C}_4 \tag{9.19}$$

with $\bar{G}_4, \bar{C}_3,$ and $\bar{C}_4$ being constants.

Before proceeding to give an expression for the variation of pressure, which does occur when the present boundary conditions are used, an additional feature of granular flow must be mentioned. This is the phenomenon of film dilatation in the normal direction due to the "thermal" effects of granular fluctuation. This expression, denoted Δh, is given by

$$\Delta h + \frac{3}{d}\int_o^h s(y)dy \qquad (9.20)$$

The expression for film pressure then contains this term and is given in dimensionless form by

$$\bar{p} = \frac{3t}{\Delta h}\bar{F} \qquad (9.21a)$$

where

$$\bar{F} = \frac{\bar{d}}{2}\sqrt{\frac{r}{\gamma}\left[\bar{B}_h^2 - 2\bar{B}_o\bar{B}_h\cos h(\bar{\lambda}) + \bar{B}_0^2\cos h(2\bar{\lambda})\right]}\frac{\cos h(\bar{\lambda})}{\sin h(\bar{\lambda})}$$

$$+\frac{1}{2}\left[\frac{2\bar{B}_o\bar{B}_h\cos h(\bar{\lambda}) - \left(\bar{B}_0^2 + \bar{B}_h^2\right)}{\sin h^2(\lambda h)}\right]$$

$$+\bar{d}\sqrt{\frac{r}{\gamma}\left(\bar{B}_o\bar{B}_h - \bar{B}_0^2\cos h(\bar{\lambda})\right)}\sin h(\bar{\lambda}) \quad \text{and} \quad \bar{\lambda} = \bar{\lambda}\bar{h} = \frac{1}{\bar{d}}\sqrt{\frac{\gamma}{r}} \quad (9.21b)$$

An additional aspect considered here is that, similar to the velocity, the solids fraction, too, varies here across the film. This variation is shown to be

$$v(y) = \frac{1}{2}\left[-\frac{1}{2.5}\left(\frac{G_s(y)}{v\max}+1\right)+\sqrt{\left(\frac{1}{2.5}\left(\frac{G_s(y)}{v\max}+1\right)\right)^2+4\frac{G_s(y)}{2.5}}\right] \qquad (9.22)$$

where $G_s(y) = \frac{d}{s(y)}\frac{\pi}{4}(1+e)$

A numerical evaluation of the foregoing equations was performed for a set of parameters, starting with a benchmark set of characteristics given in Table 9.1. From this, variations in a number of parameters were conducted to see their effects on a granular film. Some of the other fixed parameters are as follows:

$$\frac{h}{d} = 14; \quad e = 0.8; \quad \Phi = 0 \text{ to } 1; \quad \frac{\Delta H}{h} = 0.1; \quad c = 0.65$$

Some results for these benchmark quantities are given in Table 9.2 and Figure 9.6. In these results, the two surfaces were given equal boundary conditions. The plots in Figure 9.6 are, therefore, all symmetrical. The elevated granular temperatures at the boundaries imply that they are more loosely packed, and this is reflected in the

TABLE 9.1
Benchmark Simulation Parameters

Parameter	Value	Parameter	Value
A	1.000	G_0	0.80
t	2.000	A_0	0.20
q	0.036	G_h	0.80
r	1.000	A_h	0.20
a_1	1.000	h	$14 \times d$
d	5.0E-6 m	ΔH	$0.1 \times h$
ϕ_0	0.20	U	4.6 m/s
ϕ_h	0.80	C	0.649

TABLE 9.2
Dimensionless Pressure, Shear Stress, and Friction Factor Results

Figure	$\bar{p}$	$\bar{\tau}$	f
9.6 (Benchmark)	0.1746	0.0583	0.3340
9.7 ($h = 7 \times d$)	0.9439	0.2304	0.2441
9.7 ($h = 40 \times d$)	0.0331	0.0124	0.3747
9.8 ($e = 0.40$)	0.1348	0.0491	0.3644
9.8 ($e = 0.99$)	0.9256	0.1453	0.1570
9.9	0.6138	0.0910	0.1482

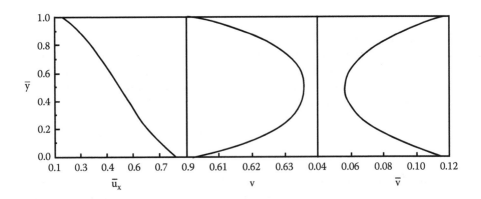

FIGURE 9.6 Performance of benchmark model—flow velocity, $\bar{u}$; granular temp, v; and solids fraction, $\bar{v}$, profiles.

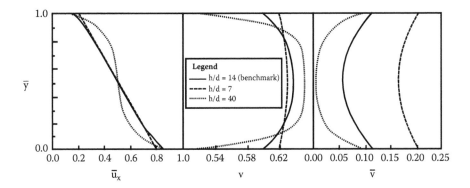

FIGURE 9.7 Effect of varying film height.

distribution of solids fraction across the film. The dimensionless pressure obtained here is $p = 0.175$, while the shear stress is $\tau = 0.058$, which yields a friction coefficient of $f = 0.34$. This is seen to be close to that obtained by Haff, as well as by other investigators. From this basic set of variables, the effects of the following parameters were plotted:

- Effect of film thickness h. Figure 9.7 shows the effects of changing the height of the granular film. The solid lines in the plots are the reference data, with the bracketing plots representing larger and smaller h's. As seen by its velocity curve, the larger film forms a sort of plug flow in the center. The solids fraction, likewise, achieves an almost constant value across the film. With a narrow gap, the velocity profile is nearly linear with little variation in the solids fraction. Also, with a decreasing gap, the generated pressure increases, as is clear from a comparison with the data of Table 9.2.
- Effect of coefficient of restitution e. Decreasing the value of e implies an increase in the inelasticity of granular collisions and, therefore, a higher loss of energy. This leads to a larger variation in granular "temperature" and solids fraction, as portrayed in Figure 9.8. For high elastic collisions,

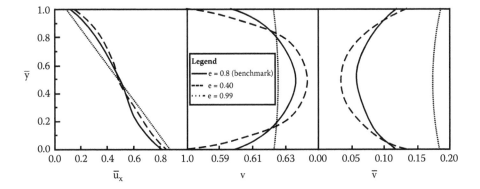

FIGURE 9.8 Effect of varying restitution constant.

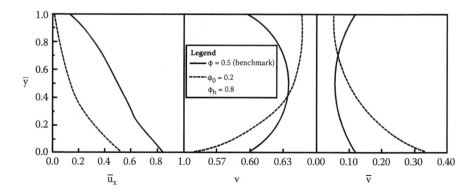

FIGURE 9.9 Effect of varying roughness on stationary surface.

there is little energy lost, and the particles retain their momentum, resulting
in nearly constant profiles for "temperatures" and solids fraction.

- Variation in surface roughness Φ. For the runs portrayed in Figure 9.9,
  the moving surface was assigned a value of Φ = 0.2, while the other one
  had Φ = 0.8. This arrangement resulted in high-velocity slip at the moving boundary and no slip at the stationary surface. Because the moving
  boundary is supplying a higher momentum to the granules through the
  slip velocity, the "temperatures" there are also higher than at the stationary surface.

9.1.4 SLIDER BEARINGS

In striving to obtain solutions for slider bearings, the formulated granular flow equations become still more cumbersome in that their velocities, pressures, etc., become
functions of distance involving, in most cases, the term x (ln p). In using numerical
methods for solving the equations, these have to employ several levels of computa-
tions and a number of iteration procedures to arrive at convergent solutions. These
ramifications are all contained in Reference 6; for relevant details, refer to that source.
Here, the results of these calculations will be given, emphasizing, in particular, the
differences in the performance of inclined sliders from parallel plates.

The evaluation of the granular flow equations for sliders was modeled after an
available set of experiments (5), employing the following parameters:

$$d = 5 \text{ microns}; \quad B = 2.6 \text{ cm}; \quad U = 4.6 \text{ m/s}; \quad \left(\frac{x_{cp}}{B}\right) = 0.6$$

$$\rho = 4260 \text{ kg/m}^3; \quad W = 155.7 \text{ N (35 lb)}$$

with equal roughness on both surfaces. Slip velocities and "temperature" effects
were included. This slider is referred to as the benchmark model.

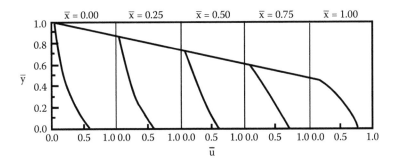

FIGURE 9.10 Velocity profiles in benchmark slider.

9.1.4.1 Results for Benchmark Slider

The velocity profiles for this case are shown in Figure 9.10. The velocity gradient along x increases at the stationary surface and decreases at the sliding surface. Correspondingly, the slip velocity along x rises at the stationary surface and decreases slightly at the moving surface. Figure 9.11 shows the "temperature" profiles at several locations along the slider. These temperatures are affected by two competing trends: the momentum supplied at the boundaries via slip and the energy sink due to inelastic grain collisions throughout the film. At the inlet, the "temperatures" at the moving surface are much higher than at the stationary surface. However, as the gap decreases, the slip velocity at the stationary surface must increase to accommodate the (constant) flow and, accordingly, the "temperature" at that surface increases and becomes greater than at the moving surface.

Figure 9.12 presents the pressure profile, as obtained from the experiments cited earlier (5) and from the present analysis (6). The experimental slider had a pivot position at 60%, which yielded a maximum pressure of 770 kPa. The present model, which naturally differed in many details from the experimental prototype, yielded a pivot position at 73% for nearly the same p_{max}. The slope of the slider from this analysis was $\left(\frac{h_2}{h_1}\right) = 0.43$, which corresponds to an inlet film height equal to 16 particle diameters.

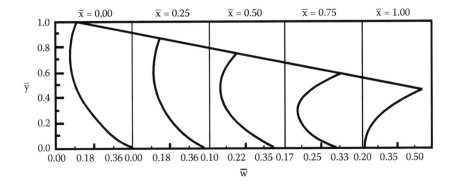

FIGURE 9.11 "Temperature" profiles in benchmark slider.

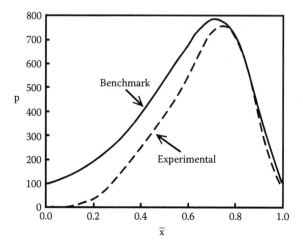

FIGURE 9.12   Experimental pressure profiles in slider.

9.1.4.2   Parametric Variations

In a departure from the basic model, the effects of varying the following parameters were investigated:

- Coefficient of restitution e. This coefficient was varied from 0.75 to 0.85, yielding the results shown in Figure 9.13. Increasing e raises the maximum pressure and moves its position slightly back.
- Particle diameter, d. This variation was performed while keeping the gap width and taper constant, while varying d from 3 to 8 microns. The latter corresponded to 10 granules at the inlet and 4 at the outlet—relatively large pellets for the film. Figure 9.14 shows the effect on both load capacity and

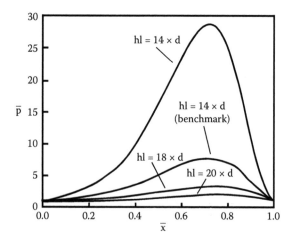

FIGURE 9.13   Effect of varying film height in slider.

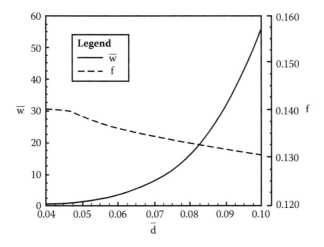

FIGURE 9.14 Effect of varying particle diameter on friction coefficient and load capacity.

friction coefficient, the main outcome being a drop in load capacity with a rise in d. Beyond the 8 micron pellets, the present equations will most likely lose their validity.

• Effect of surface roughness. Figure 9.15 shows the changes in load capacity and friction coefficient with a rise in surface roughness at the moving surface. As the roughness increases, the adjacent granules tend to assume the same surface velocity, and the wall thus supplies less collisional energy to the granules. This causes a higher solids fraction, increasing the effective viscosity near the moving wall. Both load capacity and friction rise correspondingly.

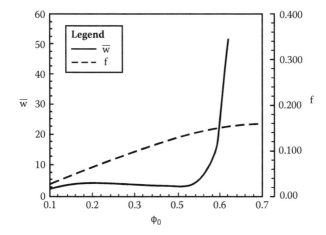

FIGURE 9.15 Effect of varying roughness on stationary surface on friction and load capacity.

9.2   DISCRETE PARTICLES APPROACH

Even though Haff's analysis contains parameters such as d and s that specify discrete particle characteristics, in the analytical approach they are all treated as if they were a continuous medium. This is evident in the forms taken by the constitutive equations of continuity and momentum, and even the energy equations all based on Newtonian models. In fact, the very usage of constitutive equations implies a parallel to fluid films. Thus, the material in the previous sections, including even the manner in which the experimental results were presented, relies on treating granular layers as a continuous medium.

An alternate approach is to treat aggregate particles on the basis of the interaction of discrete bodies followed by summing the interactions of all the granules present. This implies that the granules be relatively large and finite in number. This is the approach taken by Elrod (2,3,8). In the very first paper, Elrod emphasizes that the solutions offer no more than a qualitative portrayal of the performance of granular layers. Moreover, no analytical solutions are obtained. Whatever is achieved is the result of elaborate computer simulation runs requiring a fair amount of labor and computer time.

In outline, the analysis commences with the evaluation of the forces generated by the impact of two particles, as shown in Figure 9.16. The normal force F_N depends on

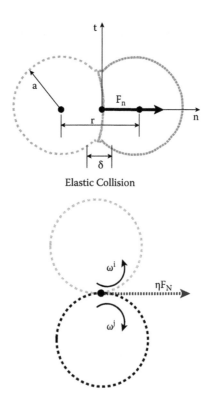

FIGURE 9.16   Interaction of two granules.

depth of penetration, which, in the present analysis, is carried to a maximum of center to center distance equal to a, the radius of the particles. This is assumed to take place without a permanent deformation occurring in either of the colliding particles. Next, a tangential force (F_T) is obtained due to the difference in the particles' angular momentum. With the forces for the individual particles determined, these are then summed over the aggregate to yield the stresses on the walls, and from these, the load and total friction are obtained.

A number of assumptions were introduced, that is,

- All particles are assumed to be identical and of circular cross section.
- Because the surfaces extend to infinity in the direction normal to motion, the configuration of the particles actually consists of cylinders of circular cross section; in the case of a large number of particles, they resemble the elements of needle bearings.
- In tapered geometries, the number of particles admitted at the inlet to the film must be no greater than can be passed through the exit, giving the upstream film a much lower solids fraction, higher particle separation, etc.
- The force of repulsion is less than that of compaction inducing dissipation in the system.
- The tangential force F_T is related to the normal force F_N via a coefficient of friction C_f.
- The effects of the carrier fluid are ignored.

9.2.1 Governing Relations

When two granules possessing the mass and inertia of

$$m = \rho \pi a^2 L \text{ and } I = \frac{ma^2}{2}$$

collide, the normal separating force exerted by granule i on granule j is

$$F_{ij} = -nG\Phi\left(\frac{r}{a}\right)H \tag{9.23}$$

with Φ a function to be determined and H the collision force. These two quantities are defined as follows:

$$H = [1+e-(1-e)\operatorname{sgn}(\delta Vo.\delta r)] \tag{9.24}$$

$$\Phi\left(\frac{r}{a}\right) = \exp[-\log_e(fr)] \tag{9.25}$$

During granule compaction, $H = 1$ and, when separating, $H = e$. A force ratio fr is now defined representing

$$fr = \left(\text{Force when } r = \frac{2a}{\text{Force}} \text{ when } r = a \right) \tag{9.26a}$$

$$\propto = \log_e(fr) \tag{9.26b}$$

The latter is a quantity that determines how rapidly the interaction force increases with particle penetration or how fast it decreases with separation.

As a measure of granule hardness, one can take the energy E required to reduce the center-to-center distance r from $d = 2a$ to a or

$$E = \int_a^{2a} (r)dr = \frac{[Ga(fr-1)]}{fr^2 \log_e(fr)} \tag{9.27}$$

If the granule is rotating with an angular velocity ω_j, the local velocity at the contact point $(r/2)$ is $(U_i - V_i + \omega_i(\frac{r}{2})t_i)$. The tangential component of granule i, relative to granule j at the point of contact, is then

$$(U_i - U_j) \bullet t_i + -\delta V \bullet t_i + \frac{(\omega_i + \omega_j)r}{2} \tag{9.28}$$

If the foregoing component is positive, granule j will reduce the angular velocity of granule i. This tangential velocity is taken to be proportional to the normal repulsive force via a friction coefficient C_f, present in the following force component:

$$F_{ij,t} = \text{sgn}[(U_i - U_j) \bullet t_i]t_i C_f G \Phi\left(\frac{r}{a}\right)H \tag{9.29}$$

In terms of the foregoing quantities, we define a hardness ratio hr as the ratio of E, the maximum particle inertia $(\frac{mU^2}{2})$ or

$$hr = \left[\frac{2Ga}{mU^2}\right]\left[\frac{(fr-1)}{fr^2 \log_e(fr)}\right] \tag{9.30}$$

The usual equations of motion are given by

$$m\left(\frac{dV_i}{dt}\right) = \Sigma_j(F_{ij,N} + F_{ij,T})$$

$$I_i\left(\frac{d\omega_i}{dt}\right) = \Sigma_j r_{ij} F_{ij,T}$$

These equations are now evaluated on a computer for the two standard cases: that of parallel surfaces and slider bearings.

9.2.2 PARALLEL PLATES

As was shown previously, the solutions for granular films depend on the physical condition of the surfaces. In the present runs, the surfaces were lined with half-particles so that the film faces stationary sets of particles similar to those in the stream. The spacing between the particles lining the surface was set equal to particle radius. The relevant equations were first applied to the standard case of a set of parallel surfaces, with one of them moving with a velocity U_o. A preliminary run investigated the dependence of the solutions on the article hardness hr. Figure 9.17 shows the dependence of the shear stress in terms of the solids fraction for three levels of hardness: 3, 5, and 10. As seen in the following Figure 9.17, a concentration of $c = 0.6$ the particle hardness has little effect on the stress; above, a concentration of some 90% of the stress is directly proportional to hardness.

The subsequent series of runs were all performed with the following set of constants:

$$hr = 5; \quad fr = 3; \quad e = 0.8; \quad C_f = 0.3$$

for three sets of film thickness, which, in terms of particle diameter, were $7d$, $14d$, and $28d$. Figures 9.18 and 9.19 give the results for the smallest film thickness: $7d$. The plot for velocity distribution across the film—$u(y)$—shows that, at low solids concentration, the flow resembles the conventional Couette distribution except for the presence of slip at the boundaries. At the higher values of c, the curve tends to steepen in the center of the film. The plots for the particle distribution, likewise, tend to form a core in the center when approaching the higher values of c and

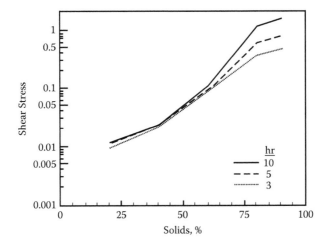

FIGURE 9.17   Shear stress dependence on hardness ratio and density.

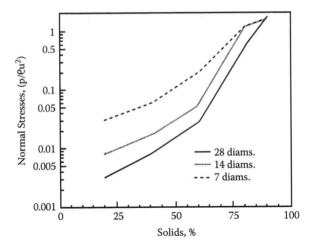

FIGURE 9.18 Normal stresses as a function of gap height and solids concentration.

the velocity slip at the boundaries tends to disappear. Figures 9.20 and 9.21 give results for the highest film thickness: $28d$. The trend here is similar. At low values of c, a near-Couette flow is obtained but at the higher concentrations, a definite stagnant portion of f or plug flow is formed in the center of film height. The plots for particle distribution are, here, more level than in Figure 9.22, the only anomaly in both figures being a slight depression in the center of particle distribution at the 90% plots.

Figures 9.23 and 9.24 provide a comparison of the results for all three film heights. As expected, the levels of solids concentrations and the pressure are inversely proportional to h. What is of interest in the plot of Figure 9.24 is that, at very high values of c, the pressure is nearly independent of gap height, the main mechanism of generating normal stresses being film compression. The friction coefficient, as given

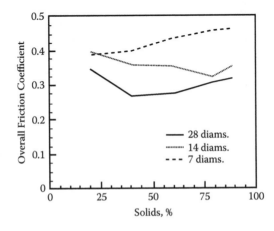

FIGURE 9.19 Overall friction coefficient as function of solids % and gap.

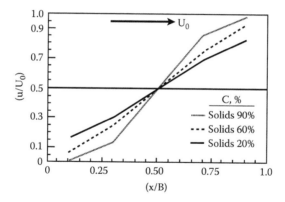

FIGURE 9.20 Velocity profiles for various solids concentrations along the surfaces.

by the overall (F/W) for the three configurations (as plotted in Figure 9.24), shows a decrease with film thickness. Of interest is also the plot in Figure 9.24, where the overall friction coefficient f is plotted for two values of particle friction constant C_f, yielding a higher f when there is high particle friction.

9.2.3 SLIDERS

A sample of the slider to be analyzed is shown in Figure 9.25, having a particular film ratio of $h = 2$. It should be noted that the height of the admitted granular layer consists of only $h_o = 4d$. This is required so that the layer can pass the exit portion of the film, which, like the rest of the surfaces, has half a granule on the top and lower surfaces; their center-to-center distance in the present case is 1.33 d. When the ratio h will be changed in the course of the parametric study, in all cases the height of h_o

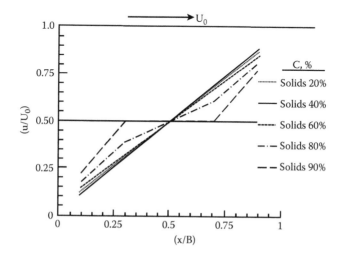

FIGURE 9.21 Velocity profiles with gap height of 28 d for various solids concentrations.

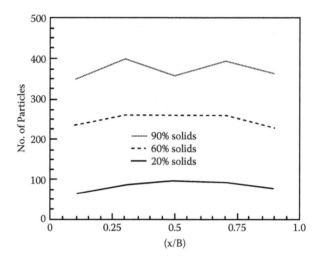

FIGURE 9.22   Distribution of particles across film for $h = 28\ d$ at various solids concentrations.

will be adjusted to the geometry of the outlet film. The parameters kept constant in the analysis are

$$hr = 5; \quad fr = 3; \quad B = 37.5d; \quad e = 1 \text{ on approach} \quad \text{and} \quad e = 0.8 \text{ at separation of granules}$$

Figures 9.26a, 9.26b, and 9.26c give a pictorial representation of the dynamics of the granular film. Part (a) shows film behavior at the inlet, where some of the admitted particles are seen being ejected backward, mimicking reverse flow in fluid films when the pressure gradient there is sufficiently large to overcome the effects of shear. On the other hand, at the exit—Figure 9.26b—the particles are

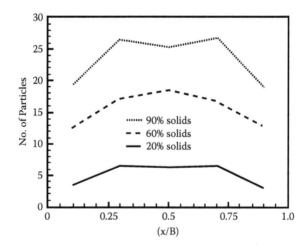

FIGURE 9.23   Distribution of particles across film for various solids concentrations.

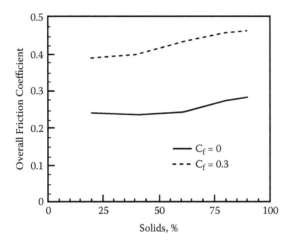

FIGURE 9.24 Dependence of overall friction coefficient on friction coefficient between granules.

propelled forward from the film with some energy due to compression in the tight gap.

Figure 9.27 gives the velocity profile $u(y)$, showing slip at the boundaries. Again, the profiles bear a striking resemblance to hydrodynamic film, including reverse flow at the inlet, which extends here over 40% of film height—all this despite the radically different physical system and different analytical tools employed. Another resemblance to hydrodynamic behavior appears in Figure 9.28, in which load and friction are plotted against different h values (8). Here, the optimum load capacity occurs at $h = 2.0$, which is close to what transpires with hydrodynamic bearings: $h = 2.2$. What is radically different here is that, outside the range of 1.5–2.0, there is little variation in the forces with a change in h. One of the consequences of this behavior is that, even at constant film thickness, there is a load

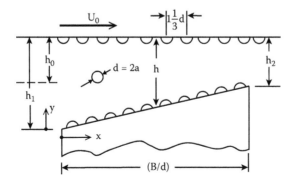

FIGURE 9.25 Geometry of investigated slider.

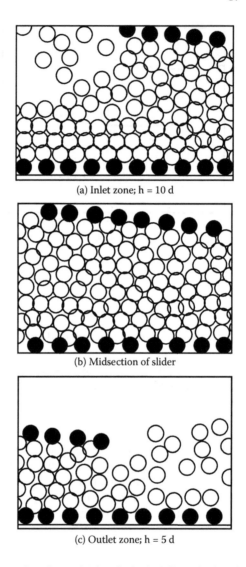

(a) Inlet zone; h = 10 d

(b) Midsection of slider

(c) Outlet zone; h = 5 d

FIGURE 9.26   Kinematics of granules in tribological film with $h_o = 4$.

capacity—about a third of the optimum value. As noted on previous occasions, this analytically confirms the experimental evidence that parallel plates using conventional lubricants can carry a load.

Figure 9.29 shows that there is, under normal sliding conditions, no great variation in film density along the slider. The peak value, occurring at about $(\frac{x}{B}) = 0.4$, differs only by some 10% from the lowest value at the extremities of the film. Figure 9.30 provides the stresses in the slider, the normal stresses representing, of course, the pressure.

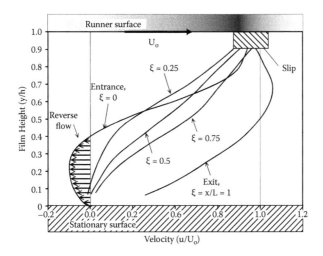

FIGURE 9.27 Velocity profiles in granular film in kinetic flow, $\bar{h} = 2$.

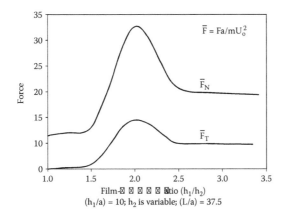

FIGURE 9.28 Effect of $\bar{h}$ on forces in kinetic flow.

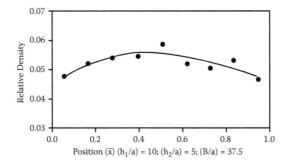

FIGURE 9.29 Density variation within film.

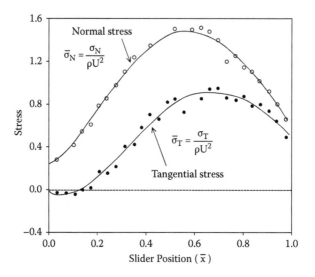

FIGURE 9.30 Stresses in kinetic flow granular film.

TABLE 9.3

Performance of Particle Assembly for Various Geometric Conditions

(h_1/d)	(h_2/d)	(B/d)	$\sigma_{N|1}$	$\sigma_{T|1}$	σ_T/B	Q	$C, \%$	F_N	F_T
				Effect of inclination					
10	3	37.5	0.110	0.935	0.97	0.74	65.5	19.5	9.81
10	4	37.5	0.280	0.550	0.50	1.06	71.5	20.5	10.30
10	5	37.5	0.150	0.620	0.56	1.72	82.7	32.7	16.90
10	7	37.5	0.423	0.175	0.138	1.73	65.6	12.3	5.44
10	9	37.5	0.530	0.130	0.09	2.29	63.9	11.8	5.24
10	10	37.5	0.564	0.140	0.08	2.61	64.5	11.6	5.06
				Effect of gap					
5	4	37.5	1.19	0.378	0.25	1.22	87.8	41.7	21.0
5	3	37.5	-.25	1.00	0.62	0.96	86.5	41.3	20.5
5	5	37.5	1.19	0.31	0.21	1.60	83.0	36.5	18.7
				Effect of length					
5	4	75	1.30	0.27	0.249	1.19	85.3	85.1	39.8
5	3	75	0.35	1.00	0.625	0.88	85.5	82.8	37.6
5	3	56	1.39	1.25	0.760	1.08	107.00	97.3	44.2
5	4	56	1.40	0.75	0.470	1.41	101.00	84.8	38.2
5	5	56	1.36	0.25	0.125	1.58	85.10	62.8	29.3

Note: $\bar{\sigma} = \sigma/(eU_0^2); \bar{F} = F/(mU_0 \, {}^2\!/_a)$.

Table 9.3 lists the detailed results of the computer simulation runs for a wide range of parameters in which the effects of slider taper, film height, and extent of slider are given in appreciable detail.

REFERENCES

1. Yu, Chih-Ming et al., Granular collision lubrication, *J. Rheology*, Vol. 38, No 4, July–August 1994.
2. Elrod, H.G., Granular flow as a tribological mechanism, Leeds-Lyon Symp., Leeds, September 1987.
3. Elrod, H.G. and Brewe, D.E., Numerical experiments with flows of elongated granules, Elsevier Tribology Series, pp. 75–88, 1992.
4. Haff, P.K., Grain flow as a fluid mechanical phenomenon, *J. Fluid Mechanics*, Vol. 134, pp. 401–430, 1983.
5. Heshmat, H., The quasi hydrodynamic mechanism of powder lubrication: Part II, *Trans. STLE*, Vol. 84, No 2, 1992.
6. Mc Keague, K.T. and Khonsari, M.M., An analysis of powder lubricated slider bearings, *ASME J. Tribology*, Vol. 118, January 1996.
7. Mc Keague, K.T. and Khonsari, M.M., Generalized boundary interactions for powder lubricated Couette flow, ibid, July 1996.
8. Elrod, H.G., Numerical experiments with flows of elongated particles—Part II, Elsevier Tribology Series, Vol. 21, pp. 347–354, 1996.
9. Heshmat, H., The quasi-hydrodynamic mechanism of powder lubrication: Part III: On theory and rheology of triboparticulates, STLE Annual Meeting, Calgary, Canada, May 17–20, 1993 (1995) *Tribology Trans.*, Vol. 38 (2), pp. 269–276.

10 The Tribological Continuum

The material presented in the previous chapters endeavored to describe the functions and characteristics of a number of interface layers capable of acting as tribological films. Table 10.1 lists the substances considered, which range from "dry" contacts to oils and discrete solid particles. While the capacity of these diverse materials to perform as lubricants holds together the topics discussed, the present chapter will attempt to extrapolate this commonality into some systematic pattern, wherein the operational functions and analytical formulations are combined into a unified matrix imparting in physical or mathematical terms an engineering order to this aspect of tribology. We shall start with a brief note on the history of previous attempts in this area and conclude by sketching a possible analytical continuum across a wide spectrum of tribological regimes.

10.1 HISTORICAL OVERVIEW

As is familiar to students of tribology, the founder of lubrication as a scientific discipline was Osborne Reynolds—an illustrious name, not only in hydrodynamic theory but also in many other branches of the physical sciences. At a meeting of the British Association for the Advancement of Science held in Montreal, Canada, in 1884, Reynolds, for the first time, discussed the governing differential equation on the hydrodynamic nature of lubrication, though no published record of it exists. The paper itself was submitted on December 29, 1885, and read to the Royal Society on February 11, 1886, in which the equation bore the following form (25):

$$\frac{\partial}{\partial x}\left(h^3 \frac{\partial p}{\partial x}\right) + \frac{\partial}{\partial z}\left(h^3 \frac{\partial p}{\partial z}\right) = 6\mu\left[(U_o + U_1)\frac{dh}{dx} + 2V\right] \qquad (10.1)$$

What is of particular interest from our present vantage point is that, at the same time (December 1885), Reynolds published another paper (26) dealing with the nature and properties of granular assemblies. The paper, entitled "On the Dilatancy of Media Composed of Rigid Particles in Contact," deals, in particular (as the title suggests), with the property of dilatancy of granular assemblies. As far as it relates to tribological layers, some of the points Reynolds stressed were

- Dilatancy is a singular property of granular media.
- Every change in the relative position of the grains is attended by a consequent change in volume.

TABLE 10.1
List of Possible Interface Layers

	Layer	Discussed in Chapter
1.	Dry metal contacts[a]	3, 4
2.	Solid lubricants	3, 4
3.	Boundary lubrication	3, 4
4.	Liquid lubricants	2, 5
	• Petroleum oils	
	• Synthetic oils	
	• Process fluids	
	• Water	
	• Cryogenics	
5.	Gaseous lubricants	5
	• Air	
	• Water vapor	
	• Process gases	
6.	Non-Newtonian substances	2, 4, 5
	• Greases	
	• Bingham plastics	
	• Rubbers	
7.	Powder films	2, 5, 6, 7, 8
8.	Granular layers	5, 9

[a] It is to be understood that no metals are "dry" but are overlaid with an inherent coating, most often an oxide.

- If arranged in a cubical formation, the density of the assembly is $\left(\frac{\pi}{6}\right)$ of particle density, or the void amounts to 48% of total volume; when tetragonal-spheroidal packing is considered, the density of the volume can go up close to 70%.
- If the granules are rigid, the relation between distortion and dilatation is independent of particle friction; the only effect of higher friction is to render the assembly more stable.

It is of particular historical cogency that the man who founded the very discipline of tribology should, at the very beginning, have spanned a whole century to delve into a tribological regime that is at the present is still in its infancy.

The most interesting next development in the tying together of the various regimes of lubrication was the discovery that the fluid film does not have to be an oil or liquid but that it could be a gas. This discovery was made by Albert Kingsbury (23), who, having built a torsion machine, discovered that when he spun the piston in its tight clearance, the slightest effort would suffice to make it spin, even when the machine was in a horizontal position when the piston was loaded. When Kingsbury instrumented the clearance space with pressure taps, he discovered a hydrodynamic pressure profile in the air film. This occurred at the turn of the nineteenth century and, a short time later in 1913, Harrison at Cambridge University formulated Reynolds equation for compressible fluids, which naturally contained a variable density term (6). By using the perfect gas equation and assuming isothermal conditions—true for most gaseous films—the density becomes proportional to pressure, eliminating the density term from the equation. Both the incompressible and compressible Reynolds equation can now be written in a single form

$$\frac{\partial}{\partial x}\left(P^n h^3 \frac{\partial p}{\partial x}\right) + \frac{\partial}{\partial z}\left(P^n h^3 \frac{\partial p}{\partial z}\right) = 6\mu\left[(U_o + U_1)\frac{dP^n h}{dx} + 2V\right] \qquad (10.2)$$

By setting $n = 0$, we get an expression for incompressible lubricants and if we set $n = 1$, the expression represents lubrication with compressible fluids.

At the same time that Kingsbury discovered gas bearings, R. Stribeck in Berlin constructed what is, from our present standpoint, probably the most important tribological diagram discipline, which is known to the present day by his name. The plot shown in Figure 10.1 encapsulates, in a single curve, three regimes of lubrication: dry friction and boundary, mixed, and the hydrodynamic regimes. The friction coefficient for all these regimes is plotted against a single dimensionless group of variables—the familiar $\left(\frac{\mu N}{P}\right)$—sometimes referred to as the Hersey number. It is the inverse of the widely used Sommerfeld number, $S = \left(\frac{R}{C}\right)^2\left[\frac{\mu N}{P}\right]$. In order to eliminate the effect of temperature in his experimental correlations, Stribeck conducted all his experiments at the constant temperature of 25°C. As seen from Figure 10.1, the different regimes also represent decreasing ratios of $\bar{h} = \left(\frac{\eta}{h}\right)$, where η is the surface roughness of the mating surfaces.

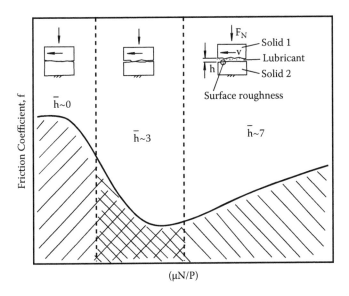

FIGURE 10.1   The Stribeck curve for three tribological regimes.

TABLE 10.2

Early Works on Non-Newtonian Substances

Year	Author	Lubricant Model
1957	Milne	Viscoelastic
1967	Hsu	Pseudoplastic
1971	Allen and Kline	Micropolar
1973	Tipei and Rohde	Directional viscosity
1978	Harnoy	Viscoelastic
1984	Bourgin and Gray	Polymer-thickened

An unfulfilled ambition of tribologists has been to have similarly extended equations, similar to Equation 10.2, to subordinate the vagaries of non-Newtonian substances to an appropriate Reynolds equation. Some of these early works, listed in Table 10.2, targeted such disparate substances as polymer-thickened oils, greases, elastomers, slurries, rubber staves, and others. A wide gap looms between the rheological expressions employed, particularly their dependence on the operational parameters; often, even the use of the same rheological model yielded different results. This has not been entirely the tribologists' fault. While with Newtonian fluids the main task is to harness known characteristics to the strictures of interface flow, with non-Newtonian substances the problem lies with the science of fluid mechanics in the first place. The tribologist is thus often forced to generate ad hoc laws for the rheology of his or her material, with unhappy results.

A radical step in unifying the disparate modes of lubrication occurred early in the 1980s. This originated under the tutelage of M. Godet and his associates at the Institut National des Sciences Appliques de Lyon in France. Their pioneering work has been classified as "third-body" effects, a term here replaced by interface layers. In their first paper, the Godet group starts from the view that, because the two basic characteristics of a tribological film are its ability to accommodate a velocity difference between the surfaces and to generate a load capacity via its normal stresses, it should be possible to generate a continuum across the disparate regimes of tribology. Such a film can exist in four different varieties (as shown in Figure 10.2) and still be perfectly capable of supporting a load. There is, with such a layer, some distinction in performance when the motion is unidirectional or reciprocating. In the latter case, the particles are trapped, providing more reliable film, whereas in the first case, the debris may be ejected at the exit, denuding the surface of the protecting layer. Experimental results shown in Figure 10.3 support this supposition in terms of the amount of wear inflicted on the contact surfaces in the two cases. This discovery of the role of wear detritus explains much of the difficulties in surface mechanics, namely, the inconsistencies in arriving at and extrapolating given material properties. There is, in fact, no such thing as an intrinsic material property, because it depends very much on the role of the "lubricant" present, namely, the layer of wear detritus, which changes with time and test conditions. The wear rate also depends on the shape and orientation of the test specimen. Thus, a rectangular sample wears least when its long axis is aligned with the direction of velocity.

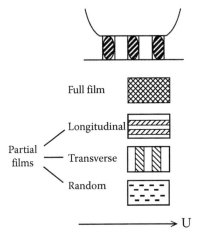

FIGURE 10.2 Several possible film shapes.

Godet's mechanism of "third-body" lubrication, due to the role that wear particles play in tribological kinematics, can be extended and boosted in efficacy by deliberately introducing such debris into the contact zone. The most energetic follow-up on this possibility is the work conducted by Heshmat who, since the early 1980s, applied it primarily to powder films (10–21). In extensive experiments with micronsized MoS_2 and TiO_2 powders, he demonstrated their capacity to support shear and generate load-carrying pressures. Assigning a quintic stress–strain relationship to the rheology of powders, he also generated computer code to yield analytical solutions in terms of the relevant operational parameters and powder properties. This work led to the first tentative applications of powder lubrication in dynamic and structural dampers and slider bearings.

The most recent step in formulating a common basis for disparate regimes of lubrication is the work of Elrod on granular films (3) that began in the early 1990s. As presented in Chapter 9, the astounding result of these analyses is that, even with such radically different materials and different postulates of their mechanics, the results yield flows, stresses, pressures, etc., that are strikingly similar to those from conventional lubricants.

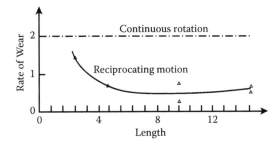

FIGURE 10.3 Wear rates for unidirectional and reciprocating motions.

10.2 OVERLAPPING TRIBOLOGICAL REGIMES

We shall take a brief look at the functions tribology is called upon to perform, with particular emphasis on the overlap of their lubrication mechanisms. This has been discussed in several previous chapters. Table 10.3 lists the major tribosystems used either in the laboratory or industry and the process involved in their operation. Tribosystems are often similar to the processes operative in machinery in general, including the phenomenon of entropy-generating losses. Figure 10.4 provides a schematic representation of such parallels to mechanical systems. One item of particular interest, indicated on the figure by a dashed line, is that of wear. Wear has, heretofore, been considered an unmitigated waste to be minimized, if not eliminated altogether. However, from the perspective of our approach to wear, it can actually be considered a boon in operating the system, much as rejected thermal energy is so often used for purposes of preheating fuels and gases before combustion.

In earlier chapters, we have dealt with the subject of overlapping tribological regimes for particular areas. They can now be brought together and summarized in the following fashion:

TABLE 10.3
Operational Mode of Tribosystems

	Tribosystem	
Boundary Lubrication	Low Speed Machinery Cycling Operation	
	Gears Cams	**Elastohydrodynamic**
	Rolling Element Bearings Clearance Seals Metalworking Cutting Tools Piston Rings Bio-tribology Clutches and Brakes	
Hydro-dynamic	Liquid Lubricant Bearings Gas Bearings Hydrostatic Bearings Dynamic Dampers	
	Rubber Staves Grease Elastomers	**Non-Newtonian**
Powder & Granular	Rubbing Seals	
	Structural Dampers Lapping	

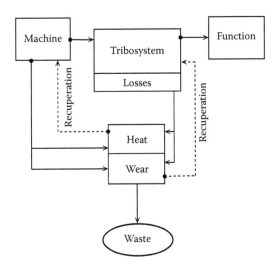

FIGURE 10.4   Tribosystem cycle.

1. Chapter 3. The three segments taken up there were boundary and hydro-dynamic lubrication and their linkage to powder films. It was noted that—from extensive experience both in the laboratory and in the field with plane sliders, flat thrust sectors, face seals, and centrally pivoted pads—they all perform satisfactorily (some even in an outstanding manner) despite the fact that hydrodynamic theory does not assign any load capacity to such geometries. Moreover, they functionally fit in with the parametric effects in full-fledged hydrodynamic devices, such as a reduction in friction with the parameter $(\frac{\mu U}{W'})$, as well as when compared to "dry" friction; the value of film thickness falls halfway between dry and hydrodynamic regimes. The generated pressure profiles are qualitatively similar to those from conventional bearings. This held, also, for powder films on the opposite end of the hydrodynamic regime, showing the presence of typical pressure profiles with all the characteristics of liquid films, such as cavitation, side leakage, and reverse flow.

2. Chapter 4. This was devoted to the commonality of four regimes, that is,
   a.   Dry surfaces
   b.   Surfaces with natural or overlaid coatings and solid lubricants
   c.   Boundary and mixed lubrication
   d.   Full or partial elastohydrodynamic operation, the latter defined by films in which the film height is less than three times the surface roughness.

   It was seen that it was impossible to separate categories (a) and (b), the only difference between them being that in (a) the coating is most likely an oxide while in (b) it could be any organic or inorganic substance. Spanning regimes (b) and (c) is the effect of temperature, which may, by melting the solid lubricants, generate conditions of boundary or even those of hydro-dynamic lubrication, as occurs intentionally in steel mill operations. The extreme of mixed lubrication easily merges into that of partial EHL where there is only a small subregion of asperity interaction. On the other hand, under high stresses and pressures, an EHL regime may convert itself into a case of solid lubricants. All these modes have been plotted in Figures 4.53 to 4.55, which provide a dependence on films ranging from monomolecular layers to full hydrodynamic condition, simply on film height or its relative $(\frac{\mu U}{W'})$.

3. Chapter 5. This chapter dealt specifically with the umbrella covering hydrodynamic and powder lubrication regimes. This showed not only that the shears, flows, and pressures in a powder film bear a strong resemblance to conventional films but that even such secondary effects such as starvation, variable viscosity and density, and elastic and inertia effects are all present in powder films, mimicking the characteristics of both liquid and gas lubrication.

The thrust of all of the foregoing observations is the proposition that, whether one employs gas, liquids, powders, or even granular assemblies, they all possess the ability to separate the mating surfaces and support an external load. At the extremities, any individual regime blends most readily into the next domain. The grouping of the parameters governing the quantitative performance of each regime varies, but only a restricted number of parameters make their appearance in all of them. These are

- Viscosity
- Rheology of interface layer, including compressibility
- Surface topography
- External loading
- Film thickness
- Configuration of system

The dimensionless groupings for each regime should, therefore, be a combination of the foregoing six parameters. While variables such as temperature, environment, and others do have an effect, they are in the nature of perturbations on the essential solutions.

10.3 THE TRIBOLOGICAL CONTINUUM

We have seen that the interrelation of tribological processes emerges when viewed from two opposite perspectives. At one extreme, it transpires that devices consisting of parallel plates using conventional lubricants produce a performance that, by standard hydrodynamic theory, should not occur at all. At the opposite extreme, it has been established, both experimentally and theoretically, that material layers such as powder beds and granular films, which in the past have not rated at all as potential lubricants, produce pressure profiles, flows, and friction, etc., that are very much in keeping with hydrodynamic behavior. This leads us to the conceptual view of tribology, which was partially broached in the preceding chapters and is summarized in the following paragraphs.

10.3.1 GENERAL CONCEPT

The essential postulate of our approach is that two generic mechanisms operate in the process of tribological interactions and, although both are present, it is their relative ratio that determines the prevalence of a particular regime. These two are the hydrodynamic and what will be here termed the *morphological elements*. These terms, along with our present definition of *lubricant*, are used in the following sense:

- Hydrodynamic elements. Effects produced in a lubricant due to the geometry of the film, most often a converging clearance in the direction of motion.
- Morphological elements. Effects due to the elastic, plastic, chemical, molecular, and other surface phenomena in the zone of contact.
- Interface layer (lubricant). Any substance—gaseous, liquid, or solid—present in the clearance space that is not integral with either surface; this includes solid matter, whether due to natural wear or deliberately introduced.

Thus, referring to Figure 10.5, in conventional hydrodynamic film, typical numbers are

	Microns	Mils
Clearance	75	3
Runner surface roughness, δ_1	4	$\left(\dfrac{1}{6}\right)$
Stationary surface roughness, δ_2	8	$\left(\dfrac{1}{3}\right)$

The hydrodynamic elements would, simplistically speaking, prevail over 86% and the morphological over 14% of the interface. In the mixed and boundary lubrication regimes, the two components would be of the same order. In "dry" operation, surface roughness and associated surface phenomena would dominate the mechanics of the process.

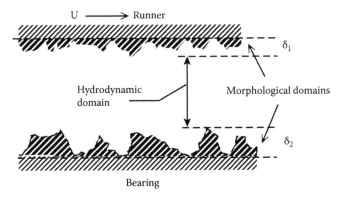

FIGURE 10.5 The two generic domains of a tribological film.

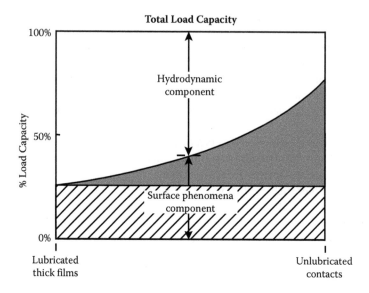

FIGURE 10.6 The weights of the two components in a tribological film.

Consequently, the entire spectrum of tribological regimes is predicated to be operating on the simultaneous presence of both elements, which, together, are responsible for generating load capacity. It is a shift in the (hydrodynamic/morphological) ratio that defines a particular regime. This is schematically portrayed in Figure 10.6, where both the percentages at the ends and the shape of the line delineating the morphological and hydrodynamic components were picked completely at random. The task of tribology is, then, to devise linked constitutive equations, which, when solved, would reflect the individual contribution of each component. Such a set of equations would perhaps be similar to Equation 10.2, valid for both gas and liquid lubrication, or to the complementary equations written by Oh (see Chapter 4) for the simultaneous representation of hydrodynamic and elastic effects in the form of

$$w_1 = Rp + U\left(\frac{dh}{dx}\right) + V\left(\frac{dh}{dt}\right) = 0$$

$$w_2 = Lp + a + h = 0$$

where the first equation contains fluid film quantities and the second equation, elastic body effects with the portions of the film where either w_1 or w_2 applies, determined as part of a solution of the two expressions.

10.3.2 THE CONTINUUM CURVE

In setting up a progression of lubrication processes from dry contacts to granular films, it is possible to say that this hierarchy corresponds to a progressive rise in lubricant viscosity, film thickness, and in the ratio $\lambda = (\frac{h}{\sigma})$, though the latter makes

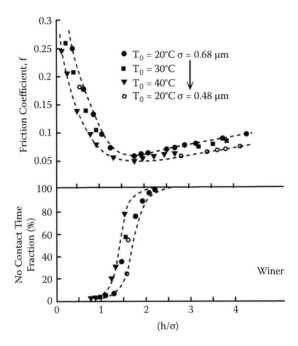

FIGURE 10.7   Friction coefficient in terms of (h/σ).

an impact only in the mixed and early EHL regimes. The effect of the last parameter, not discussed much earlier, is shown in Figure 10.7. Surface roughness is here
defined as the square root of the sum of the squares of the two rms surface rough-
nesses. At $\lambda < 1$ we have boundary lubrication, at $\lambda = 1$ the mixed regime, and at
$\lambda = 2.5$ a hydrodynamic film makes an appearance. The temperature effects are seen
to be small.

We can now, based on the theory and experimental evidence (10–21), group
together the local correlations postulated in previous chapters. The four parts of
Figure 10.8 (a–d), taken from the previous chapters, represent continua over several
local regimes of operation:

1. Figure 10.8a: Dependence of the four regimes—boundary, mixed, EHL,
   and hydrodynamic on film thickness; with a rise in film thickness, friction
   and rate of wear consistently decrease.
2. Figure 10.8b: This diagram, too, brings together the four regimes listed
   earlier, but now plotted against the dimensionless quantity $(\frac{\mu U}{W'})$, which,
   essentially, reproduces the Stribeck curve. One refinement contained in
   the present plot is the distinction between several modes of "dry" friction.
   Thus, friction decreases if we replace a natural surface oxide with a chosen
   coating and replace a coating by a solid lubricant.
3. Figure 10.8c: Here, the extended Stribeck curve, plotted against the
   Sommerfeld number—essentially equivalent to the $(\frac{\mu U}{W'})$ group—is

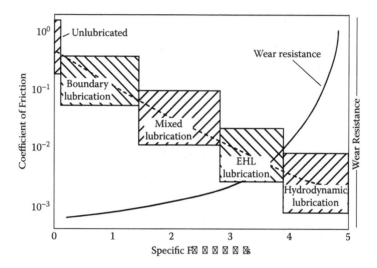

FIGURE 10.8a Coefficient of friction and wear rate for several tribological regimes in terms of film thickness.

accompanied by a corresponding variation in f and rate of wear. As expected, f rises continuously, while at very high values of $(\frac{\mu U}{W})$, wear, which originally declined, starts rising again due to the powder reaching its limiting shear stress level. Unlike the conventional Stribeck curve, as depicted in Figure 10.8c, the frictional curve ultimately

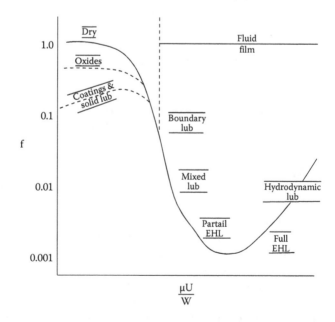

FIGURE 10.8b Coefficient of friction for several tribological regimes in terms of $(\mu U/W)$.

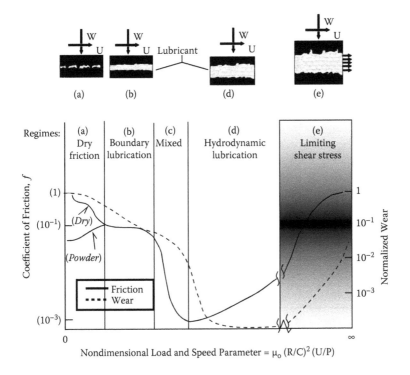

FIGURE 10.8c Extension of Stribeck curve to include powder lubrication and limiting shear stress.

attains the limitation in coefficient of friction as the Sommerfeld number approaches its limits.

4. Figure 10.8d: This part brings us close to our ultimate purpose in this volume, that is, to be able to construct an equivalent Stribeck curve in terms of appropriate dimensionless quantities for all lubrication regimes. In part c, we

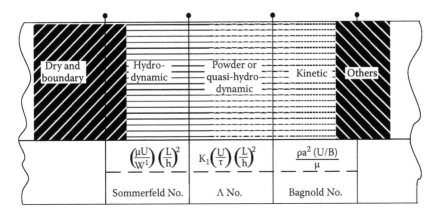

FIGURE 10.8d Contiguity of various tribological regimes.

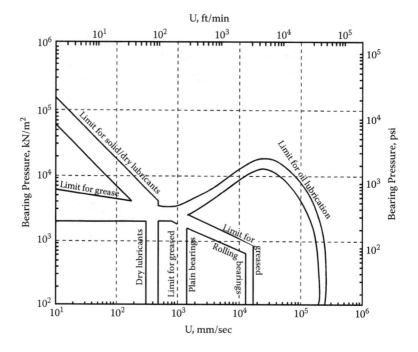

FIGURE 10.9   Operational envelopes for several tribosystems (from Winer, W.O., 1982).

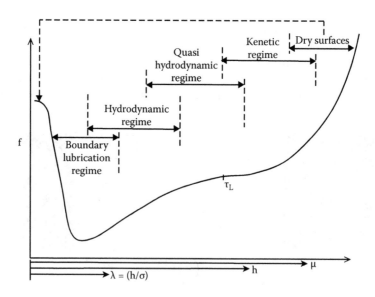

FIGURE 10.10   Tribological continuum curve in terms of increasing μ, h, and λ.

have sketched the compressible, incompressible, and kinetic regimes, each in terms of an appropriate dimensionless set of parameters: Sommerfeld number for incompressible and compressible hydrodynamic operation, a quantity $\Lambda = K_1(\frac{U}{\tau})(\frac{L}{h})^2$ for the compressible mode, and a Bagnold number for powder and granular flows.

A somewhat similar attempt by Winer (28) placed several tribological processes on a common plane, as shown in Figure 10.9. These are configured in terms of speed and applied pressure with the proviso that these tribosystems be operated with an "adequate" lubricant viscosity. If one takes this to mean the same average or effective viscosity, than this would be equivalent to plotting the various envelopes in terms of the familiar $(\frac{\mu U}{W})$ group of variables.

We conclude by presenting here three continuum curves embracing tribological regimes from "dry" to granular films, the latter looping at its extreme end back to dry contacts. On the curve of Figure 10.10, we see that the continuum curve can be plotted in terms of constantly rising parameters of viscosity, film thickness, and λ. In Figure 10.11, we have expressed the various regimes in terms of $(\frac{\mu U}{W\sigma})$ dimensionless grouping, as they have been deduced in various chapters. These have been replotted in Figure 10.12 against the continuum curve spanning all the tribological regimes where they overlap neighboring lubrication modes, as has been postulated all along. In scanning these parameters, one finds that the following variables keep reappearing:

Viscosity, μ
Rheology of layer, ρ, τ_L, U, a
Surface topography, σ
External loading, W
Film thickness, h
System configuration, B, h

	Incompressible		Compressible			Incompressible
	Mixed	Liquids	Gases	Powders	Granular	Solids
$(h/\sigma)=$ $\left[\frac{\mu U}{W\sigma}\right]$		$\frac{\mu U}{W'}\left(\frac{R}{C}\right)^2$	$\frac{\mu}{\sigma a}\left(\frac{R}{C}\right)^2$	$\frac{U}{\sigma L}\left(\frac{L}{h}\right)^2$	$\frac{\sigma a^2}{\mu}\left(\frac{U}{B}\right)^2$	

FIGURE 10.11.   Contiguity of various tribological regimes.

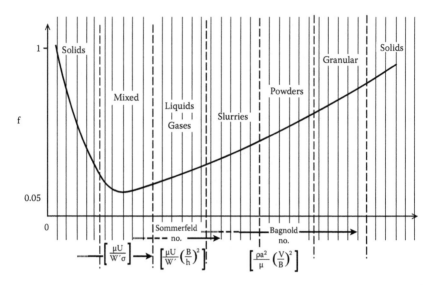

FIGURE 10.12   Tribological continuum curve.

There are the very variables forecast in the previous section to be the basic determinants of tribological operation.

REFERENCES

1. Allen, S.J. and Kline, R.A., Lubrication theory for micropolar fluids, *ASME J. Applied Mechanics*, September 1971.
2. Bourgin, P. and Gay, B., Determination of the load capacity of a finite width journal bearing by a finite element method in the case of a non-Newtonian fluid, *ASME J. Lubrication Technology*, April 1984.
3. Elrod, H.G., Granular flow as a tribological mechanism—A first look, Interface Dynamics Symposium, Leeds-Lyon, BHRA, pp. 75–102, 1988.
4. Godet, M, Play, D., and Berthe, D., An attempt to provide a unified treatment of tribology etc., *ASME J. Lubrication Technology*, Vol. 102, pp. 153–164, 1980.
5. Godet, M., The Third body approach: A mechanical view of wear, *Wear*, Vol. 100, pp. 437–452, 1984.
6. Harrison, W.J., The hydrodynamic theory of lubrication with special reference to air as a lubricant, *Trans. Camb. Phil. Soc.* Vol. 22, No 3, 1913.
7. Harnoy, A., An analysis of stress relaxation in elastico-viscous fluid lubrication of journal bearings, *ASME J. Lubrication Technology*, April 1978.
8. Heshmat, H., The quasi-hydrodynamic mechanism of powder lubrication, Part II: Lubricant film pressure profile, *Lubrication Eng.*, Vol. 48 pp. 373–383, 1992.
9. Heshmat, H., Pinkus, O., and Godet, M., On a common tribological mechanism between interacting surfaces, *STLE Tribolgy Trans.* pp. 32–41, May 1988.
10. Heshmat, H. and Heshmat, C.A., The effect of slider geometry on the performance of a powder lubricated bearing—experimental illustrations, 52nd Annual Meeting of STLE, Kansas City, May 18–22, 1997 (1999), *Tribology Trans.*, Vol. 42 (3), pp. 640–646.

11. Higgs, C.F. III, Heshmat, C.A., and Heshmat, H., Comparative evaluation of MoS_2 and WS_2 as powder lubricants in high speed, multi-pad journal bearings, Paper 98-TRIB-61, ASME/STLE Tribology Conference, October 1998 (1998), American Society of Mechanical Engineers (Paper), (98-TRIB-1-61), pp. 1–6 (1999) *J. Tribology*, Vol. 121 (3), pp. 625–630.

12. Higgs, C.F. III and Heshmat, H., Characterization of pelletized MoS_2 powder particle detachment, 2000-TRIB-46, 1999 ASME/STLE Joint Tribology Conference (2001) *J. Tribology*, Vol. 123 (3), pp. 455–461.

13. Heshmat, H. and Heshmat, C.A., On the rheodynamics of powder lubricated journal bearing: Theory and experiment, *25th Leeds-Lyon Tribology Symposium*, September 8–11, 1998 Elsevier Science B.V., ed. D. Dowson (1999) *Tribology Series*, Vol. 36, pp. 537–544.

14. Heshmat, H., Experimental determination of powder film shear and damping characteristics, ASME Paper 2000-GT-0368, *ASME Turbo Expo*, IGTI 2000, Germany, May 2000.

15. Kaur, R.G. and Heshmat, H., On the development of self-acting and self-contained powder lubricated auxiliary bearing, 2000 ASME/STLE Joint Conference, Nashville, Tennessee.

16. Kaur, R.G., Heshmat, H., and Higgs, C.F. III, Pin-on-disc tests of pelletized molybdenum disulfide (2001), *Tribology Trans.*, Vol. 44 (1), pp. 79–87.

17. Heshmat, H., The effect of slider geometry on the performance of a powder lubricated bearing—Theoretical considerations, 54th Annual Meeting of STLE, May 1999 (2000), *Tribology Trans.*, Vol. 43 (2), pp. 213–220.

18. Kaur, R.G. and Heshmat, H., 100 mm diameter self-contained solid/powder lubricated auxiliary bearing operated at 30,000 rpm (2002), *Tribology Trans.*, Vol. 45 (1), pp. 76–84 (2002) *Lubrication Eng.*, Vol. 58 (6), pp. 13–20.

19. Heshmat, H. and Kaur, R.G., Powder lubricated bearing operation to PV of half a million, 2nd World Tribology Congress, Vienna, Sept. 3–7, 2001 (2002) *Tribology Series*, Vol. 40, pp. 477–480.

20. Heshmat, H., On the theory of quasi-hydrodynamic lubrication with dry powder: application to development of high speed journal bearing for hostile environment, *20th Leeds-Lyon Symposium on Dissipative Processes in Tribology*, Lyon, France, September 7–10, 1993, Dowson et al., eds. (1994) Elsevier Science B.V., *Tribology Series* 27, pp. 45–64.

21. Iordanoff, I., Berthier, Y., Descartes, S., and Heshmat, H., A review of recent approaches for modeling solid third bodies, ASME/STLE Joint Tribology Conference, October 2001, San Francisco, CA (2002) *J. Tribology*, Vol. 124 (4), pp. 725–735.

22. Hsu, Y.C., Non-Newtonian flow in an infinite length full journal bearing, *ASME J. Lubrication Technology*, July 1967.

23. Kingsbury, A., Experiments with an air lubricated bearing, *J. Am. Soc. Naval Engineers*, Vol. 9, 1897.

24. Milne, A.A., A contribution to the theory of lubrication, *Wear*, Vol. 1, No 1, 1957.

25. Reynolds, O., On the theory of lubrication and its application to Mr. Beauchamp Tower's experiments including an experimental determination of viscosity of olive oil, *Phil. Trans.* Vol. 177 (i), pp. 157–234, 1886.

26. Reynolds, O., On the dilatancy of media composed of rigid particles in contact—with experimental illustrations, *Philosophical Magazine and J. Sci.*, London, Vol. 55, 20, pp. 469–481, 1885.

27. Stribeck, R. Die wesentliche Eigenschaften der Gleit—and Rollenlager, *VDI Zeitschrift*, 46, 143–163, 1902.

28. Winer, W.O. Lubricants, *Tribological Technology*, Vol. II, p. 409, ed. P.B. Senholzi, Martinus Nijhoff Publication, Netherlands, 1982.

Index

9 780367 452339